Proceedings of the 8th International Symposium on Insect-Plant Relationships

SERIES ENTOMOLOGICA

VOLUME 49

Proceedings of the
8th International
Symposium on Insect-Plant
Relationships

Edited by

S. B. J. Menken, J. H. Visser and P. Harrewijn

SPRINGER SCIENCE+BUSINESS MEDIA, B.V.

ISBN 978-94-010-4723-4   ISBN 978-94-011-1654-1 (eBook)
DOI 10.1007/978-94-011-1654-1

Cover design by F.J.J. von Planta, Wageningen Agricultural University

Printed on acid-free paper

The 8th International Symposium on Insect-Plant Relationships was held at the International Agricultural Centre in Wageningen, The Netherlands, on 9-13 March 1992

Organizing Committee:

- Prof.dr. L.M. Schoonhoven (*Chairman*),
  Department of Entomology, Wageningen Agricultural University
- Dr. J.H. Visser (*Secretary*),
  Research Institute for Plant Protection IPO-DLO, Wageningen
- Dr. J.J.A. van Loon (*Treasurer*),
  Department of Entomology, Wageningen Agricultural University
- Prof.dr. S.B.J. Menken,
  Institute of Taxonomic Zoology, University of Amsterdam
- Dr. P. Harrewijn,
  Research Institute for Plant Protection IPO-DLO, Wageningen
- Prof.dr. E. van der Meijden,
  Department of Population Biology, University of Leiden
- Dr. C. Mollema,
  Centre for Plant Breeding and Reproduction Research CPRO-DLO, Wageningen
- Prof.dr. M.W. Sabelis,
  Department of Pure and Applied Ecology, University of Amsterdam
- Ms. A.A. Greeven & Ms. E.J.L. Hotke (*IAC Congress Organization*),
  International Agricultural Centre, Wageningen
- Mr. R.P.L.A. de Rooij (*Second Secretary for Proceedings*),
  Research Institute for Plant Protection IPO-DLO, Wageningen

Acknowledgements

The Organizing Committee wishes to acknowledge the support of several institutions and private organizations for their contributions, sponsorships and cooperation, particularly the following:
- Koninklijke Nederlandse Akademie van Wetenschappen, Amsterdam
- Ministerie van Economische Zaken, Den Haag
- European Science Foundation, Strasbourg
- Landbouwuniversiteit Wageningen
- Dienst Landbouwkundig Onderzoek DLO-NL, Wageningen
- Instituut voor Planteziektenkundig Onderzoek IPO-DLO, Wageningen
- Nederlandse Vereniging voor het Tuinzaadbedrijfsleven, Wassenaar
- Internationaal Agrarisch Centrum, Wageningen

The Second International Symposium on Iron in Soil Relationships was held at the International Agricultural Centre in Wageningen, The Netherlands on 1-3 May 1984.

Contents

Host-Plant Selection

Genetics and Evolution

Host-Plant Resistance and Application of Transgenic Plants

Multitrophic Interactions

Proc. 8th Int. Symp. Insect-Plant Relationships, Dordrecht: Kluwer Acad. Publ.
S.B.J. Menken, J.H. Visser & P. Harrewijn (eds), 1992

Introduction

Initially a mainly European affair in 1958, the Symposia on Insect-Plant Interactions have developed into an international forum where researchers in both fundamental and applied entomology can meet with other scientists. It was a great pleasure for the Organizing Committee of this 8th Symposium to welcome over 180 participants from 26 countries from all over the world: unfortunately, contributions from the developing countries fell short of expectations.

The proceedings show the progress this field has made since the previous symposium in 1989 in Budapest. This volume follows the symposium program quite strictly. Papers are organized along the five major symposium topics: insect-plant communities, host-plant selection, genetics and evolution, host-plant resistance and application of transgenic plants, and multitrophic interactions. Besides seven invited papers and a paper with concluding remarks, this volume contains the short communications of all 115 oral presentations and posters. Included are also the summaries of four European Science Foundation workshops held over the past two years where European scientists discussed the state of the art and the future of major topics in insect-plant interactions in order to develop better integrated research programs. This occurred concomitant with a further political and economical integration of Europe. It is just 150 km from Maastricht to Wageningen.

The field of insect-plant interactions nowadays includes almost all of biology, as well as parts of chemistry and physics. It takes a central position in biology because insects are the most abundant animal group, half of them are herbivores, and they dominate all terrestrial ecosystems. Knowledge of insect-plant interactions is thus fundamental to an understanding of the evolution of life on Earth.

Two major topics of world-wide concern give this field an extra dimension. First, large amounts of food crops are still lost due to insect pests. With the increasing concern for environmental pollution and the subsequent plans to drastically reduce pesticides, integrated pest management and development of resistant crops become a major focus in agriculture. The importance of the study of insect-plant relationships is thus continuously augmented. Clearly, successful pest control demands sufficient fundamental knowledge of pest-host interactions. Second, our work can contribute towards stopping or even counterbalancing the threatening biodiversity crisis thanks to an understanding of how the interaction of insects and plants has influenced and still influences the diversification and speciation (evolution) of both groups. These problems should, of course, be approached at a multitrophic level.

The editors would like to thank Raph de Rooij and Paul Piron for their extensive help in preparing the camera-ready manuscript, Albert Koedam for photographic procedures and Ninette de Zylva for correcting part of the English text. The assistance of Sandrine Ulenberg, Diny Winthagen, Marco Roos, Kees van Achterberg, and Sybren de Hoog is gratefully acknowledged.

The participants wish to dedicate this volume to Louis Schoonhoven, who recently retired from Wageningen Agricultural University, for his stimulation of the field of insect-plant relations in general and that of physiological entomology in particular, and for being chairman of this 8th Symposium on Insect-Plant Interactions.

The editors

Insect-Plant Communities

Proc. 8th Int. Symp. Insect-Plant Relationships, Dordrecht: Kluwer Acad. Publ.
S.B.J. Menken, J.H. Visser & P. Harrewijn (eds), 1992

The importance of herbivore population density in multitrophic interactions in natural and agricultural ecosystems

Donald R. Strong[1] and Stig Larsson[2]
[1] *Bodega Marine Laboratory, University of California, Bodega Bay, USA*
[2] *Dept of Plant and Forest Protection, Swedish Univ. of Agricultural Sciences, Uppsala, Sweden*

Key words: Herbivory, heritability, host plant resistance, insect outbreaks

Summary

Multitrophic interactions are an important aspect of evolution for insects on plants, with carnivorous insects comprising a large part of plant defense against herbivores. We explore a conflict between ecological and evolutionary aspects of tritrophic interactions. We reason that, by greatly lowering the densities of herbivores, carnivorous natural enemies can thwart the opportunities for selection of resistance to herbivory. With effective natural enemies, these opportunities should be fairly rare and restricted to odd periods of herbivore outbreak.

An example of a gallmidge outbreak and consequent population decline in a genetic experiment with willow saplings illustrates this idea. At the height of the outbreak, great additive genetic variance and heritability in resistance to the midge occurred among willow genotypes. With increasing mortality from natural enemies, gall densities decreased over several midge generations, and additive variance and heritability for willow resistance to the midge decreased in kind. At lowest gall densities heritability for resistance equalled zero. Endemic densities of the midge are usually much lower than even the lowest in the experiment, so heritability for resistance in nature is normally nil. Since no plants died, no selection took place among these saplings. However, seedlings can probably suffer heavy mortality from these galls. This is an example of an interspecific genotype by environment interaction, with herbivore density being the environmental factor that influences the expression of plant genotype. Without heritability, no response to selection and no evolution of resistance can take place. Thus, carnivorous natural enemies that suppress herbivore populations can protect plants ecologically but at the same time thwart opportunities for selection of herbivore resistance.

Introduction

The interactions of plants with herbivores do not proceed independently of other trophic levels. Carnivores and microbes have important roles in virtually every natural system of plants and herbivores, and in a very large fraction of agricultural systems as well. Over ten years ago, this theme was explored by Price *et al.* (1980), under the rubric of "tritrophic interactions". These authors observed with dismay that contemporary general theory was primarily made for two trophic levels, and was oblivious to the influences of carnivores upon herbivores and plants. Their paper came two decades after the much discussed hypothesis of Hairston *et al.* (1960), which carried a similarly multi-trophic message, *viz.* the remarkable fact that the "earth is green" is owing to the suppression of insect herbivore populations by carnivores. Hairston *et al.* (1960) asserted that these natural enemies limited herbivore populations, in general, to levels sufficiently low that the terrestrial world was saved from being grazed bare. The tritrophic argument of Price *et al.* (1980) was more evolutionary than that of Hairston *et al.* (1960), and envisioned plants to have evolved adaptations enlisting parasitoids, predators, and parasites into the "battery of defenses against herbivores." The prime example of this evolutionary encouragement is recruitment of ants, those "pugnacious bodyguards" of plants (Bentley, 1977; Vrieling *et al.*, 1991), but parasitoids can also fit this mold (Weis & Abrahamson, 1986; Price & Clancy 1986).

In this paper we will explore the prospect that, in some cases, natural enemies so reduce insect herbivore density that selection pressure by these herbivores upon the plant's internal defenses are virtually eliminated. Thus, the effectiveness of defenses external to the plant in suppressing herbivores can thwart the evolution of internal resistance to herbivory. When the rare outbreak of these herbivores occurs, there would be an abundance of susceptible genotypes in the plant population.

As an example, we will discuss resistance to herbivory by a gall midge in *Salix viminalis* L., the basket willow. This plant is cultivated for biomass fuel in parts of Europe, where it also grows wild. The data are the results of a serendipitous experiment, in which the midge unexpectedly colonized an elaborate common garden of willow saplings (Strong *et al.*, in press). The garden was planted with offspring of a partial factorial cross of willows meant to study wood productivity differences among genotypes; parents of the plant were feral. The gall midge *Dasineura marginemtorquens* Brem. (Diptera, Cecidomyiidae), quickly grew to outbreak densities then declined over four generations under heavy mortality from natural enemies, populations of which grew rapidly in tandem to the gall midge outbreak.

The midge oviposits on very young leaves of the terminal bud. Eggs hatch, and first instars search for galling sites as the leaves expand (Larsson & Strong, in press). *D. marginemtorquens* is a "pocket galler," causing the leaf to fold around the developing larvae in the fashion of a taco. Parasitoids and predators come and go through the ends of the folded leaf. The midge has from three to four discrete generations in central Sweden, each of which leaves a band of galled leaves interspersed between bands of ungalled leaves on the continuously growing willow shoots.

Methods

We defined resistance operationally, as the absence of galling by the midge. Genetic variation for resistance was determined for paternal and maternal half-sib families and for full sib families of willows. The garden was planted in the spring of 1988 with saplings grown indoors in late winter from stem cuttings of 240 willow clones. Each clone was replicated 10 times to yield 2400 separate plants in the garden. Eight seed parents ("mothers") and eight pollen parents ("fathers") were used in the partial factorial cross that produced the clones. The parents were feral plants from Sweden and Holland that had been transplanted to the Department of Forest Genetics for the crosses. Six full sibs from each of 40 (of the 64 possible) crosses were in the garden. Thirty offspring for each of the eight willow fathers and each of the willow mothers comprised the 240 clones, with each paternal half-sib family representing five mothers and each half-sib maternal family representing five fathers. Each father and mother were represented equally among the 240 offspring clones. The garden was divided into two halves, corresponding to high and low nutrients for the willows. Each half was divided into five parallel replicate areas, with one plant from each of the 240 clones planted at a random grid position in each replicate. Thus, there were five complete plot replicates for each nutrient treatment.

Results

The plot was in prime habitat for both basket willow and the insects. Even though in a "common garden," the environment was certainly not artificial or strange to these organisms. The outbreak of *D. marginemtorquens* occurred spontaneously in the plot in 1988, either from a few hitchhiking immatures on the cultivated plants or from adult immigrants flying in from nearby fields of the plant (Fig. 1). After 1988, midge densities declined through the three generations of 1989. By the first generation of 1990, gall

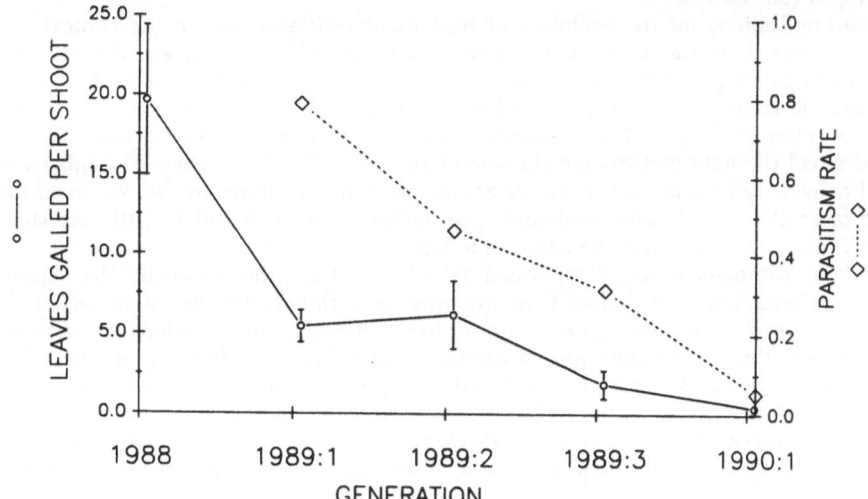

Figure 1. Level of galling of basket willow, *Salix viminalis* plants by the midge *Dasineura marginemtorquens* and combined parasitism rate of galls as the outbreak of the midge declined from 1988 through 1991. Substantial additional mortality due to predation by the bug *Anthocoris nemoralis* was suffered both by larval midges and larval parasitoids. Endemic gall densities are at least an order of magnitude lower than those observed at the end of the outbreak.

density had decreased to quite low levels. However, the level of 0.43 galled leaves per shoot in the spring of 1990 was at least an order of magnitude higher than endemic levels of the midge in the Uppsala area. Endemic densities of galls on feral and cultivated plants range from about 0.01 downward to 0.001 galled leaves per shoot (Larsson and Strong, pers. obs.), so natural levels of attack are usually much lower than those studied in the outbreak. The 1988 galls occurred mostly in the third generation of midges. The first generation had passed before the planting of the plot, and gall density of the second generation was fairly low.

Combined parasitism of individual *D. marginemtorquens* larvae and pupae by three hymenopteran species: *Aprostocetus abydenus* Walker, *A. torquentis* Graham (Eulophidae) and *Synopeas myles* (Wallner) (Platygastridae) had risen to approximately 80% in the first generation of 1989, when we began to study mortality from natural enemies. We measured neither predation nor parasitism rate in 1988. All parasitized individuals of the midge died. Parasitism decreased in a roughly linear fashion from the high level in the first generation of 1989 to approximately 45% in the second and 32% in the third generation of that year. Finally, combined parasitism was but 4% in the first generation of 1990 (Fig. 1). In addition to parasitism, larvae and pupae were also killed by the bug *Anthochoris nemoralis* L. (Anthocoridae), which had become quite abundant in the galls by the third generation of 1989. This predator also ate larvae and pupae of the parasitoids that were developing upon the midge larvae in the folded leaves. The high mortality from parasitoids and the predator is consistent with the inference that the outbreak of midges was quenched by natural enemies.

The genetic constitution of the willows in the plot did not change over the experiment. No plants died. The stem cuttings were all quite vigorous when planted, with well developed root masses.

The implications for the evolution of host plant resistance within the context of tritrophic interactions are shown, within generations, by differences in gall density among the willow genotypes during the course of the outbreak (Fig. 2). There were 13 extremely resistant clones that hardly experienced any galling at all (Fig. 2, bottom). Two full-sib families -- out of the total of 40 -- contained all of these 13 clones (Fig. 2, top). These two families had different mothers but the same father, from Växjö, Sweden. The full array of genetic diversity in herbivore resistance among all clones is shown by Fig. 2b. In addition, three more clones had substantial resistance, falling between 5 and 10 galls per stem in 1988. These three clones had the father from Uppsala.

In brief summary of the complicated set of data from this outbreak, the extremely resistant clones were not "induced" in any way by herbivory because stem sections that had been stored from plants grown with no insect attack produced foliage as resistant to the midge as those in the outbreak. In laboratory experiments, ovipositing midges did not discriminate among extremely resistant and susceptible clones. Eggs hatched abundantly on both, but first instar larvae failed, by and large, to initiate galls and died on the resistants (Larsson & Strong, in press). However, we have no evidence either in favor or against induction of the remaining clones which define the upper mode of resistance (Fig. 2).

Galling differences were not a function of vigor, so the variable "leaves galled per shoot" does reflect resistance rather than leaf production differences among genotypes. The paternal, maternal, and interaction genetic variances for leaves galled per shoot were significant with very low p values in generations 1988, 1989:1, 1989:2 for both high and low nutrients, and in generation 1989:3 for high nutrients, but were not significant in generations 1989:3, low nutrients or in 1990:1, high nutrients (no data for low nutrient replicates were available in 1990:1). Maternal inheritance did not figure substantially in the patterns of resistance, and maternal half-sib heritabilities were roughly commensurate with paternal ones. Though interaction variances of fathers x mothers were significant in most instances, in no instance did this interaction contribute more than 5% of explained variance, in components of variance analysis. So, the additive variance for resistance was a very large part of the genetic variability for this character among the plants in the garden. The extreme resistance has indications of being caused by a single gene or set of closely-linked genes. There were females among the extremely resistant willows, so sex of the host plant did not appear to restrict expression of this trait.

Substantial quantitative genetic variability in resistance, independent of the extreme resistance, occurred among the plants of the experiment. This is suggested both in 1988 among full-sib families (Fig. 2, top) and among clones (Fig. 2, bottom), where an upper mode is clearly defined above the 13 extremely resistant plants, below. Between about 40% and 60% of the heritability in the entire sample remains when the offspring with the Växjö father are excluded from the analysis; paternal and maternal variances are mostly

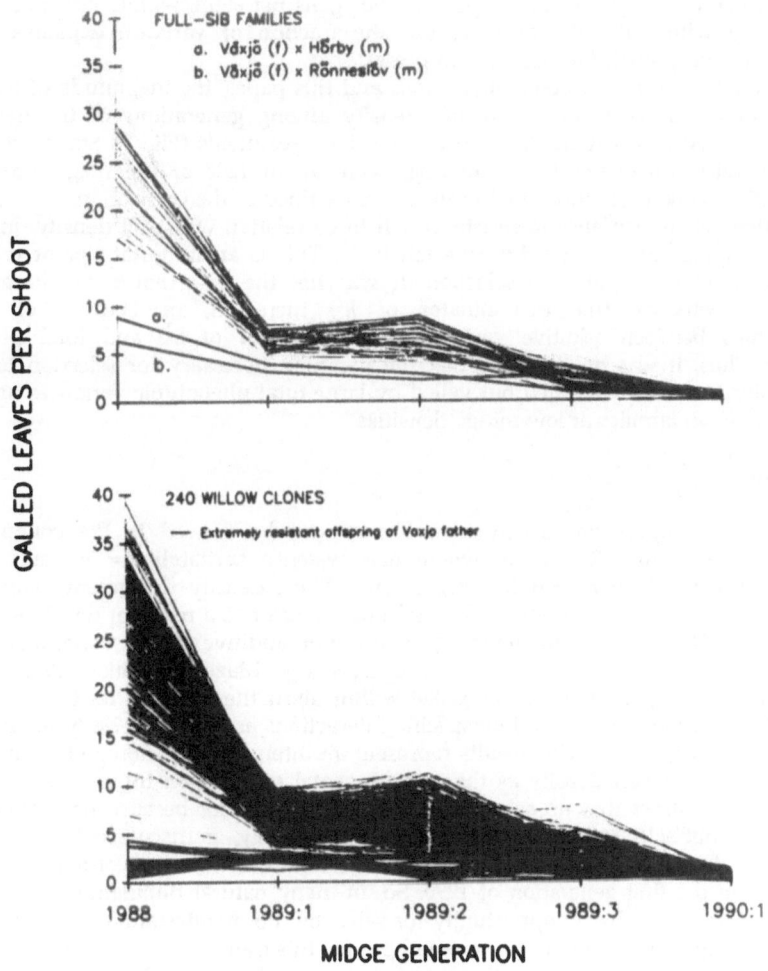

Figure 2. Galled leaves per shoot of the 40 basket willow full-sib families (top) and 240 clones comprising these families (bottom) during the midge generations occurring from 1988 through the beginning of the growing season in 1990; high nutrient treatment only (low nutrients gave a similar picture). The full-sib families with the father from Växjö contained all of the extremely resistant clones (a top, father from Växjö and mother from Hörby; b top, father from Växjö and mother from Rönneslöv). None of the other seven fathers contributed any extremely resistant offspring. Genetic differences in resistance among the genotypes were most apparent in 1988 at the height of the outbreak. The progressive reduction in distinction among willow genotypes that developed as mean gall density declined was not due to plant mortality; none died over the course of the study.

significant even without the families with the Växjö father. Nutrient treatment did not much affect the resistance of these willows to *D. marginemtorquens*. Nutrients greatly affected growth rates of the plants but they did not affect the fraction of leaves galled per stem or the number of galls per leaf. Willow fathers do not interact statistically with nutrient treatment in terms of galls per leaf. For galls per stem, willow fathers do interact significantly with nutrient treatment, but the fraction of variance explained by this interaction is very small (Strong *et al.*, in press).

For the issues of multitrophic interactions and this paper, the magnitude of heritability for resistance was correlated with gall density among generations of the insect. This pattern occurred similarly in high and low nutrient treatments (Fig. 3). Standard errors of the heritability, calculated by jackknifing, were about half of the magnitudes of the heritabilities, so no exact functional form can be ascribed to the correlation.

Additive genetic variance in resistance fell in correlation with gall density, just as did total phenotypic variance, yielding the fall in h^2. This is an important point relevant to density-dependent response to selection. It was not the case that h^2 fell because total phenotypic variance (the denominator of h^2) increased, or because of complex relationships between additive variance (the numerator of h^2) and total phenotypic variance. Thus, it was not the case that the variation necessary for selection to operate upon resistance was manifested but veiled by large total phenotypic variances among the paternal half-sib families at low midge densities.

Discussion

The effects of population density in evolution can be viewed in the context of the response to selection, $R = h^2 S$, where narrow-sense heritability $= h^2$, and selection differential $= S$ (Maynard Smith 1989, p. 110). Thus, density dependent evolution can proceed as a result of density-dependent selection, **and/or** as a result of density-dependent heritabilities. The effects of population densities upon additive genetic variation and h^2 are just beginning to be explored for single species (*e.g.*, Mazer & Schick, 1992), and our results with *D. marginemtorquens* on basket willow show the potential for both h^2 and S to be affected in evolution due to interspecific interactions as well, especially for interaction of herbivores with plants. Our results represent an interspecific genotype by environment interaction, with insect density as the environmental effect upon the expression of plant genotype. (We stress that no selection took place in the plot because no plants died or reproduced; but selection of this resistance is not unlikely, as discussed below). Endemic densities of *D. marginemtorquens* are much lower than even those recorded at the end of our study in the first generation of 1990. So, in many natural populations of the midge, there would be virtually no opportunity for selection, no manifestation of heritability and no selection differential among plant genotypes for this trait.

For multitrophic interactions, the result that natural enemies lower herbivore attack enough that resistance variation in the host population is not expressed adds another possibility to the list provided by Simms & Fritz (1992, p.358) for evolution of tritrophic interactions. First in the list is that "plant resistance may enhance the impact of natural enemies either through chemical cues that elicit parasitoid search or by prolonging herbivore development and increasing the chance of parasitism (synergism)." Second in the list is that "density-dependent effects could increase parasitism and/or predation on susceptible plants, which are expected to have higher herbivore densities." Third, is "enemies may not respond to variation in plant resistance, making enemy impact independent of and additive to, plant resistance." Indeed, any of these are possible for our system during the unusually high midge densities of the outbreak, although we have no data pertinent to them. On the surface, our results seem closest to the third possibility.

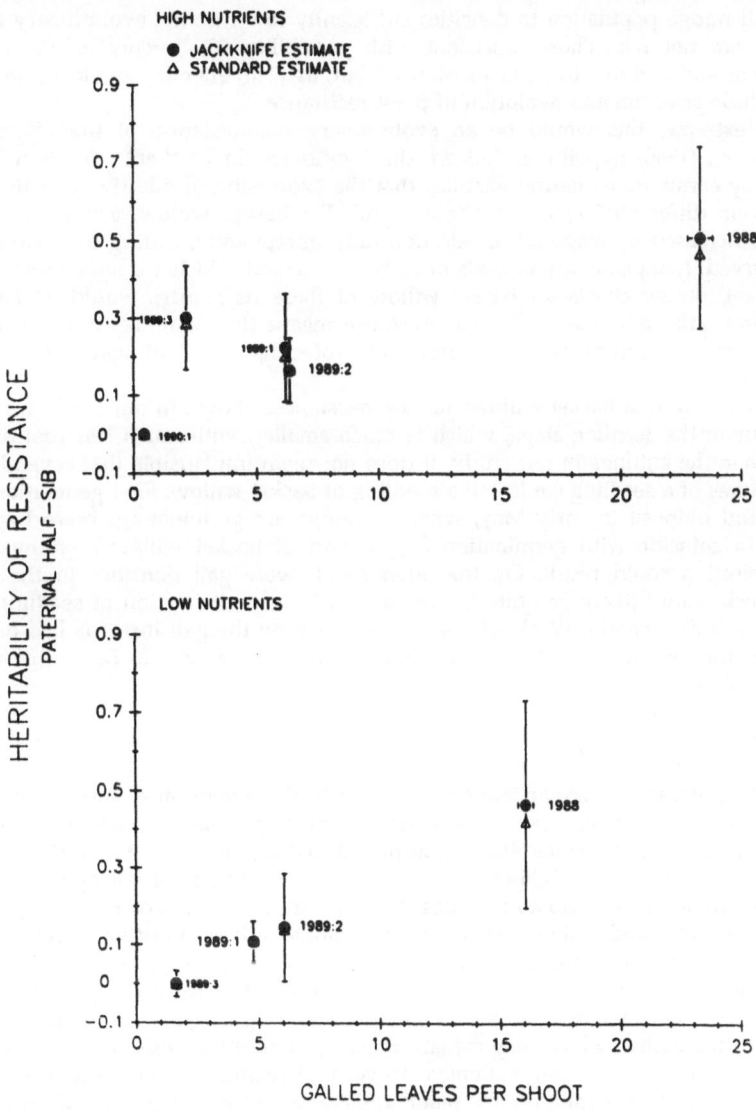

Figure 3. Heritability of resistance to *Dasineura marginemtorquens* in the full population of basket willow clones, as a function of gall density. Top, high nutrients. Bottom, low nutrients. Standard errors of heritability are calculated by jackknifing paternal half-sib families. All additive genetic variances for these data are statistically significant except for generation 1990:1 in high nutrients and generation 1989:3 in low nutrients. From Strong *et al.* (in press).

11

However, there is a key distinction in the fact that most of the time in most places, densities of the midge are so much lower than in the plot; we infer that natural enemies reduce gall midge population to densities sufficiently low that the evolutionary effects of herbivory are not felt. Thus, consistent with possibility #3, "enemy effects would be independent and additive to plant resistance," but, as well, enemies would be so effective as to preclude selection and evolution of plant resistance.

In the extreme, this would be an evolutionary manifestation of the effects of the Hairston et al. (1960) hypothesis. Indeed, the "world would be (kept so) green," most of the time, by carnivorous natural enemies that the expression of additive genetic variance and selection differential could be slight or nil. For basket willow, genes for resistance would be increased by response to selection only infrequently, during outbreaks like we have observed. Independently of costs of resistance (about which we have no information in this case), fitness effects for basket willow of these resistances would be temporally variable and quite infrequent. This circumstance means that some resistance variation is usually selectively neutral, at least in terms of protecting the plant against the herbivore in question.

Natural selection in basket willows for the resistances shown in our study would most likely occur in the seedling stage, which is much smaller, with much less root and shoot biomass than the cuttings in our study. It does not seem implausible that several galls on the few leaves of a seedling could kill a seedling of basket willow. First generation midges disperse and oviposit in early May, when seedlings are germinating. Were high midge densities to coincide with germination of a cohort of basket willow seedlings, intense natural selection could result. On the other hand, were gall densities in the endemic range, attack would likely be quite limited and only a small fraction of seedlings would be galled, in this scenario. Work with seedling attack by the gall insect is just beginning, and at this time we have no data on selection of these resistances to D. marginemtorquens in basket willow.

Conclusions

Multitrophic interactions are an important aspect in the evolution of insects on plants in both agricultural and natural ecosystems. Carnivorous natural enemies, predators, parasitoids, and diseases of herbivores can be enlisted in various ways by the plant in its protection. The list of possibilities for a scenario of evolution of tritrophic interactions offered by Simms & Fritz (1990) includes direct positive correlations between protections offered by enemies and endogenous plant protections, indirect positive correlations (with the higher herbivore densities on susceptible plant genotypes receiving greater proportional attack from enemies, due to some form of density dependent parasitism alone), and enemy attack independent of plant resistance genotype. We argue that there is yet a fourth possibility, which may explain the large amount of resistance variation found in natural and agricultural populations; carnivorous enemies are so effective in reducing herbivore attack that evolution of plant resistance occurs only sporadically, during infrequent outbreaks of the herbivore. This is an extension both of the venerable proposition of Hairston et al. (1960) and that of tritrophic interaction theory.

By way of example, we describe genetic work with basket willows during the collapse of a gallmidge outbreak on cultivated willows that was probably suppressed by parasitoids and predators. Great variance in resistance to the midge occurred among willow genotypes, and the outbreak effectively occurred only among a susceptible fraction of these plants. Heritability in resistance, quite high at the zenith of the outbreak, declined to zero as the outbreak waned. Thus, density-dependent evolution of resistance can occur as a result of **both** heritability and selection differential being density dependent in time.

Potentially potent selective pressures upon willow populations generated by midge galling are likely to be rare. We speculate that these selection pressures could happen during the coincident events of a mass seedling germination and an outbreak of the gall midge.

Acknowledgements

Göran Nordlander was of great help in the identification of the parasitoids.

References

Bentley, B.L. (1977). Extrafloral nectaries and protection by pugnacious bodyguards. *Annu. Rev. Ecol. Syst.* **8**: 407-427.

Hairston, N.G., F.E. Smith & L.B. Slobodkin (1960). Community structure, population control, and competition. *Am. Nat.* **44**: 421-425.

Larsson, S. & D.R. Strong (1992). Oviposition choice and larval survival of *Dasineura marginemtorquens* (Diptera: Cecidomyiidae) on resistant and susceptible *Salix viminalis. Ecol. Entomol..* (in press).

Maynard Smith, J. (1989). *Evolutionary Genetics.* Oxford: Oxford Univ. Press.

Mazer, S.J. & C.T. Schick (1991). Constancy of population parameters for life history and floral traits in *Raphanus sativus* L. II. Effects of planting density on phenotype and heritability estimates. *Evolution* **45**: 1888-1907.

Price, P.W., C.E. Bouton, P. Gross, B.A. McPheron, J.N. Thompson & A.E. Weis (1980). Interactions among three trophic levels: influence of plants on interactions between insect herbivores and natural enemies. *Annu. Rev. Ecol. Syst.* **11**: 41-65.

Price, P.W. & K.M. Clancy (1986). Interactions among three trophic levels: gall size and parasitoid attack. *Ecology* **67**: 1593-1600.

Simms, E.L. & R.S. Fritz (1992). The ecology and evolution of host-plant resistance to insects. *Trends Ecol. Evol.* **5**: 356-360.

Strong, D.R., S. Larsson & U. Gullberg (1992). Heritability of host plant resistance to herbivory changes with gallmidge density during an outbreak on willow. *Evolution* (in press).

Vrieling, K., W. Smit & E. van der Meijden (1991). Tritrophic interactions between aphids (*Aphis jacobaeae* Schrank), ant species, *Tyria jacobaea* L. and *Senecio jacobaea* L. lead to maintenance of genetic variation in pyrrolizidine alkaloid concentration. *Oecologia* **86**: 177-182.

Weis, A. & W.G. Abrahamson (1986). Evolution of host-plant manipulation by gall makers: ecological and genetic factors in the *Eurosta-Solidago* system. *Am. Nat.* **127**: 681-695.

Proc. 8th Int. Symp. Insect-Plant Relationships, Dordrecht: Kluwer Acad. Publ.
S.B.J. Menken, J.H. Visser & P. Harrewijn (eds), 1992

The evolution of plant resistance and correlated characters

Ellen L. Simms
Dept of Ecology and Evolution, University of Chicago, Chicago, Illinois, USA

Key words: Allocation trade-offs, compensatory growth, genetic correlation, herbivory, tolerance

Summary

The paradigm of plant herbivore coevolution suggests that determining the evolutionary response of plants to insect herbivory will improve our understanding of many aspects of plant and insect ecology. There are two potential evolutionary responses of plants to herbivory: resistance and tolerance. This paper examines problems inherent in measuring one factor involved in the evolution of plant resistance (the allocation cost of resistance) and then briefly considers the relationship between the evolution of tolerance and resistance in the common morning glory, *Ipomoea purpurea* Roth.

Introduction

Almost thirty years ago Ehrlich and Raven (1964) proposed that plants and the insects that feed on them coevolve. Subsequent work has tentatively supported this hypothesis (Berenbaum & Feeny, 1981; Berenbaum, 1983; Farrell *et al.*, 1991, but see Thompson, 1986, Miller, 1987, and Mitter *et al.*, 1988, for a critique of Berenbaum's thesis), suggesting that understanding how plants evolve resistance to herbivory may illuminate one of the causes of both plant and insect diversity. Furthermore, because insect herbivory affects plant growth and reproduction (Marquis, 1992), and because variation among plants in resistance to herbivory can influence the composition of consumer communities (Fritz, 1992; Karban, 1992; Simms & Fritz, 1990), understanding how plants evolve resistance to herbivory can enhance our understanding of plant and animal ecology as well. At the end of the paper, however, I will question the continued focus on plant resistance and suggest that other plant responses to herbivory, such as the evolution of tolerance, might have very different implications for the diversity and ecology of plants and insects.

In most models of the evolution of plant resistance to herbivores, costs of resistance are important determinants of equilibrium levels of resistance (Fagerström *et al.*, 1987; Simms & Rausher, 1987, 1989; Fagerström, 1989; Rausher & Simms, 1989). These models show that under certain circumstances stabilizing selection can maintain in a population an intermediate optimal level of resistance determined by a balance between the costs and benefits of resistance (Fig. 1). Consequently, measuring the cost of resistance is an important step towards predicting the level of resistance.

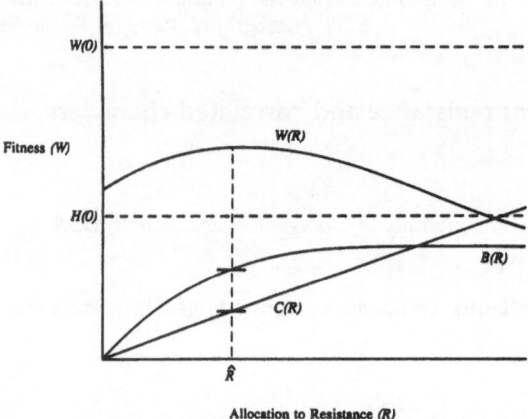

Figure 1. Optimality model for the calculation of cost and benefit of resistance. The fitness of the most susceptible genotype in the absence of herbivory is designated $W(0)$. The amount of fitness lost to herbivory by the most susceptible genotype is designated $H(0)$. The fitness cost of resistance, as a function of the resources allocated to resistance, is $C(R)$. $B(R)$ is the fitness benefit of resistance as a function of allocation. Fitness in the presence of herbivory, as a function of allocation to resistance, is described by $W(R)$. The point $\hat{R}$, corresponds to the genotype with maximum fitness. (Reprinted from Simms, 1992.)

Mechanisms of costs and methods for measuring them

A number of mechanisms may produce fitness costs of resistance to herbivory (Simms, 1992), but most verbal presentations of plant herbivore theory have focused on allocation costs, which are postulated to arise from the competitive allocation of resources to resistance and other fitness-enhancing functions (growth, maintenance, reproduction) (Chew & Rodman, 1979; Rhoades, 1979). This type of resistance cost is analogous to the allocation trade-offs believed to constrain the evolution of life-history traits (*e.g.*, Reznick, 1992).

Simms & Rausher (1987) argued that an allocation cost of resistance should be detected as a negative genetic correlation between fitness in the absence of herbivory and resistance. They reasoned that, if there is an allocation cost of resistance, this correlation should be negative because, in the absence of herbivory, plants allocating resources to resistance will not benefit from that resistance and should thus have lower fitness than plants not allocating resources to resistance (Fig. 2).

Evidence regarding allocation costs

In one study using this method, Berenbaum *et al.* (1986) reported large negative genetic correlations in wild parsnip (*Pastinaca sativa* L.) between inflorescence size (a component of fitness) in an herbivore-free greenhouse and production of several types of furanocoumarins that provide resistance to the parsnip webworm, *Depressaria pastinacella* (Dup.). In another pair of studies, however, Simms and Rausher (1987, 1989) eliminated insect herbivores with pesticides and found no genetic correlation between fitness of *Ipomoea purpurea* Roth (Convolvulaceae) in the absence of herbivory and resistances to four types of insect herbivores. This result also agreed with their measurements of

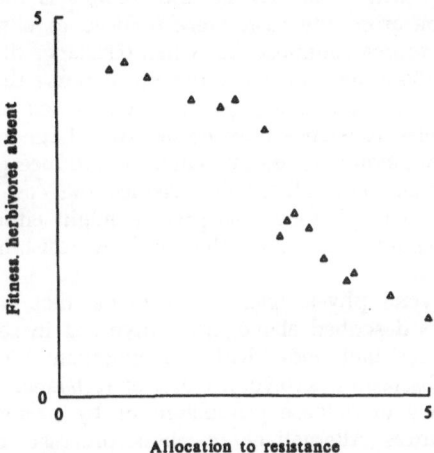

Figure 2. Genetic correlation method for detecting an allocation cost of resistance. The y-axis represents fitness in the absence of insect herbivores, w. The x-axis represents the resource allocation to resistance, R. An allocation cost is detected as $r_{R,w} \leq 0$.

selection on resistance to herbivory in the presence of insects (Rausher & Simms, 1989). Instead of finding stabilizing selection, they found directional selection favoring resistance to fruit damage and detected no evidence of selection on resistance to folivory.

In a review of the literature (Simms, 1992) concluded that costs of resistance exist in some circumstances but are either nonexistent or unmeasurable in other cases. This finding raises the interesting issue of why such an intuitively appealing assumption is sometimes unsupported empirically. Most of the remainder of this paper is devoted to examining this question. There are two possible answers: Either (1) resistance is not costly, or (2) the genetic correlation method may not allow us to detect costs. I will address each answer in turn.

Why costs might be absent

There are two potential reasons that resistance may involve little or no cost. First, traits involved in resistance may have benefits other than those associated with reducing herbivory. For example, lupine alkaloids defend plants against insect herbivory but also reduce herbivory by molluscs and vertebrates (Wink, 1984b), inhibit pathogen infection (Wink, 1984a; Wippich & Wink, 1985) and growth of neighboring plants (Wink, 1983), and function in nitrogen transport and storage (Wink & Witte, 1985). If these alternative benefits outweigh the allocation costs of traits involved in herbivore resistance, then there may be no allocation costs associated with the function of these traits in herbivore resistance. For testing the validity of models of coevolution between plants and insect herbivores, measures of allocation costs of resistance traits should be restricted to those directly due to the function in plant defense against insect herbivory. Detecting costs as the negative genetic correlation between fitness in the absence of insect herbivores and resistance includes only costs that can be charged directly to the resistance function of these traits.

Another explanation for why costs may be undetectable is that the traits may have been costly when they first arose, but costs were reduced or eliminated by subsequent selection. To illustrate this process, suppose that when resistance first arises as a new trait, it involves a substantial allocation cost. Nevertheless, because there is a net benefit to resistance, the trait will spread and the population will reach an equilibrium level of resistance. However, because resistance now occurs in a large number of individuals, there is opportunity for mutation and recombination to produce genetic variation in the cost. Selection will favor any new alleles that reduce the cost of resistance without reducing its benefit. Given enough time, this process might either reduce or eliminate costs of resistance, and could even allow the level of resistance to increase in the population.

One could imagine several physiological mechanisms that could mitigate costs of resistance. For example, as described above, traits involved in resistance could acquire new functions, causing resistant individuals to maintain high fitness relative to susceptibles even in the absence of herbivory. Costs of resistance might also be reduced by increasing the efficiency of defense production or by performing other necessary functions with fewer resources. Alternatively, catabolic processes could arise that permit recycling of resources from resistance into other fitness-enhancing functions, potentially allowing resistance compounds to function as storage molecules for important nutrients such as nitrogen or carbon. Finally, if resources are allocated to resistance only when defenses are mounted, then inducible resistance could reduce costs. This mechanism will also complicate attempts to measure the allocation cost of resistance, as I will describe below.

Although we have no evidence for evolution in the magnitude of trade-offs involved in resistance to herbivory, two examples from other systems are particularly compelling. McKenzie and colleagues (1980) reported evolutionary change in the cost of single-gene resistance to the pesticide diazinon in the Australian sheep blowfly, *Lucilia cuprina* (Wied.). In 1969-1970 the resistance allele reduced blowfly fitness in the absence of diazinon (unpublished results by Arnold & Whitten, reported in McKenzie *et al.*, 1980), suggesting that the resistance allele should decline in frequency after discontinuing diazinon. By 1977-1978, however, field surveys found that most *L. cuprina* populations were close to fixation for the resistance allele (McKenzie *et al.*, 1980). McKenzie and colleagues (1982) solved this paradox by discovering that most populations no longer exhibited a cost of resistance to diazinon. In a series of elegant backcross experiments, they demonstrated that the cost of resistance had been reduced by the evolution of modifiers at other loci.

Lenski (1988a, 1988b) also showed that the cost to *Escherichia coli* (Migula) Castellani & Chalmers of resistance to virus T4 can evolve. He discovered three distinct genetic mechanisms whereby costs of virus resistance were reduced. In one case he found several different resistance alleles at the same locus, each with different accompanying costs (1988a). In the same experiment, he also found other loci with resistance mutations that involved lower costs. Finally, Lenski (1988b) found that fitness modifiers could reduce the costs of viral resistance in *E. coli* in much the same manner as occurred for costs of resistance to diazinon in *L. cuprina*.

Why costs might go undetected

Clearly, there are plausible explanations for why costs of resistance to herbivory might be undetectably small or absent. However, as is often the case for negative results, such an unexpected outcome inevitably leads one to question the experimental methodology used to produce it.

There are several reasons that the correlation method might fail to detect existing allocations costs of resistance. I will discuss two that have interesting biological implications, beginning with inducible resistance. In addition to reducing the actual cost of resistance, inducible resistance also violates a critical assumption of the correlational method for detecting allocation costs. Specifically, the fitness of plants in the absence of herbivory may not reflect allocation costs of resistance that is mustered only in response to herbivory. If one can artificially induce resistance without damaging the plant, as was elegantly accomplished by Baldwin *et al.* (1990), then an allocation cost of resistance may be detected as the genetic correlation between the fitness of induced but undamaged plants and the resistance level of induced plants. However, inducible resistance presents a host of new complexities. For example, if resistance is induced only once, then its cost may be underestimated, but inducing resistance continuously could overestimate costs. Thus, to measure the ecologically relevant allocation cost of inducible resistance, one must determine the mean frequency of defense induction. Inducible resistance also involves cost due to mechanisms not relevant to constitutive resistance. For example, the plant must gather information about when to mount resistance and this is most likely done by incurring damage. Information gathering is a cost attributed to many switchable or plastic phenotypes (Van Tienderen, 1991), which suggests that the evolution of inducible resistance should be examined using the theoretical framework erected to understand the evolution of phenotypic plasticity (Sultan, 1987; Via & Lande, 1985; Van Noordwijk, 1989; Van Tienderen, 1991).

Another interesting biological reason that the correlational method for detecting costs may fail is that the genetic correlation between fitness in the absence of herbivory and resistance depends critically on two factors: (1) variation in allocation to resistance and, (2) variation in the resource pool from which allocations are made. This assertion can be understood by recasting the cost-benefit model more explicitly as a resource-allocation model of the type used for studying trade-offs in life-history traits (Riska, 1986, Van Noordwijk & de Jong, 1986). Suppose that the total resources available to a plant, called T, can be partitioned between two functions (Fig. 3). Resources can be allocated either to

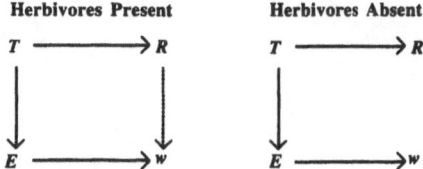

Figure 3. Resource allocation model for the allocation trade-off between fitness and resistance. The total resources available, T, may be allocated either to resistance, R, or to other fitness-enhancing functions, E. The proportion of resources allocated to R is c. The remaining proportion, $1-c$, is allocated to E. Hence, $R = cT$ and $E = (1-c)T$. In the presence of herbivores, both R and E contribute to fitness, w. In the absence of herbivores, resources allocated to R do not contribute to fitness, w.

resistance, R, or to other fitness-enhancing functions such as growth or seed production, which I shall call E. Suppose that the proportion of T allocated to R is determined by a proportionality term, c. Presumably, in the absence of herbivory the portion of resources allocated to R does not contribute to fitness, w, and w is therefore determined directly by E. The allocation trade-off embodied in this model can be detected as $r_{R,w}$, the genetic correlation between resources allocated to fitness in the absence of herbivory and resources allocated to resistance. To avoid charging to resistance all allocation costs of traits with multiple functions, the level of allocation to resistance should be estimated by bioassay. Thus, this correlation is equivalent to the one recommended by Simms & Rausher (1987) for detecting allocation costs of resistance.

From the allocation model (Figure 3), $\mathrm{cov}(R,w) = c(c\text{-}1)\,\mathrm{var}(T) - \mathrm{var}(c)[T^2 - \mathrm{var}(T)]$ where T is the average resource availability (Van Noordwijk & de Jong, 1986; Riska, 1986). When most variation in R is due to variation in c, then a negative correlation between fitness in the absence of herbivory and resistance will result (Fig. 4a). In contrast, when most variation in R is due to variation in T, the correlation will be positive (Fig. 4b).

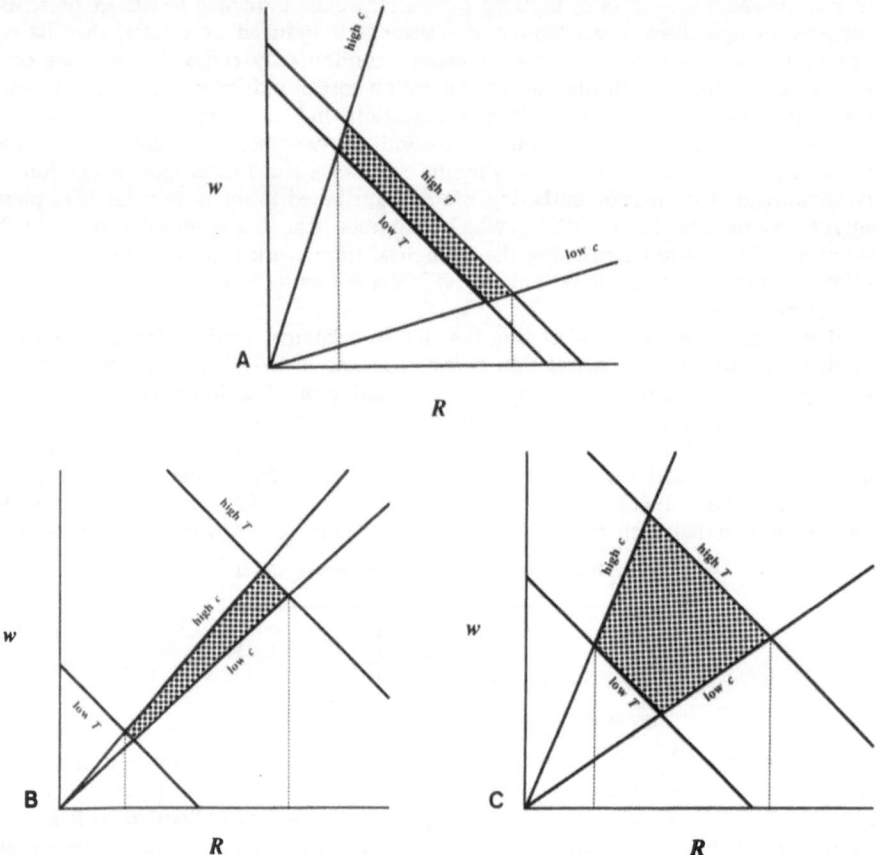

Figure 4. Variation in allocation to resistance, R, can be caused by variation in either the total resources available for allocation, T, or the proportion allocated to resistance, c. The line representing c has positive slope and passes through the origin, T has a slope of negative one because the total amount of resource available for both R and w sums to one. In each panel, R has equal range and variance. The shaded region represents potential genotypic values in a population with the given variation in c and T. A. If variance in R is due mainly to variance in c, then $r_{R,w}$ the genetic correlation between fitness in the absence of herbivores, w, and allocation to resistance, R, will be negative. B. If variance in R is due mainly to variance in T, then $r_{R,w}$ will be positive. C. If variance in R is due to similar amounts of variance in both c and T, then $r_{R,w}$ will be close to zero. (Redrawn from Van Noordijk & de Jong, 1986.).

20

Finally, when R varies due to similar amounts of variation in both c and T, then the correlation may be zero (Fig. 4c). One interesting and non-intuitive prediction of this model is that when c and T both vary, for a given level of variance in c and T, the correlation is more likely to be negative when T is high (Fig. 5). Thus, the trade-off is more likely to be detected under high resource conditions.

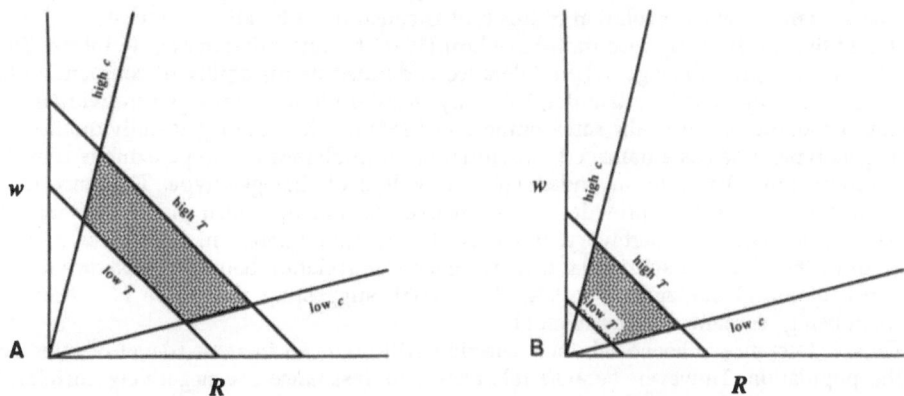

Figure 5. When both the total resources available for allocation, T, and the proportion allocated to resistance, c, vary, then for a given level of variation in T and c, the correlation $r_{R,w}$ is more likely to be negative when the mean amount of resource available for allocation is high (panel A) than when the mean is low (panel B).

Houle (1991) developed a model that assumes that all three traits (c, T, and R) are polygenically controlled and at equilibrium due to mutation-selection balance. It suggests that if the number of loci contributing to variance in T is large relative to the number contributing to variance in c, then $r_{R,w}$ may be positive. Houle used information on biochemical pathways to argue that resource acquisition might commonly be influenced by more loci than allocation. The model also suggests that stabilizing selection for an optimal allocation of resources to resistance will reduce the number of loci contributing variance to c and thus lead to near zero genetic covariances between fitness in the absence of herbivores and resistance. Thus, the very types of selection pressures predicted by the optimality model of evolution of plant resistance may lead to an equilibrium situation in which $r_{R,w}$ is either zero or positive.

How can this problem be dealt with? One possibility would be to measure variation in acquisition ability and "factor" it out of the correlation analysis (*e.g.*, Futuyma & Phillipi, 1987). For example, one could use plant size to estimate acquisition ability, and then calculate the correlation between fitness in the absence of herbivory and resistance while controlling for plant size. This method might reveal a previously undetected negative correlation, although J. Fry (pers. comm.) has argued that it still does not permit using the lack of a correlation as evidence for the absence of a trade-off. This argument holds because, as outlined by Pease and Bull (1988) and elaborated upon by Charlesworth (1990), if trade-offs responsible for the cost of a trait such as resistance to herbivory involve more than two traits, then particular pairs of traits may be positively correlated without alleviating the constraint imposed by the trade-off. Instead, the trade-off may be manifested by negative genetic correlations between other traits involved in the trade-off.

We can look to *Ipomoea purpurea* for an example where recent information suggests the presence of a complex functional constraint that may limit the evolution of resistance. In addition to the insect folivory and fruit damage on *I. purpurea* studied by Simms & Rausher (1989), Fineblum (1991) reported damage to apical meristems by generalist insect herbivores. Resistance to this damage varies among inbred lines and thus is genetically variable. Furthermore, Fineblum did not detect a negative correlation between fitness in the absence of herbivores and resistance to apical damage, even though she measured fitness in the absence of herbivores in both the greenhouse and in the field. Thus, there was no evidence that the evolution of this trait is constrained by allocation costs.

In addition to this resistance trait, Fineblum (1991) found variation among inbred lines in tolerance to apical damage, where tolerance is defined as the ability to compensate for damage caused by herbivory (Painter, 1958). By this definition, a damaged individual of a tolerant genotype produces the same number of seeds as an undamaged individual of the same genotype, whereas a damaged individual of an intolerant genotype exhibits reduced seed production relative to an undamaged individual of that genotype. Tolerance is an interesting trait because it provides an alternative method by which plants can mitigate the deleterious effects of herbivory. But it is of particular interest in *I. purpurea* because Fineblum (1991) found a substantial negative genetic correlation between resistance to and tolerance of apical damage ($r = -0.94$, $P < 0.005$), suggesting that there is a trade-off between being resistant or being tolerant.

Thus, if tolerance is beneficial, then selection will favor an increased level of tolerance in the population. However, because tolerance and resistance are negatively correlated, selection favoring increased tolerance will cause a reduction in the population mean level of resistance, which selection on resistance is likely to oppose. Thus, although there is no evidence for direct allocation trade-offs of either resistance or tolerance, it appears that in the *I. purpurea* population from which these inbred lines were established, both traits may be maintained at intermediate optima by stabilizing selection due to another trade-off, the opposing benefits of increased tolerance and increased resistance. It is not yet known whether this trade-off is rooted in competitive resource allocation or is due to another mechanism (Fineblum, 1991).

Conclusions

The issues I have discussed here lead to several conclusions and suggestions for directions of future research. First, more empirical evidence is needed to determine whether allocation costs associated with defensive function can constrain the evolution of resistance to herbivory. Although several studies have detected evidence for allocation costs of resistance, such trade-offs are not ubiquitous. Given that there are clear theoretical reasons for why allocation costs of resistance might be small or non-existent, we must broaden our focus to elucidate the circumstances under which allocation costs can limit the evolution of plant resistance and when other factors may be important.

Thus, the question should not be whether there is an allocation trade-off; each plant has a certain pool of resources available and these must be allocated to various competing functions, including resistance. The appropriate question is instead: How often does this allocation trade-off constrain the evolution of increased resistance in a plant population? The genetic correlation is a statistical abstraction that describes the average effects of large numbers of factors contributing to the mean phenotype in a population (Riska, 1989). As such, it can encompass the allocation trade-off but also many other aspects of the phenotype that may be constraining the evolution of plant resistance. In a broad sense, evolutionary constraints are created by the physical limitations imposed on the phenotype by both the environment and the configuration of the genetic information encoding the phenotype. Therefore, they involve the entire phenotype and are unlikely to be restricted to allocation trade-offs. The genetic correlation method will be most successful, therefore, when it is used in conjuction with functional analysis (of physiology, morphology, insect behavior, and even genetics) to determine the mechanisms of particular constraints.

In addition to broadening our views about trade-off mechanisms, the last example I provided also emphasizes the importance of moving beyond the narrow confines of plant resistance to ask broader questions about how plants respond to herbivory. As Fineblum's (1991) work shows, plant response can include resistance, tolerance, or a combination of the two traits.

Recognizing the diversity of plant responses to herbivory will allow us to extend the paradigms used to understand plant-insect coevolution. For example, a focus on plant resistance has centered attention on the arms race paradigm, leading to expectations of ever increasing plant defense and herbivore offense. How would Ehrlich and Raven's view of plant-insect coevolution have differed if they had instead viewed the evolution of tolerance as the dominant mode of plant response to herbivory? The evolution of tolerance may allow a plant lineage to enter a new adaptive zone (one in which it is not affected by herbivory), but may not produce selection for host shifts in the herbivore, thereby reducing its chances of entering a new adaptive zone and subsequent diversification. What affect would this response have on rates of diversification and extinction in insects? Finally, tolerance in its extreme form produces a commensal relationship between plant and herbivore, and if plants overcompensate for herbivory, plant-herbivore mutualisms might even develop (Owen & Wiegert, 1976; McNaughton, 1979; Owen, 1980; Paige & Whitham, 1987; Vail, 1992).

Currently, tolerance appears to be rare in plants, but that perception may be due to the relative difficulty of detecting it. Resistance variation is easy to see in the field as variation in damage, whereas detecting tolerance requires experimental manipulations. To eliminate this methodological bias, future work must focus on the evolution of all potential plant evolutionary responses to herbivory.

Acknowledgements

This work was supported in part by BSR-9196188 from the U. S. National Science Foundation. The clarity of the manuscript was improved by the comments of an anonymous reviewer.

References

Baldwin, I.T., C.L. Sims & S.E. Kean (1990). The reproductive consequences associated with inducible alkaloidal responses in wild tobacco. *Ecology* 71: 252-262.

Berenbaum, M. (1983). Coumarins and caterpillars: a case for coevolution. *Evolution* 37: 163-179.

Berenbaum, M.R., A.R. Zangerl & J.K. Nitao (1986). Constraints on chemical coevolution: wild parsnips and the parsnip webworm. *Evolution* 40: 1215-1228.

Berenbaum, M. & P.P. Feeny (1981). Toxicity of furanocoumarins to swallowtail butterflies: escalation in a coevolutionary arms race? *Science* 212: 927-929.

Charlesworth, B. (1990). Optimization models, quantitative genetics, and mutation. *Evolution* 44: 520-538.

Chew, F.S. & J.E. Rodman (1979). Plant resources for chemical defense. In: G.A. Rosenthal & D.H. Janzen (eds), *Herbivores: Their Interactions with Secondary Plant Metabolites*, pp. 271-307. New York: Academic Press.

Ehrlich, P.R. & P.H. Raven (1964). Butterflies and plants: a study in coevolution. *Evolution* 18: 586-608.

Fagerström, T. (1989). Anti-herbivory chemical defense in plants: a note on the concept of cost. *Am. Nat.* **133**: 281-287.

Fagerström, T., S. Larsson & O. Tenow (1987). On optimal defense in plants. *Funct. Ecol.* **1**: 73-81.

Farrell, B.D., D.E. Dussourd & C. Mitter (1991). Escalation of plant defense: do latex and resin canals spur plant diversification? *Am. Nat.* **138**: 881-900.

Fineblum, W.L. (1991). Genetic constraints on the evolution of resistance to host plant enemies. Ph.D. dissertation. Duke University: Durham, N.C.

Fritz, R.S. (1992). Community structure and species interactions of phytophagous insects on resistant and susceptible host plants. In: R.S. Fritz & E. L. Simms (eds), *Plant Resistance to Herbivores and Pathogens: Ecology, Evolution, and Genetics*, pp. 240-277. Chicago: University of Chicago Press.

Futuyma, D.J. & T.E. Philippi (1987). Genetic variation and covariation in responses to host plants by *Alsophila pometaria* (Lepidoptera: Geometridae). *Evolution* **41**: 269-279.

Houle, D. (1991). Genetic covariance of fitness correlates: what genetic correlations are made of and why it matters. *Evolution* **45**: 630-648.

Karban, R. (1992). Plant variation: its effects on populations of herbivorous insects. In: R.S. Fritz & E.L. Simms (eds), *Plant Resistance to Herbivores and Pathogens: Ecology, Evolution, and Genetics*, pp. 195-215. Chicago: University of Chicago Press.

Lenski, R.E. (1988a). Experimental studies of pleiotropy and epistasis in *Escherichia coli*. I. Variation in competitive fitness among mutants resistant to virus T4. *Evolution* **42**: 425-432.

Lenski, R.E. (1988b). Experimental studies of pleiotropy and epistasis in *Escherichia coli*. II. Compensation for maladaptive effects associated with resistance to virus T4. *Evolution* **42**: 433-440.

Marquis, R.J. (1992). The selective impact of herbivores. In: R.S. Fritz & E.L. Simms (eds), *Plant Resistance to Herbivores and Pathogens: Ecology, Evolution, and Genetics*, pp. 301-325. Chicago: University of Chicago Press.

McKenzie, J.A., J.M. Dearn & M.J. Whitten (1980). Genetic basis of resistance to diazinon in Victorian populations of the Australian sheep blowfly, *Lucilia cuprina*. *Aust. J. Biol. Sci.* **33**: 85-95.

McKenzie, J.A., M.J. Whitten & M.A. Adena (1982). The effect of genetic background on the fitness of diazinon resistance genotypes of the Australian sheep blowfly, *Lucilia cuprina*. *Heredity* **49**: 1-9.

McNaughton, S.J. (1979). Grazing as an optimization process: grass-ungulate relationships in the Serengeti. *Am. Nat.* **113**: 691-703.

Miller, J.S. (1987). Host-plant relationships in the Papilionidae (Lepidoptera): parallel cladogenesis or colonization? *Cladistics* **3**: 105-120.

Mitter, C., B. Farrel & B. Weigmann (1988). The phylogenetic study of adaptive zones: has phytophagy promoted insect diversification? *Am. Nat.* **132**: 107-128.

Owen, D.F. (1980). How plants may benefit from the animals that eat them. *Oikos* **35**: 230-235.

Owen, D.F. & R.G. Wiegert (1976). Do consumers maximize plant fitness? *Oikos* **27**: 488-492.

Paige, K.N. & T.G. Whitham (1987). Overcompensation in response to mammalian herbivory: the advantage of being eaten. *Am. Nat.* **129**: 407-416.

Painter, R.H. (1958). Resistance of plants to insects. *Annu. Rev. Entomol.* **3**: 267-290.

Pease, C.M. & J.J. Bull (1988). A critique of methods for measuring life history trade-offs. *J. Evol. Biol.* **1**: 293-303.

Rausher, M.D. & E.L. Simms (1989). The evolution of resistance to herbivory in *Ipomoea purpurea*. I. Attempting to detect selection. *Evolution* **43**: 563-572.

Reznick, D. (1992). Measuring the costs of reproduction. *Trends Ecol. Evol.* **7**: 42-45.

Rhoades, D.F. (1979). Evolution of plant chemical defenses against herbivory. In: G.A. Rosenthal & D.H. Janzen (eds), *Herbivores: Their Interaction with Secondary Plant Metabolites*, pp. 3-54. New York: Academic Press.

Riska, B. (1986). Some models for development, growth, and morphometric correlation. *Evolution* **40**: 1303-1311.

Riska, B. (1989). Composite traits, selection response, and evolution. *Evolution* **43**: 1172-1191.

Simms, E.L. (1992). Costs of plant resistance to herbivory. In: R.S. Fritz & E.L. Simms (eds), *Plant Resistance to Herbivores and Pathogens: Ecology, Evolution, and Genetics*, pp. 392-425. Chicago: University of Chicago Press.

Simms, E.L. & R.S. Fritz (1990). The ecology and evolution of host-plant resistance to insects. *Trends Ecol. Evol.* **5**: 356-360.

Simms, E.L. & M.D. Rausher (1987). Costs and benefits of plant defense to herbivory. *Am. Nat.* **130**: 570-581.

Simms, E.L. & M.D. Rausher (1989). The evolution of resistance to herbivory in *Ipomoea purpurea*. II. Natural selection by insects and costs of resistance. *Evolution* **43**: 573-585.

Sultan, S.E. (1987). Evolutionary implications of phenotypic plasticity in plants. *Evol. Biol.* **21**: 127-178.

Thompson, J.N. (1986). Patterns in coevolution. In: A.R. Stone & D.L. Hawkworth (eds), *Coevolution and Systematics*, pp. 119-143. Systematics Association Special Volume 32. Oxford: Clarendon.

Vail, S.G. (1992). Selection for overcompensatory plant resonses to herbivory: a mechanism for the evolution of plant-herbivore mutualism. *Am. Nat.* **139**: 1-8.

Van Noordwijk, A.J. (1989). Reaction norms in genetical ecology. Studies of the great tit exemplify the combination of ecophysiology and quantitative genetics. *BioScience* **39**: 453-458.

Van Noordwijk, A.J. & G. de Jong (1986). Acquisition and allocation of resources: their influence on variation in life history tactics. *Am. Nat.* **128**: 137-142.

Van Tienderen, P.H. (1991). Evolution of generalists and specialists in spatially heterogeneous environments. *Evolution* **45**: 1317-1331.

Via, S. & R. Lande (1985). Genotype-environment interaction and the evolution of phenotypic plasticity. *Evolution* **39**: 505-522.

Wink, M. (1983). Inhibition of seed germination by quinolizidine alkaloids. *Planta* **158**: 365-368.

Wink, M. (1984a). Chemical defense of Leguminosae. Are quinolizidine alkaloids part of the antimicrobial defense system of lupins? *Z. Naturforsch.* **39C**: 548-552.

Wink, M. (1984b). Chemical defense of lupins. Mollusc-repellent properties of quinolizidine alkaloids. *Z. Naturforsch.* **39C**: 553-558.

Wink, M. & L. Witte (1985). Quinolizidine alkaloids as nitrogen source for lupin seedlings and cell cultures. *Z. Naturforsch.* **40C**: 767-775.

Wippich, C. & M. Wink (1985). Biological properties of alkaloids: influence of quinolizidine alkaloids and gramine on the germination and development of powdery mildew, *Erysiphe graminis* f. sp. *hordei*. *Experientia* **41**: 1477-1479.

Proc. 8th Int. Symp. Insect-Plant Relationships, Dordrecht: Kluwer Acad. Publ.
S.B.J. Menken, J.H. Visser & P. Harrewijn (eds), 1992

Habitat impact on insect communities of annual and perennial grasses

H.-J. Greiler and T. Tscharntke
Zoologisches Institut I, Universität Karlsruhe, Karlsruhe, Germany

Key words: Fallows, herbivores, life-cycle strategy, meadows, species richness

Variability of insect species richness is largely explained by major determinants like size of geographic host range of the plant and plant architecture, but a residual variation still remains unexplained (Strong *et al.*, 1984). Habitat characteristics play an important role, since annual grasses dominate in early succesional habitats and perennials in late succesional habitats (Brown & Southwood, 1987). We analyzed determinants of the variability of insect numbers on annual and perennial grasses and tested the hypotheses that variability of species richness of insect communities of grasses can be explained by (1) plant chemistry, (2) plant architecture, (3) life-cycle of grasses (annuals compared with perennials) or (4) by habitat characteristics.

Material and methods

Plant chemistry was characterized by analyzing contents of nitrogen, silicate, sugar, and starch of 4 annual grass species (*Alopecurus myosuroides* Huds., *Apera-spica venti* (L.), *Avena fatua* L. and *Bromus sterilis* L.) and 4 perennial grass species (*Agrostis stolonifera* L., *Elymus repens* L., *Dactylis glomerata* L. and *Lolium multiflorum* Lam.). Plant architecture was characterized by measuring the length, number of nodes and diameter of shoots of these grasses. The impact of habitat and life-cycle was tested experimentally in an agricultural region in southern Germany using the annual *A. myosuroides* and the perennial *D. glomerata* and using two habitat types: one year old fallows of previously cultivated arable land, and old meadows. Eight pots with the annual grass and 8 pots with the perennial grass were planted into each of five fallows and five meadows. In May and June, ectophytic pests of half of the grass pots were sampled by a suction sampler. Endophytic insects of the other half of the grass pots were sampled by dissection of stems.

Results and discussion

The content of nitrogen, minerals, silicate, sugar and starch of the shoots of annuals and perennials did not differ significantly nor did the plant architecture. In fact, the length, number of nodes and basal diameter of shoots of annual and perennial grass species showed great similarity.

In the case of ectophytes, appreciable variability in species richness of insects colonizing annuals and perennials was not caused by habitat characteristics or life-cycle. The mean number of ectophytic species infesting the annual *A. myosuroides* did not differ from the mean number of insects attacking *D. glomerata*, regardless of whether the habitat was a young fallow or an old meadow (Fig. 1a).

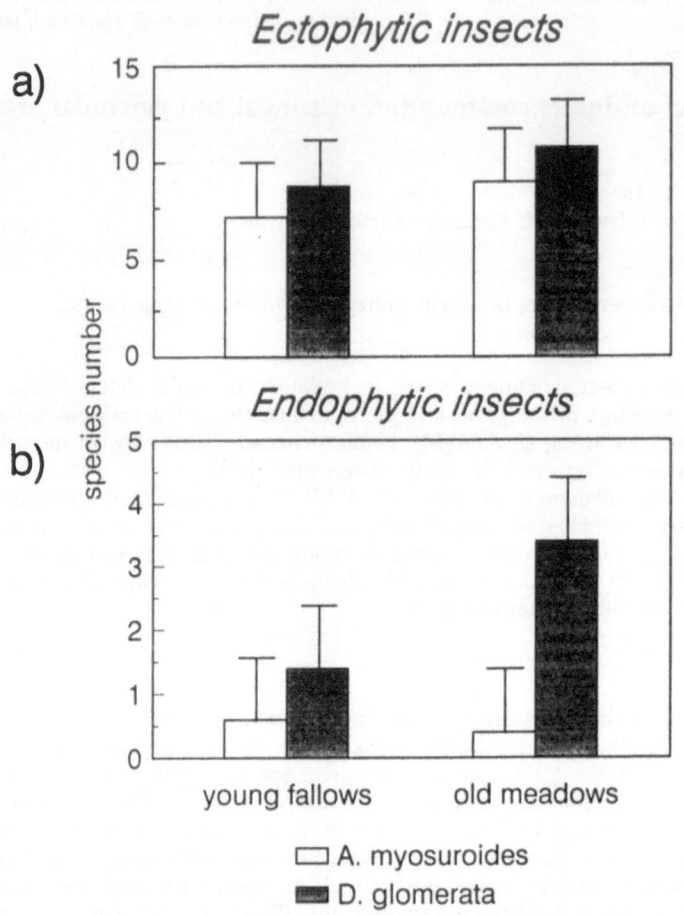

Figure 1. Species numbers of (a) ectophytic and (b) endophytic insects on *A. myosuroides* and *D. glomerata* in relation to habitat type (young fallow-old meadow) and life-cycle strategy (annual-perennial). Arithmetic mean and 95% confidence limits are given.
a) $F_{3,16} = 1.2$, $P = 0.34$; b) $F_{3,16} = 8.0$, $P = 0.0018$.

In contrast, the impact of habitat and life-cycle proved to be important for endophytes. Stems of the annual grass were infested by only one endophytic gall-midge, whereas stems of the perennial grass had a total of 10 stem-feeding species (Greiler & Tscharntke, 1991). The mean number of insect species attacking *D. glomerata* increased with the age of the habitat and was always greater than the mean number of insect species attacking *A. myosuroides*, irrespective of the habitat. In contrast, the incidence of the gall-midge infesting the annual grass was not influenced by habitat (Fig. 1b).

The impoverished community of endophytes on *A. myosuroides* could be explained by the scanty appearance or predictability of annuals as a result of their life-cycle, whereas habitat characteristics explained differences in endophytes' colonization of the perennial grass (the "*D. glomerata* on meadows" mean is significantly different from the three other groups: LSD-test).

References

Brown, V.K. & T.R.E. Southwood (1987). Secondary succession: Patterns and strategies. In: A.J. Gray, M.J. Crawley & D.J. Edwards (eds), *Colonization, Succession and Stability*, pp. 315-337. Oxford: Blackwell Sientific Publications.
Greiler, H.-J. & T. Tscharntke (1991). Artenreichtum von Pflanzen und Grasinsekten auf gemähten und ungemähten Rotationsbrachen. *Verh. Ges. Ökol.* 20/1: 429-434.
Strong, D.R., H.J. Lawton & T.R.E. Southwood (1984). *Insects on Plants*. Oxford: Blackwell Scientific Publications.

Proc. 8th Int. Symp. Insect-Plant Relationships, Dordrecht: Kluwer Acad. Publ.
S.B.J. Menken, J.H. Visser & P. Harrewijn (eds), 1992

Predispersal seed predation in the limitation of native thistle

Svata M. Louda, Martha A. Potvin and Sharon K. Collinge
University of Nebraska, Lincoln, Nebraska, USA

Key words: *Cirsium canescens*, insect herbivores, plant competition, vertebrates

Invertebrate herbivores, such as insects, have been hypothesized to be less important than vertebrates in plant population dynamics (Crawley, 1989). This generalization, if true, is an important one. However, few experimental data are available on the role of invertebrates in the recruitment and density of native plants. Also, few studies analyze both interactions simultaneously. The purpose of this study was to concurrently test the importance of three biological interactions that may limit plant densities. In three concurrent experiments, we excluded insects or vertebrates, or altered the level of seedling competition with established grasses.

The study plant, Platte thistle (*Cirsium canescens* Nutt.), is a native monocarpic species of sand prairie in the middle U.S.A. This plant and its interactions lend themselves to generalization for several reasons. First, the study builds on detailed studies of *Cirsium* biology and interactions. Second, short-lived perennial plants with fugitive life history strategies, such as thistles, are important in the maintenance of biological diversity in grasslands; yet, we know little about consumer effects on their population dynamics. And, third, many species in the genus are problematic weeds, so the results should have immediate application.

Results

In the first experiment, we studied the impact of flower- and seed-feeding insects by reducing damage to developing flowerheads with insecticide. We compared the performance of insecticide-treated plants with control plants treated with water or with nothing (N = 11, 12, 8 replicates/treatment). Predispersal consumption of flowers and developing seed caused a significant decrease in the seed matured (Fig. 1). A three-fold reduction in seed by insects led to a six-fold decrease in seedling density. No compensation in seedling survival occurred; the insect-induced difference in numbers persisted and, in fact, grew during succeeding stages (Fig. 1). Reduction of insects eventually led to a three- to 37-fold increase in the number of progeny that reached maturity per test plant (Fig. 1; Louda & Potvin, unpubl.). The differences in recruitment and ultimate reproduction were established by the variation in insect damage. Note that the increases in seedling and subsequent adult densities in this experiment occurred in spite of ambient levels of both postdispersal seed predation and plant competition.

In the second experiment, postdispersal seed consumers were excluded with full cages, or allowed access in partial cage or no cage control treatments (N = 30 seeds/ treatment, 30 x 30 cm area/treatment, cage mesh = 1.27 cm). Replicates were placed in two habitats: open and amid established grasses (N = 5/habitat). Seedling establishment and survival were compared over two years. Germination was low in general (6.7%, full cage).

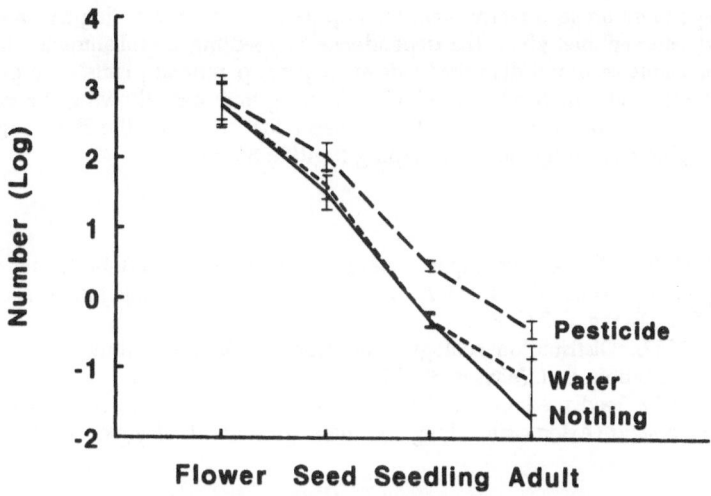

Figure 1. Number (log) of Platte thistle individuals in three treatments (insecticide spray, water-only spray, no spray; 1984-85 experiments combined) over sequential life history stages: Flower = total number of flowers initiated per plant; Seed = total number of viable seeds released per plant, after flower- and seed-predation by insects; Seedling = number of seedlings established per plant; Adult = number of progeny matured per plant by test plants as of May 1991.

Vertebrates did not have a definitive, statistically significant effect on seedling establishment overall (ANOVA: treatment effect $F_{2,10}$ = 2.53, P < 0.10). However, 80% of seedlings were in cages that excluded vertebrates. Postdispersal predators reduced seedling recruitment in the open, but not in the grass (ANOVA: location effect $F_{1,10}$ = 5.52, P < 0.02, treatment x location interaction $F_{1,10}$ = 4.95, P < 0.03). Postdispersal seed loss thus amplified the negative impact of insects in the open habitat.

In the third experiment, we transplanted seedlings into two treatments to test the effect of competition on seedling establishment. The treatments were: in open vs within openings in clones of Switchgrass, Panicum virgatum L. (N = 21 seedlings/treatment). Within 9 weeks, established grasses significantly reduced seedling survival to 4.8%, compared with 42.9% in the open (ANOVA, $F_{1,20}$ = 7.23, P < 0.02). After 2 years, 9.5% of the seedlings survived in open vs none in the vicinity of grasses. Thus, competition with established grasses supplemented the negative effect of insect herbivory and postdispersal seed predation by vertebrates.

Conclusions

These results have several important implications. First, the data suggest that flower- and seed-feeding insects had a greater impact than did vertebrates on the population dynamics of this plant, contrary to expectation. These results are consistent, however, with the only other set of experiments that directly test the effect of seed destruction on seedling establishment in situ (Louda, 1982, 1983). In both studies, seed was found to be more limiting to seedling establishment than safe sites. Second, a generalization suggested by these studies is that short-lived perennial plants with fugitive life history strategies are highly vulnerable to insect impact on their dynamics. Why might this be so? Short-lived

31

plants are dependent on seed recruitment for population persistence. Fugitive species may be particularly susceptible, given the dependence of seedling establishment on the joint probability of viable seed and disturbed safe-sites. Also, perennial plants may persist long enough to be relatively predictable to adapted insect granivores, allowing the cumulative development and maintenance of significant insect loads. Third, the data suggest that predation complements competition in shaping fugitive life histories.

References

Crawley, M.J. (1989). The relative importance of vertebrate and invertebrates herbivores in plant population dynamics. In: E.A. Bernays (ed), *Insect-Plant Interactions*, Vol. I, pp. 45-71. Boca Raton: C.R.C. Press.

Louda, S.M. (1982). Distribution ecology: Variation in plant recruitment in relation to insect seed predation. *Ecol. Monogr.* **52**: 25-41.

Louda, S.M. (1983). Seed predation and seedling mortality in the recruitment of a shrub, *Haplopappus venetus* (Asteraceae), along a climatic gradient. *Ecology* **62**: 511-521.

Proc. 8th Int. Symp. Insect-Plant Relationships, Dordrecht: Kluwer Acad. Publ.
S.B.J. Menken, J.H. Visser & P. Harrewijn (eds), 1992

Preference, tree resistance, or chance: how to interpret differences in gall density among trees?

David Wool and Moshe Burstein
Dept of Zoology, Tel Aviv University, Tel Aviv, Israel

Key words: Gall aphids, gall density, performance, *Pistacia*

The ability of insects to select between different species of host plants (preference) is well documented, and insect-plant coevolution is often assumed to have occurred through a positive covariance between preference and performance (Thompson, 1988). In gall-forming insects, high gall-density on one of several host species may indicate preference. But it is unclear whether this argument can be extended to differences in density among different individuals of the same host species.

It is often observed that some *Pistacia* trees are heavily colonized by galling aphids (Fordinae, Aphididae, Homoptera), while neighbouring trees carry few galls or none at all. Fifteen species of galling aphids colonize *Pistacia* hosts (Anarcardiaceae) in Israel (Koach & Wool, 1977).

We sought answers to two questions: (1) Are there persistent differences in gall density among trees, over and above the considerable, temporal variation (Wool, 1990)? (2) If so, do they result from preference or are there other causes? In this paper, we describe the results of long-term observations conducted at three sites on two host species, *viz.*, *Baizongia pistaciae* (L.) which colonizes *Pistacia palaestina* Boiss. and *Smynthurodes betae* Westw. which colonizes *Pistacia atlantica* Desf. We claim that this widespread phenomenon involves no preference.

Table 1. Colonization of recovering trees at the CAR site in 1991 by *Baizongia pistaciae*

Growth status on April 1	Colonized > 1 gall	Not colonized 0 galls
Growing	15	7
Dormant	7	10

Mean number of galls on colonized trees:

Growing	16.7 (range 2-59)	n = 15
Dormant	2.1 (range 1-8)	n = 7

Results and discussion

Gall densities of *B. pistaciae* were recorded for 13 years on 64 marked trees at three sites in Israel. The gall densities of *S. betae*, however, are known for three years only, on a few trees in each of two sites. In addition, 39 trees of *P. palaestina*, recovering after complete destruction by forest fire, provided data (in 1991) on the early stages of recolonization by *B. pistaciae*. Two-way ANOVA with no replication on square-root-transformed gall densities, revealed highly significant differences among trees, apart from the very large temporal fluctuations. Thus, differences in gall density among trees are persistent (question 1). At one site, 3 out of 33 trees were almost never colonized (with 1, 2 and 0 galls observed in 13 years) while neighbouring trees were almost always colonized (Wool, unpubl. results).

Table 2. Average density of fundatrix and final galls of *S. betae* on different trees at two sites in 1988 and 1991 ("bad years") and in 1989 ("good year")

Year	Galls per shoot			Replacement rate
	Tree	Fundatrix	Final	(Final / fundatrix)
1988	1-GB	6.0	5.7	0.950
	4-GB	3.2	2.0	0.620
	9-GB	14.4	2.9	0.203
1989	1-GB	2.0	7.3	3.650
	3-GB	2.9	6.0	2.070
	4-GB	1.1	1.2	1.091
	5-GB	2.2	6.9	3.136
	9-GB(a)	2.0	9.0	4.500
	9-GB(b)	5.2	8.5	1.634
	12-GB	2.1	0.89	0.423
1991	1-GB	4.8	5.3	1.100
	4-GB	5.4	1.4	0.264
	9-GB	7.9	3.3	0.427
1989	2-TAU	3.0	11.7	3.900
	6-TAU	5.6	9.8	1.741
	7-TAU	5.0	16.9	3.378
1991	6-TAU	7.5	17.2	2.299
	7-TAU	5.0	20.6	4.161

Gall density distribution among the 39 trees recovering after the fire indicated significant departure from Poisson expectations and clumping. Thus, colonization success was apparently not a random process. However, colonization on trees which began growing before April 1 was significantly heavier than on trees that were still dormant at that time (Table 1) (t-test, $P < 0.05$). In the complex life cycle of the Fordinae, the only stage when colonizers may select hosts is during the spring migration of sexuparae from the secondary hosts, one year before gall formation (see also Moran & Whitham, 1990). Clearly, the sexuparae distinguish between different species of *Pistacia* (Fordinae are host specific in Israel), although they do make mistakes (Wool, unpubl.). We have no data on the ability of sexuparae of *B. pistaciae* to choose among individual trees. But in *S. betae* the two processes - colonization and fundatrix success (performance) - are partially separable due to the extreme complexity of its life cycle (Wool & Burstein, 1991). The evidence (Table 2) shows that fundatrices colonized all trees every year, but in 1988, and again in 1991, trees at GB hardly produced any final galls (< 1 per fundatrix). Table 2 illustrates that fundatrix gall density (preference) is unrelated to or negatively correlated with fundatrix performance. In 1989, fundatrix performance was good on all trees.

Rohfritsch (1981) described a true case of tree resistance to galling adelgids. In our case, a chemical defence mechanism is unlikely because 2 of the resistant trees did carry 1-2 galls over the years, and the third was heavily galled each year by a different species of Fordinae (*Geoica utricularia* (Pass.)). We suspect that the clue lies in tree phenology and the synchronization of aphid attack and tree condition, which is heavily dependent on the environment.

Two *P. atlantica* trees grow side-by-side on the slope of Mt. Carmel, with their branches intertwined. One, a male tree, is heavily colonized every year by *Slavum wertheimae* HRL, another species of the Fordinae. The other tree, a female, carries no galls of this species. Our observations (in 1991) show that both trees commence flowering at the same time in spring, but the male tree begins to break its colonizable shoot buds 10-15 days earlier than the female tree! Differences in the timing of bud break - which may or may not be related to the sex of the tree but may indeed be modified by the microenvironment - can be decisive in the short time window available to aphid colonizers.

References

Koach, J. & D. Wool (1977). Geographic distribution and host specificity of gall-forming aphids (Homoptera, Fordinae) on *Pistacia* trees in Israel. *Marcellia* 40: 207-216.

Moran, N.A. & T.G. Whitham (1990). Differential colonization of resistant and susceptible host plants: *Pemphigus* and *Populus*. *Ecology* 71: 1059-1067.

Rohfritsch, O. (1981). A "defense" mechanism of *Picea excelsa* L. against the gall-former *Chermes abietis* L. (Homoptera, Adelgidae). *Z. Angew. Entomol.* 92: 18-26.

Thompson, J.N. (1988). Evolutionary ecology of the relationship between oviposition preference and performance in phytophagous insects. *Entomol. exp. appl.* 47: 3-14.

Wool, D. (1990). Regular alternation of high and low population size of gall-forming aphids: analysis of ten years of data. *Oikos* 57: 73-79.

Wool, D. & M. Burstein (1991). A galling aphid with extra life cycle complexity: population ecology and evolutionary considerations. *Res. Popul. Ecol.* 33: 307-322.

Proc. 8th Int. Symp. Insect-Plant Relationships, Dordrecht: Kluwer Acad. Publ.
S.B.J. Menken, J.H. Visser & P. Harrewijn (eds), 1992

Aggregation of aphid galls at 'preferred' sites within trees: do colonizers have a choice?

Moshe Burstein and David Wool
Dept of Zoology, Tel Aviv University, Tel Aviv, Israel

Key words: Aphididae, performance, preference, *Pistacia*, *Smynthurodes betae*

Many studies have been published on the ability of phytophagous insects to select an optimal host for larval or adult development, survival and reproduction. Moreover, the ability to select leaves or shoots within a plant is also very important since these may differ in quality at different sites on the same plant. This ability is known as the covariance of preference and performance (review in Thompson, 1988).

Most aphidologists agree that aphids can exercise preference both among and within host plants (*e.g.*, Whitham, 1978). Gall-forming aphids are particularly suitable for preference studies because of their host specificity and their restricted colonization sites; colonization being limited to a short period during leaf expansion. In this study we aimed at finding out what determines the intraplant colonization by the gall-forming aphid *Smynthurodes betae* West. (Homoptera, Pemphigidae), and the relationship of site selection to performance.

Material and methods

The aphid *S. betae* induces galls on *Pistacia atlantica* Desf. trees in Israel. The ecology of this aphid has been described in detail by Wool & Burstein (1991). The study was carried out in 1991 on trees in Tel Aviv University Botanical garden (TAU) and in Givat Brener (GB) 35 km south of Tel Aviv. We monitored the trees twice a week during the gall initiation stage, when we measured the shoot length and counted the number of leaves, fundatrix and final galls and recorded their location on the shoot. The galls were collected and examined in the laboratory at the peak of aphid clone size. Gall survival and clone size were used as performance parameters. We also transferred aphids from an early-budding tree in TAU to another tree which started growing about 15 days later. Since pruning causes an extension of shoot growth, in 1990 we pruned several shoots on a resistant tree in GB, and counted the galls on it in 1991.

Results

The fundatrix galls of *S. betae* were formed on leaves 1-8 on the shoot, while the final galls were formed on leaves 7-17. Final gall density was positively correlated with shoot length (average $r = 0.8$, $P < 0.01$) and with the number of leaves on the shoot ($r = 0.75$, $P < 0.01$). Survival and clone size decreased with increasing leaf position whereas final gall density (preference), had a symmetrical distribution with a peak on leaves 11-13 (Fig. 1). Consequently, correlations between preference and performance variables were low and non-significant. F_2 aphids from tree 6 were transferred to tree 10 about 15 days before the native F_2 aphids were born. The transferred aphids colonized the lower leaves (3-6) of the shoot (Fig. 2), but the average clone size of galls they produced was not significantly different from the mean of a random sample of native galls on leaves 8-11 on the same tree (t = 1.079, df = 20, NS).

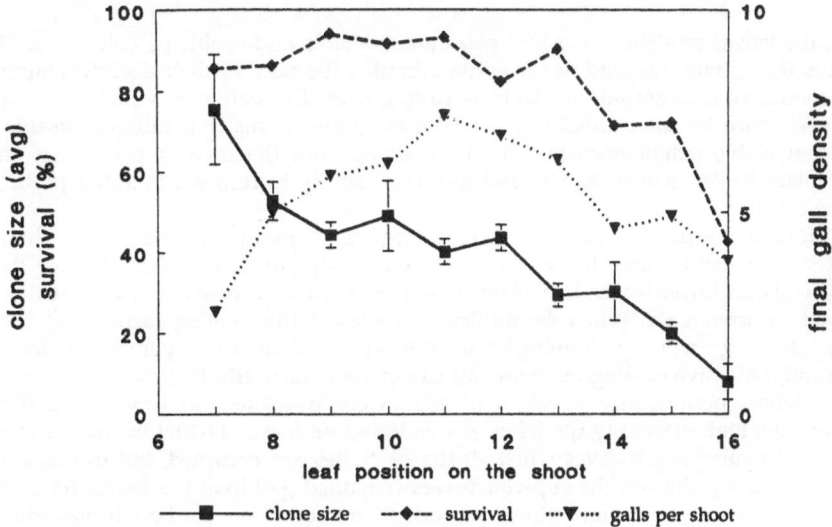

Figure 1. Final gall density ("preference") and performance variables (clone size and gall survival) in relation to leaf position on the shoot on tree 7 (TAU).

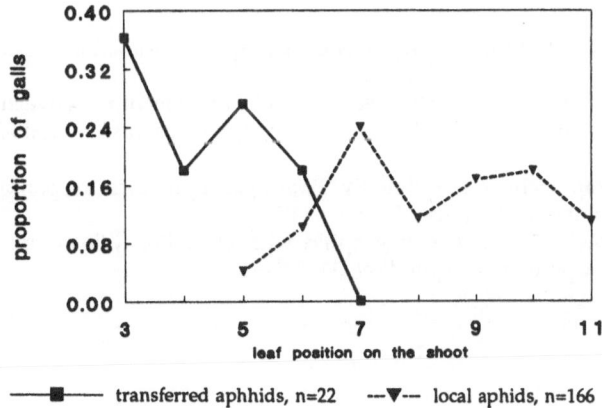

Figure 2. Location of final galls induced by transferred and local aphids on tree 10 (TAU). Transferred aphids were introduced to the tree 15 days before local aphids infestation.

37

Ten pruning cuts were made on one tree in 1990, but produced only three shoots at the cuts in the spring of 1991. Shoots that grew near the pruning cuts were more than three times longer, and bore more leaves (t = 5.83, df = 20, P < 0.001). They also carried six times more final galls per shoot than normal shoots (average: 19 vs 3 galls/shoot).

Discussion

Not all the leaves or shoots on a host plant are available and usable by colonizers. Shoots or leaves that cannot respond to the aphid stimulus (because of their age, developmental stage, chemistry etc.) are not available as galling sites. Evaluation of whether the aphids show preference for an optimal leaf or shoot can only be made if different usable sites become available simultaneously. This is obviously not the case in elongating shoots, where lower leaves mature rapidly and lose their ability to respond to aphid probing by forming a gall.

Gall formers exclusively exploit immature organs, therefore leaf position on the shoot indicates a time scale, since higher leaves are obviously younger. Hence, the low density of final galls on lower leaves is explained by the absence of available leaves at the right time, not by preference. When we artificially removed the limiting factor, and induced further shoot growth by pruning, the number of final galls per shoot increased significantly. Massive grazing by mammals can cause similar effects.

The highest performance variables in galls were measured on leaves 7-12 (Fig. 2). However, the highest density (preference) was found on leaves 11-13. One may argue that aphids go to suboptimal leaves when all the best sites are occupied, but our data show that many galling sites on the superior leaves remained gall-free. We found no evidence that intra-specific competition reduces reproductive success on the best leaves (data not shown).

Insects can indeed discriminate between different host plant taxa (*e.g.*, Thompson, 1988). However, insects must find it much more difficult to detect differences among individual trees of the same species or among leaves on the same shoot. As our data show, there is no need in involving active preference to explain distribution patterns within hosts (Rhomberg, 1984).

References

Rhomberg, L. (1984). Inferring habitat selection by aphids from the dispersion of their galls over the tree. *Am. Nat.* **124**: 751-756.

Thompson, J.N. (1988). Evolutionary ecology of relationship between oviposition preference and performance of offspring in phytophagous insects. *Entomol. exp. appl.* **47**: 3-14.

Whitham, T.G. (1978). Habitat selection by *Pemphigus* aphids in response to resource limitations and competition. *Ecology* **59**: 1164-1176.

Wool, D. & M. Burstein (1991). A galling aphid with extra life-cycle complexity: ecology and evolutionary aspects. *Res. Popul. Ecol.* **33**: 307-322.

Proc. 8th Int. Symp. Insect-Plant Relationships, Dordrecht: Kluwer Acad. Publ.
S.B.J. Menken, J.H. Visser & P. Harrewijn (eds), 1992

Abundance and mortality of a specialist leaf miner in response to shading and fertilization of American holly

Daniel A. Potter
University of Kentucky, Dept of Entomology, Lexington, Kentucky, USA

Key words: Herbivory, *Ilex opaca, Phytomyza ilicicola*, plant defense

Phytomyza ilicicola Loew (Diptera: Agromyzidae), a univoltine specialist leaf miner, typically reaches higher densities on cultivated *Ilex opaca* Aiton planted in sunny, urban sites than it does on native trees in the forest understory (Potter & Kimmerer, 1986; Potter, 1992). I manipulated shading and fertilization to study their effects on holly leaf morphology, nutritional quality, early leaf abscission, and incidence and survival of leaf miners. I hypothesized that one or more of the following factors could explain the higher leaf miner densities on cultivated trees:

Structural constraints associated with mining thinner, shaded leaves could reduce survival of leaf miners on forest trees. The palisade mesophyll consists of 3-4 cell layers in unshaded holly leaves, but is only 2 cell layers thick in shaded leaves. The abaxial cell layer contains abundant crystals, probably of calcium oxalate. Leaf miners feed mainly in the middle layer of unshaded leaves, avoiding the crystals and leaving the adaxial layer as the mine roof. Shaded leaves do not have a middle palisade layer free of crystal-containing cells.

Relatively higher levels of foliar N in cultivated trees could enhance suitability for leaf miners. Mature holly leaves are very low in nitrogen and water. Urban trees are often planted in fertilized lawns, or may themselves be fertilized. Sun leaves are higher in total nitrogen than shaded leaves.

Differential mortality from early leaf abscission may favor leaf miners on cultivated trees. Greater incidence of abscission of mined leaves on cultivated trees, induced by direct sunlight and earlier senescence, could enhance leaf miner survival by reducing mortality from pupal parasitoids, which do not search fallen leaves (Kahn & Cornell, 1989).

Methods

Clonal trees were planted at a common site and experimentally shaded and fertilized (0, 52%, or 93% shade; 0 or 1.5 kg N/100 m^2 in April & Nov.; 3x2 factorial, 6 replicates). Low wire enclosures surrounded each tree to confine abscised leaves. Trees were grown for 14 months under these conditions before being exposed to oviposition by adult leaf miners, and then for 12 additional months to allow leaf miners to complete one generation. Differences in leaf size and number/tree, foliar N, water content, and leaf thickness and morphology were determined by standard assays and examination of leaf cross sections. Incidence of aborted and completed leaf mines, mine area, pupation date, weight of puparia, and percentage survival of leaf miners was determined for each tree. Rates of abscission of mined and unmined leaves were monitored, and abscission-related mortality of leaf miners assessed by dissecting abscised leaves after the emergence period. Experimental details are reported elsewhere (Potter, 1992).

Results

Leaves from shaded trees were larger and about 20% thinner than sun leaves. However, there was little or no difference in leaf miner abundance, developmental rate, survival to pupation, area of finished mines, or pupal weight between shaded and unshaded trees. Leaf miners compensated for feeding within shaded leaves by consuming portions of the abaxial and adaxial cell layers and leaving a thinner roof on the mine. Furthermore, there was no difference in thickness between leaves with successful or aborted mines in either sun or shade. Leaves from fertilized trees contained 37% higher nitrogen than controls. Fertilization did not significantly affect leaf miner abundance, developmental rate, mine area or pupal weight. Survival to pupation was slightly *lower* on fertilized trees. Interaction between shading and fertilization was nonsignificant for all tree and leaf miner parameters measured.

Abscission rates of mined leaves were nearly twice as high for mined leaves than for unmined leaves. Rates of abscission were slightly higher for fertilized trees than for unfertilized trees, and much higher in full sun than in shade. Abscission-related mortality of leaf miners was 44% in full sun, 31% in partial, and 26% in deep shade, a trend opposite that which would be expected if early abscission *per se* were responsible for observed variation between urban and woods trees. Most mortality was from flies being unable to emerge from dried fallen leaves. Pupal parasitism was negligible.

Discussion

These results indicate that structural constraints on leaf miner larvae within thinner, shaded leaves, differences in leaf nitrogen related to variation in soil fertility, and differential mortality resulting from early leaf abscission are probably *not* the proximate causes of density variation of this leaf miner between woods and urban habitats. Other factors, including effects of habitat structure on behavior and survival of adult flies, differential pressure from natural enemies, and genetic variation between cultivated and forest trees may be important instead.

References

Kahn, D.M. & H.V. Cornell (1989). Leafminers, early abscission, and parasitoids: a tritrophic interaction. *Ecology* 70: 1219-1226.

Kimmerer, T.W. & D.A. Potter (1987). Nutritional quality of specific leaf tissues and selective feeding by a specialist leafminer. *Oecologia* 71: 548-551.

Potter, D.A. (1992). Abundance and mortality of a specialist leafminer in response to experimental shading and fertilization of American holly. *Oecologia* 89: (in press).

Proc. 8th Int. Symp. Insect-Plant Relationships, Dordrecht: Kluwer Acad. Publ.
S.B.J. Menken, J.H. Visser & P. Harrewijn (eds), 1992

Within-population variation in demography of a herbivorous lady beetle

Takayuki Ohgushi
Faculty of Agriculture, Shiga Prefectural Jr. Coll., Kusatsu, Shiga 525, Japan

Key words: Cohort life table, *Epilachna niponica*, fitness, population dynamics, survivorship curve

The traditional view of insect population ecology has long assumed that individuals within a population are identical to each other in terms of survival and reproduction, handling populations as homogenized averages. This is, however, a biologically unrealistic assumption, because in any population there is a great variety of individuals of different ages, sizes, and degrees of fitness, thus exhibiting different demography. Recent studies on herbivorous insects have revealed a considerable variation in resource-use tactics within a population (Ohgushi, 1992). Nevertheless, few insect population studies have aimed at illustrating within-population variation in demography, although theoretical studies have recently emphasized individual differences as an essential source in determining population dynamics in natural populations.

I studied the population biology of a thistle-feeding lady beetle, *Epilachna niponica* (Lewis), at six localities (A-F) in the northwestern part of Shiga Prefecture, central Japan (Ohgushi, 1986, 1991). I investigated the average demography of each local population and the cohort demography within a population at two study sites A and F, by constructing life tables for populations and cohorts. My particular aim was to determine: (1) how demography varies from one cohort to another over a reproductive season, and (2) how variation in demography on different ecological scales of locality, year, and cohort occurs.

The lady beetle is a univoltine specialist herbivore feeding on leaves of thistle plants. Overwintering adult females emerge from hibernation in early May and begin to lay eggs in clusters on the undersurface of thistle leaves. Larvae pass through four instars. New adults emerge from early July to early September, feeding on thistle leaves through the autumn. By early November they enter hibernation. The present results are based on five years of mark-recapture experiments for individual beetles and life table statistics.

Survivorship curves clearly demonstrated a large variation in demography among cohorts within the populations at the two study sites, compared to that among six other localities in the same year, and that among five different years at site A and site F. In fact, the coefficient of variation of survival from the egg to the reproductive stage (overall survival) was much higher among cohorts than that among the localities. The overall survival differed on different ecological scales. First, there was a decreasing tendency from downstream to upstream areas. Second, the temporal pattern of changes in the cohort survival differed considerably between site A and site F. At the downstream site A, early cohorts had higher survival than later ones. At the upstream site F, however, late cohorts undoubtedly enjoyed a higher survival. Components of survival contributed differently to the variation in overall survival. Egg and larval survival largely explained among-cohort variation, adult survival contributed little. Nonetheless, adult survival was the most important factor in producing variation between different years in the respective localities. The major cause of the differential pattern of cohort survival between the two sites was specific seasonal variation in the intensity of mortality during the egg and larval periods. More specifically, arthropod predation acting early in the season and host plant deterioration advancing late in the season were essential in generating the among-cohort variation within a population at both sites.

This study emphasizes that contrary to the traditional population ecology viewpoint, no cohort has identical demography with an averaged demography of the overall population. Thus, to clearly understand dynamic features in insect herbivore populations, we should pay more attention to within-population variation in demography, derived from individual differences of different phenotypes or genotypes.

References

Ohgushi, T. (1986). Population dynamics of an herbivorous lady beetle, *Henosepilachna niponica* in a seasonal environment. *J. Anim. Ecol.* 55: 861-879.
Ohgushi, T. (1991). Lifetime fitness and evolution of reproductive pattern in the herbivorous lady beetle. *Ecology* 72: 2110-2122.
Ohgushi, T. (1992). Resource limitation on insect herbivore populations. In: M.D. Hunter, T. Ohgushi & P.W. Price (eds), *Effects of Resource Distribution on Animal-Plant Interactions*, San Diego: Academic Press.

Proc. 8th Int. Symp. Insect-Plant Relationships, Dordrecht: Kluwer Acad. Publ.
S.B.J. Menken, J.H. Visser & P. Harrewijn (eds), 1992

Continental-scale host plant use by a specialist insect herbivore: milkweeds, cardenolides and the monarch butterfly

Stephen B. Malcolm[1], Barbara J. Cockrell[1] and Lincoln P. Brower[2]
[1] *Dept of Biological Sciences, Western Michigan University, Kalamazoo, Michigan, USA*
[2] *Dept of Zoology, University of Florida, Gainesville, Florida, USA*

Key words: *Asclepias*, chemical fingerprint, *Danaus plexippus*, defence, migration, sequestration

Distributions of the host plants of mobile herbivorous insects form complex mosaics in space and time. Thus it is difficult to track large scale patterns of herbivorous insect movement without time- and labour-intensive trapping, tracking, and marking techniques that are subject to misinterpretation. Here we summarize the use of a chemical fingerprinting technique to describe the continental scale pattern of migration by the monarch butterfly (*Danaus plexippus* (L.)) in relation to the spatial and temporal distributions of its milkweed larval host plants (*Asclepias* spp.) in North America.

We distinguish between two alternative hypotheses: (1) single sweep migration in which overwintered adults migrate from overwintering sites and distribute eggs on milkweed plants across their entire breeding range, or (2) successive brood migration in which overwintered adults lay their eggs on the first abundant early spring milkweeds they encounter, and their offspring continue the migration.

Each March, monarch butterflies remigrate from Mexican overwintering sites and by June they are distributed across eastern North America as far north as southern Canada. Larvae produced by these butterflies feed on the milkweed genus *Asclepias* in North America. Almost all of the 108 species are perennial and contain toxic, steroidal cardenolides as their characteristic chemical defence against herbivores (Malcolm, 1992). Nevertheless, monarch larvae feed as specialists on these plants and sequester cardenolide amounts and arrays for their own chemical defences against natural enemies that are characteristic of each exploited species of *Asclepias* (Malcolm *et al.*, 1989). Thus these cardenolide "fingerprints" can be used to assign wild-caught butterflies to a geographical and temporal origin within that of their larval food plant species.

On a coarse scale, the most abundant and frequently used monarch host plants east of the Rocky Mountains include the southern USA species *Asclepias viridis* Walt. and *Asclepias humistrata* Walt. and the northern species *Asclepias syriaca* L. (Malcolm *et al.*, 1992). The two southern species appear in early spring (March) and the extremely abundant northern species appears in late spring (May). Each species produces monarchs with a characteristic quantitative and qualitative cardenolide fingerprint (Malcolm *et al.*, 1992). These fingerprints were derived from quantitative assays of cardenolide concentrations by spectrophotometry (Malcolm *et al.*, 1989), as well as qualitative separations of cardenolides by thin layer chromatography (TLC) in overwintering and migrant monarch butterflies (Malcolm *et al.*, 1992; Table 1).

Results and discussion

Using these milkweed-derived cardenolide fingerprints in adult, field-collected migrant butterflies we found that monarchs migrate by a *successive brood* strategy each spring. Thus almost all monarchs overwintering in Mexico, and early spring remigrants across the southern USA, have the low cardenolide concentrations and TLC patterns characteristic of butterflies that fed as larvae on the northern milkweed *A. syriaca* (Malcolm *et al.*, 1989; Table 1). These butterflies lay all of their eggs on milkweeds like *A. viridis* and *A. humistrata* in the southern USA and the ensuing generation continues the migration north to account for the later spring arrivals in the northern USA with significantly higher cardenolide concentrations (Wilcoxon z = 12.21, P < 0.0001) and a TLC pattern dominated by monarchs that fed as larvae on the southern *A. viridis* (Table 1).

Table 1. Quantitative and qualitative cardenolide finger-prints of monarch butterflies collected overwintering in Mexico (January-March) or migrating in the south (April-May) and north (May-June) of their North American breeding range, east of the Rocky Mountains.

	Overwintering	Migrating	
	Mexico	South	North
Quantitative cardenolide fingerprint (cardenolide concentration in µg/0.1 g dry lean weight)			
Mean	79	62	157
SD	80	65	81
N	562	133	646
Qualitative cardenolide fingerprint (% TLC cardenolide pattern derived from larval host plant)			
Asclepias viridis	0	3	84
Asclepias humistrata	0	4	6
Asclepias syriaca	92	84	6
Indeterminate	8	9	5
N	386	133	629

Chemical fingerprints help us to show the influence of variation in the distributions of different milkweed host plant species on the life history of the monarch butterfly and its chemically-based protection against natural enemies. Thus chemical defence during the monarch's annual cycle of up to five successive generations varies from poorly protected overwintering migrants to highly protected late spring migrants. Such life history variation in host plant use has significant implications for our understanding of the evolution of plant-herbivore interactions within at least three interacting trophic levels.

References

Malcolm, S.B., B.J. Cockrell & L.P. Brower (1989). The cardenolide fingerprint of monarch butterflies reared on the common milkweed, *Asclepias syriaca*. *J. Chem. Ecol.* **15**: 819-853.

Malcolm, S.B. (1992). Cardenolide-mediated interactions between plants and herbivores. In: G.A. Rosenthal & M.R. Berenbaum (eds), *Herbivores: Their Interaction with Secondary Plant Metabolites*, Second Edition. San Diego: Academic Press.

Malcolm, S.B., B.J. Cockrell & L.P. Brower (1992). Spring recolonization of eastern North America by the monarch butterfly: successive brood or single sweep migration? In: S.B. Malcolm & M.P. Zalucki (eds), *Biology and Conservation of the Monarch Butterfly*. Los Angeles: Natural History Museum of Los Angeles County.

Proc. 8th Int. Symp. Insect-Plant Relationships, Dordrecht: Kluwer Acad. Publ.
S.B.J. Menken, J.H. Visser & P. Harrewijn (eds), 1992

Resource partitioning of host plants by insects on a geographic scale

F. Kozár
Plant Protection Institute, Hungarian Academy of Sciences, Budapest, Hungary

Key words: *Coccoidea*, ecological theories, evolutionary theories, scale insects, species richness

Studies on geographic scale resource partitioning in insect-plant relationships are scarce (Strong *et al.*, 1984). This may be due to the absence of sufficient taxonomic knowledge about various groups of insects, and insufficient data concerning distribution of insects in different regions as well as on different plants.

In this paper, resource partitioning on plants differing in evolutionary age, species richness, distribution, chemistry, etc., is discussed taking the very well explored Diaspididae family as an example, on a global and regional scale.

The plant genus *Buxus* L. is found everywhere, but large numbers of local insect species were only found in the Palearctic region. On *Celtis* L. the greatest insect richness was observed in the Nearctic region followed by the Palearctic region. *Cornus* L. has an equally large number of insects both in the Nearctic and the Palearctic region. Despite the abundant species richness of *Crataegus* L. in the Nearctic region, the largest number of insects was found in the Palearctic region. The number of insects on *Elaeagnus* in the Palearctic region and the Oriental region is equally high. *Euonymus* L., on the other hand, has a larger number of insects in the Palearctic region than in the Oriental region. Conversely, the species richness of Gramineae is greater in the Oriental region. *Loranthus* L. has a large number of insect species in the Oriental region, followed by the Ethiopian region. *Melia* has only 1-2 local, specialized insects in all regions, but numerous widely distributed, polyphagous species. *Pinus* L. and *Quercus* L. have large numbers of insect species in the Palearctic region, followed by the Nearctic region (Fig. 1).

In the Palearctic region, I found a North-South and similarly a West-East partitioning of host plants by local groups of scale insects on fruit trees. The maps also show, that wide distribution ranges of host plants are not occupied by scale insects. Often, regions rich in plant species do not coincide with regions rich in scale insects. Thus, speciose plant genera often appear to harbour low numbers of scale insect species and *vice versa*. Finally, the average number of insect species per plant species (based upon 517 species of scale insects on 10,771 plant species) amounted to 0.05; within a plant genus, this figure varied from 0.02 to 1.6. The highest ratio of species was found on *Melia*, which represents a very specific chemistry. In this case the number of widely distributed, polyphagous insects was also very high. However, on *Pinus* and *Quercus*, also very chemically specific, the number of locally distributed, specialized insects were much higher, which underlines the opportunistic character of speciation of these insects to host plants (Eastop, 1979).

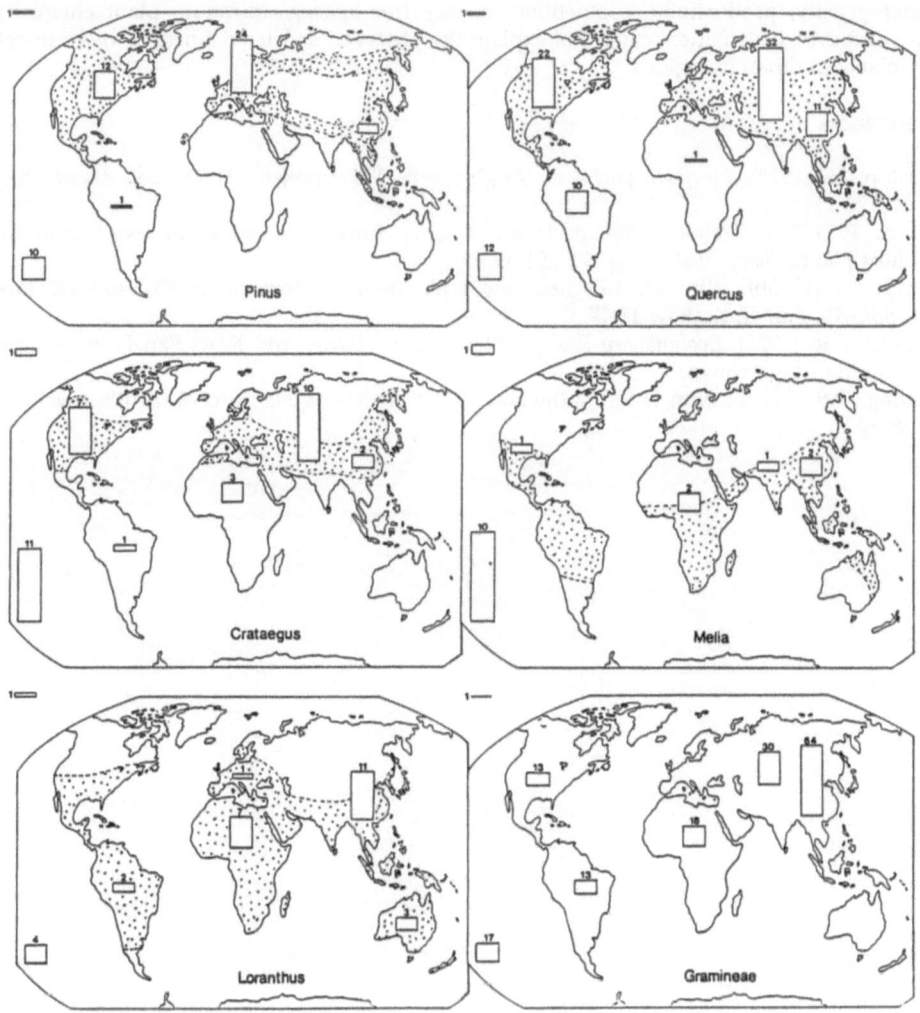

Figure 1. The number of local scale insect species on six different host plant genera (the separate vertical bar shows the number of scale insect species with world-wide distribution on that particular plant genus).

Conclusions

1. The geographic scale resource partitioning of host plants by scale insects is based on evolutionary patterns, and could only be explained by evolutionary (historical age, genetic capabilities) and by biogeographic (historical events, plate tectonics, isolation, glaciation, climatic zonation, etc.; Kozár, 1990a, b) processes.

2. Ecologically based theories and hypotheses (Pianka, 1978), like ecological age, heterogeneity, productivity, competition, enemy free space, saturation, plant chemistry, etc., and even coevolution, could not explain the observed species richness of these insects on plants in different regions.

References

Eastop, V. (1978). Sternorrhyncha as Angiosperm Taxonomists. *Symp. Bot. Upsal. XXII* **4**: 120-134.

Kozár, F. (1990a). Trends in the speciation of some Homoptera groups in association with host plants. *Symp. Biol. Hung.* **39**: 491-493.

Kozár, F. (1990b). Why are there so few scale insects (Homoptera: Coccoidea)? *Proc. ISSIS-VI, Part II*. Krakow 13-17.

Pianka, E.R. (1978). *Evolutionary Ecology*. New York: Harper and Row; San Francisco and London: Hagerstown.

Strong, D.R., J.H. Lawton & R. Southwood (1984). *Insects on Plants*. Oxford: Blackwell Sci. Publ.

Proc. 8th Int. Symp. Insect-Plant Relationships, Dordrecht: Kluwer Acad. Publ.
S.B.J. Menken, J.H. Visser & P. Harrewijn (eds), 1992

Why do droughts often result in devastating insect epidemics? The African armyworm, *Spodoptera exempta*, as an example

Jan Janssen
Dept of Entomology, Wageningen Agricultural University, Wageningen, The Netherlands

Key words: Development, nitrogen, nutrition, reproduction

Insect outbreaks can either be caused by a sudden change in the environment of an insect or by changes in the intrinsic genetic or physiological properties of individual organisms in a population (Berryman, 1987). Drought has long been recognized as an important factor in causing outbreaks. Two different situations can be distinguished in the drought-outbreak relationship: one in which outbreaks develop on plants that are being stressed by drought, and one in which outbreaks develop on unstressed plants that have recently been stressed by drought.

A number of explanations has pointed out that severe droughts are frequently followed by severe pest outbreaks (Brown, 1962; Pedgley *et al.*, 1989; McDonald *et al.*, 1990). In particular: a) the effectiveness of predators and parasites in regulating population growth may be significantly reduced by droughts (Marcovitch, 1957); b) plants that have survived drought are thought to be superior hosts because of their greater nutritional value (White, 1984) or because drought stress may cause their defence mechanisms to be temporarily compromised (Rhoades, 1985), or the increased production of secondary allelochemicals in many stressed plants may enhance the development of the insect's detoxification system thus increasing survival (Mattson & Haack, 1987). Good experimental proof for these explanations is scant. Their relative importance is as yet unclear and possibly varies according to species, time and location. However, the coincidence of droughts and subsequent outbreaks seems to point to for a general underlying mechanism. A thorough knowledge of the drought-outbreak relationship is a prerequisite for improving forecasts of severe pest outbreaks as well as their control.

Brown (1962) and Pedgley *et al.* (1989) observed the drought-outbreak relationship of the African armyworm (*Spodoptera exempta* (Walker)). This species feeds almost exclusively on plants of the families Gramineae and Cyperaceae and periodically reaches high population densities (up to 1000 larvae per m^2) in Eastern Africa. Population development was found to differ enormously between years, suggesting that variation in drought intensity during the long dry season might be responsible. Severe droughts induce severe armyworm outbreaks because of higher nitrate levels in the soil upon wetting (Birch, 1959). The higher nitrogen levels in the host plants result in an increased development rate and survival of the larvae (especially the very young; White, 1984) and an increased fecundity of the subsequent adults.

Results

Comparing field experiments carried out for the years 1988 to 1991 in Kitui District, probably the most important primary outbreak area in Kenya, reveal major differences in organic nitrogen levels in host plants of the African armyworm during the first month after the onset of the short rainy season - the start of the armyworm season. The organic nitrogen levels seem to be positively correlated with soil nitrate levels, and with soil temperatures during the preceding long dry season (Table 1).

Table 1. Average soil temperatures during the long dry season at the Makindu weather station (Kenya), and average soil nitrate and plant organic nitrogen levels during the first month after the onset of the short rainy seasons at two sampling sites in Kitui District for the years 1988-1991

Year	1988	1989	1990	1991
Soil temperature at 12.00 hr (°C)	28.5	26.4	27.1	28.1
Soil nitrate level (mg/kg)	5.3	2.9	2.3	5.7
Organic nitrogen content in wild grasses (g/kg)	38.4	29.6	28.3	36.9

However, laboratory experiments with a water-culture system carried out in a greenhouse with maize plants showed only minor effects of the organic nitrogen content on armyworm fitness: only preoviposition period and fecundity were affected. Many caterpillars survived and their development was only slightly delayed in by nitrogen levels as low as 1.5%. An estimation of the amount of leaf material eaten by the larvae indicates that the absence of a nitrogen effect was due to compensatory feeding (Table 2). The relevance of these laboratory results to the field situation is not clear.

Table 2. Effect of organic nitrogen in maize plants, grown on four different nitrate concentrations in a water-culture system, on armyworm fitness

		Nitrate concentration (nM) of water culture			
		3.75	2.50	1.25	0.50
Maize plants	organic nitrogen (g/kg)	37	33	23	13
Caterpillars	survival (%)	92	91	94	89
	fresh weight, day 8 (mg)	223	230	209	131
	larval-pupal duration (days)	20.5	21.3	20.8	21.5
	leaf fresh weight eaten (g/caterpillar)	1.9	2.6	3.2	3.9
Moths	longevity (days)	8.3	7.9	8.5	8.0
	preoviposition period (days)	2.6	2.8	3.0	3.5
	fecundity (eggs)	1003	954	841	596
	(eggs/mg pharate pupal weight)	6.0	5.3	5.1	4.2

Conclusion

Though there is evidence of substantial differences between years in the quality of the flush of new growth in primary outbreak areas, possibly as a result of differences in drought intensity during the long dry season, this fact alone is not sufficient to explain the differences observed between years in outbreak development of the African armyworm.

References

Berryman, A.A. (1987). The theory and classification of outbreaks. In: P. Barbosa & J.C. Schultz (eds), *Insect Outbreaks*, pp. 3-30. San Diego: Academic Press.

Birch, H.F. (1959). Further observations on humus decomposition and nitrification. *Pl. Soil* **11**: 262-286.

Brown, E.S. (1962). The African army worm *Spodoptera exempta* (Walker) (Lepidoptera, Noctuidae): a review of the literature. London: Commonwealth Institute of Entomology.

Marcovitch, S. (1957). Forecasting armyworm outbreaks - a possibility. *J. Econ. Entomol.* **50**: 112-113.

Mattson, W.J. & R.A. Haack (1987). The role of drought stress in provoking outbreaks of phytophagous insects. In: P. Barbosa & J.C. Schultz (eds), *Insect Outbreaks*, pp. 365-407. San Diego: Academic Press.

McDonald, G., K.P. Bryceson & R.A. Farrow (1990). The development of the 1983 outbreak of the common armyworm, *Mythimna convecta*, in Eastern Australia. *J. appl. Ecol.* **27**: 1001-1019.

Pedgley, D.E., W.W. Page, A. Mushi, P. Odiyo, J. Amisi, C.F. Dewhurst, W.R. Dunstan, L.D.C. Fishpool, A.W. Harvey, T. Megenasa & D.J.W. Rose (1989). Onset and spread of an African armyworm upsurge. *Ecol. Entomol.* **14**: 311-333.

Rhoades, D.F. (1985). Offensive-defensive interactions between herbivores and plants: their relevance in herbivore population dynamics and ecological theory. *Am. Nat.* **125**: 205-238.

White, T.C.R. (1984). The abundance of invertebrate herbivores in relation to the availability of nitrogen in stressed food plants. *Oecologia* **63**: 90-105.

Proc. 8th Int. Symp. Insect-Plant Relationships, Dordrecht: Kluwer Acad. Publ.
S.B.J. Menken, J.H. Visser & P. Harrewijn (eds), 1992

The impact of water and nutrient stress on oak leaf quality and gypsy moth performance

Karl W. Kleiner[1], Marc. D. Abrams[2] and Jack C. Schultz[1]
[1] *Dept of Entomology, Pesticide Research Laboratory, Pennsylvania, USA*
[2] *School of Forestry, Penn State Univ., Univ. Park, Pennsylvania, USA*

Key words: Baculovirus, insect outbreaks, phenolics, photosynthesis

It is widely accepted that abiotic stress can increase the suitability of plants to insect herbivores and subsequently improve herbivore performance and increase their abundance (Mattson & Haack, 1987; Larsson, 1989; Jones & Coleman, 1991). In the eastern United States, gypsy moth (*Lymantria dispar* L.) populations exhibit outbreaks periodically every 8-12 years and are most likely to develop in stands growing on sandy or thin rocky soils. In central Pennsylvania, these forest stands occur on steep upper slopes and ridgetops which have been characterized as having extremely well drained, poor quality soils (Doane & McManus, 1981).

We tested the hypothesis that abiotic stresses (water and nutrients) on chestnut oak (*Quercus prinus* L., a xeric species) and red oak (*Quercus rubra* L., a more mesic species) would result in 1) physiological plant stress, 2) changes in foliar chemical constituents and 3) subsequent impacts on gypsy moth life history traits. In addition, we predicted that a species adapted to low soil resource availabilities (chestnut oak), would exhibit less response to variation in water and nutrients than a species adapted to greater resource availabilities (red oak).

Materials and methods

Red oak and chestnut oak seedlings were grown for two years in 8" x 16" pots with a 1:1:1 mixture of peat:perlite:forest soil. One half of the plants received the equivalent of 201/130/165 lbs (90/60/75 kg) per acre of NPK supplied weekly as NH_4NO_3 and KH_2PO_4. In the second year, water was withheld from half the plants two weeks after budbreak. Two weeks after water was withheld, (4 weeks post budbreak), diurnal measures of physiology and destructive harvests for measures of growth and foliar chemistry were made on 5 plants in each of the four possible treatments. These measures continued every other week for 10 weeks.

Dry weight gains were determined for second instar gypsy moth larvae that were reared for approximately two weeks on a cohort of plants not used for physiology. Viral mortality was recorded for third instar larvae that were given a 50 mm^2 leaf disc on which was placed a 2 µl aliquot of NPV (nuclear polyhedrosis virus) containing approximately 50,000 PIBs (polyhedral inclusion bodies). Third instar is the larval stage in which population influencing epizootics begin. Foliage was analyzed for chemical traits which have been correlated with the growth performance and viral resistance of the gypsy moth.

Results and conclusions

Reductions in photosynthetic rates and leaf water potentials were not evident until week 8 of the drydown for red oak and week 10 for chestnut oak. In addition, photosynthetic rates were more likely to be reduced in plants with low water and high nutrient availability-conditions which are not representative of xeric ridges. Fertilized plants accumulated significantly more shoot and root dry matter than unfertilized plants, particularly in the more mesic species, red oak. Low water availability resulted in significantly lower shoot and root dry weight gains in red oak only.

Low nutrient availability increased measures of foliar phenolics and decreased protein concentrations while low water availability had no effect on measures of leaf chemistry. Low nutrient availability significantly reduced protein concentrations in both red oak and chestnut oak. Significant increases in the measures of total phenolics (Folin-Denis), hydrolyzable tannins (Potassium iodate) and protein binding capacity (Radial diffusion assay) were limited to red oak. There was no effect of differential fertilization on condensed tannin measures (HCL-Butanol) in either species.

Larval dry weight gains were influenced by the differential nutrient treatment but not by differential water treatment. Dry weight gains were greater for larvae reared on plants of both oak species in the fertilizer treatment than those in the unfertilized treatment. There was no effect of differential watering or fertilization on the mortality of larvae ingesting NPV and oak foliage.

Larval dry weight gains were related to variation in foliar chemistry of red oak only. None of the chestnut oak leaf chemistry measures were correlated with larval dry weight gains. Dry weight gains were positively correlated with red oak protein concentration and negatively correlated with hydrolyzable tannins, total phenolics and protein binding capacity.

Expressed in terms of plant fitness, low nutrient availability had a negative impact on plant growth, and low water availability had a negative impact on net photosynthesis and water status. However, relative to the implied (foliar chemistry) and actual impact on herbivore fitness, only low nutrient availability had a significant effect. Low nutrient availability decreased foliar protein concentrations and increased measures of phenolics and phenolic activity. These measures were associated with reduced growth of larvae on chestnut oak (the xeric species) and correlated with reduced growth of larvae on red oak (the more mesic species).

It does not appear that water stress is likely to have a significant impact on the biology of an early season feeder such as the gypsy moth. By the time photosynthetic rates and leaf water potentials were reduced in this study, the gypsy moth feeding period would have been finished. Moreover, neither foliar chemistry traits nor gypsy moth performance were affected by withholding water during this study. In a separate study, chestnut oaks and red oaks on a ridgetop in central Pennsylvania exhibited no change in foliar chemistry associated with a severe drought (Second driest month of June in 88 years, June precipitation was 76% below normal).

Of the two oak species, chestnut oak exhibited less response to variation in water and nutrient availability than red oak. Moreover, chestnut oak supported better gypsy moth growth than red oak and provided a modicum of resistance to the virus. Chestnut oak is the most abundant tree species on xeric ridges. Although water and nutrient stress are not likely to increase the suitability of individual trees to the gypsy moth, they are likely to increase the abundance of trees that are more suitable.

References

Doane, C.C. & M.L. McManus (1981). The gypsy moth: research toward integrated pest management. *USDA Forest Service Technical Bulletin* No. 1584.

Jones, C.G. & J.S. Coleman (1991). Plant stress and insect herbivory: toward an integrated perspective. In: H.A. Mooney, W.E. Winner & E.J. Pell (eds), *Response of Plants to Multiple Stresses*, pp. 249-280. San Diego: Academic Press.

Larsson, S. (1989). Stressful times for the plant stress-insect performance hypothesis. *Oikos* 56: 277-283.

Mattson, W.J. & R.A. Haack (1987). The role of plant water deficits in provoking outbreaks of phytophagous insects. In: P. Barbosa & J.C. Schultz (eds), *Insect Outbreaks*, pp. 365-407. San Diego: Academic Press.

Proc. 8th Int. Symp. Insect-Plant Relationships, Dordrecht: Kluwer Acad. Publ.
S.B.J. Menken, J.H. Visser & P. Harrewijn (eds), 1992

Forest insect trends along an acidic deposition gradient in the central United States

Robert A. Haack
USDA Forest Service, North Central Forest Experiment Station, East Lansing, Michigan, USA

Key words: Cerambycidae, Cossidae, gypsy moth, oak, soil Ca:Al ratio

The Ohio Corridor Study (OCS) was designed to detect possible effects of acidic deposition on oak-hickory forests in the central United States (Haack & Blank, 1991). The hypothesis tested was that for analogous forest stand conditions and soil types, differences in forest response along a geographic acidic-dose gradient could be explained by differences in pollutant dose.

Several parameters were measured by a team of researchers in 7 oak-hickory (*Quercus-Carya*) forest sites along a 1200-km-long gradient from Arkansas to Ohio during the period 1987-1990. Total sulfate deposition (wet + dry) increases 2-3 fold from west (Arkansas) to east (Ohio). Similarly, total nitrogen deposition increases from west to east. Growing-season ozone concentrations were not significantly different along the gradient during the decade of the 1980's (Loucks, 1990).

From west to east, there was 1 site in Arkansas, and 2 each in Illinois, Indiana, and Ohio. The 7 sites were broadly analogous in stand composition and age, aspect, mean annual temperature and rainfall, and soil type (poorly buffered, sandstone-derived soils). However, subsequent detailed soil analyses indicated that the Arkansas site was sufficiently different from the other sites, and therefore it was withdrawn from analyses.

Since many insects show improved performance on stressed plants (Mattson & Haack, 1987), several studies were conducted to determine if certain insect parameters changed across the gradient. The main objectives were to determine if variation existed across the gradient in regards to (a) attack density of trunk-boring insects, (b) the density of tree-defoliating Lepidoptera, and (c) the preference of lepidopterous larvae to foliage collected along the gradient.

Results from each of the above insect studies were compared with the soil Ca:Al ratio in the upper 50 cm. Soil Ca:Al ratio is a good indicator of the degree of acidification that has occurred on otherwise analogous sites, with lower values indicating greater acidification. In the OCS, individual-tree growth decline of oaks was higher at sites with low soil Ca:Al ratios: <0.25 (Loucks, 1990).

Results and conclusions

Experiment 1. Attack densities of living-oak borers, which are primarily wood-boring insects in the families Cerambycidae and Cossidae that do not kill their hosts, were determined on black oaks (*Quercus velutina* Lami.) and white oaks (*Quercus alba* L.) along the gradient (about 100 trees/site). Individual attacks were quantified and expressed on a per-unit-area of bark basis. Combining data from all oaks within each State, mean attack densities increased as soil Ca:Al ratios fell, being highest in the eastern sites (Indiana and Ohio).

Experiment 2. Population densities of defoliating Lepidoptera were estimated from the number of larval head-capsules collected in ground-based traps (90 traps/site). Sampling occurred over a 2-yr period. Densities of spring defoliators, expressed as the number of head-capsules/m^2 of ground surface area during the period of April to June, were highest in Indiana, which had the sites with the lowest Ca:Al ratios.

Experiment 3. Feeding preference studies were conducted in May (early-season foliage) and August (late-season foliage), using one site per State. At each site, foliage was collected from 5 white oaks and 5 black oaks, and returned on ice to the laboratory. Leaf discs were cut from foliage and provided to gypsy moth (*Lymantria dispar* L.) larvae in choice-test arenas. In both tests, larvae preferred foliage (*i.e.,* consumed more leaf-disc area) from the sites with the lowest Ca:Al ratios (Indiana and Ohio). The results were similar for both black oaks and white oaks. Leaf water content did not appear related to the feeding pattern. However, total leaf nitrogen was greater for leaves collected in Indiana and Ohio. Increased foliar nitrogen may reflect tree responses to stress (Mattson & Haack, 1987), and/or to greater nitrogen deposition in the eastern sites of Indiana and Ohio.

In addition to the above studies, other co-investigators noted strong correlations between low soil Ca:Al ratios and reduced tree growth, reduced soil pH, reduced soil invertebrate densities, and increased soil carbon levels (Loucks, 1990). Overall, the pattern of change observed in the Ohio Corridor Study is consistent with hypotheses relating long-term acidic deposition to these changes.

References

Haack, R.A. & R.W. Blank (1991). Incidence of twolined chestnut borer and *Hypoxylon atropunctatum* on dead oaks along an acidic deposition gradient from Arkansas to Ohio. In: *Proc. 8th Central Hardwood Forest Conference.* U.S. Dep. Agric. Forest Service, Gen. Tech. Rept. NE-148. pp. 373-387.

Loucks, O.L., ed. (1990). *Air Pollutants and Forest Response: The Ohio Corridor Study.* Final Report. Miami University, Department of Zoology, Oxford, Ohio.

Mattson, W.J. & R.A. Haack (1987). The role of drought in outbreaks of plant-eating insects. *BioScience* **37**: 110-118.

Proc. 8th Int. Symp. Insect-Plant Relationships, Dordrecht: Kluwer Acad. Publ.
S.B.J. Menken, J.H. Visser & P. Harrewijn (eds), 1992

Birch foliage quality and population density of *Eriocrania* miners in a pollution-affected area

Julia Koricheva and Erkki Haukioja
Laboratory of Ecological Zoology, Dept of Biology, University of Turku, Turku, Finland

Key words: Aerial pollution, *Betula*, heavy metals, leaf-mining insects, Lepidoptera

Aerial pollution frequently modifies population densities of herbivorous insects, and this is often attributed to changes in host-plant quality. The role of indirect pollution mediated by changes in host plant quality can be elucidated only by monitoring both insect density and plant quality in natural environments. A major problem with this approach is how to identify parameters of plant quality which directly estimate its effects on insects, and which can be evaluated in the field.

We estimated birch foliage quality as food for *Eriocrania* miners by monitoring an index of the efficiency of conversion of leaf material to larval body mass. Our aim was to examine the effects of aerial pollution on this index, and to determine whether the changes in plant quality modified the performance and population density of *Eriocrania* miners. In addition, we attempted to estimate the direct impacts of pollutants (heavy metals) on miners.

Material and methods

The field work was carried out around a large copper smelter in Harjavalta, Southwestern Finland, in 1991. The main pollutants present were sulphuric oxides and heavy metals. Insect sampling was conducted at 13 sites situated at different directions and varying distances (0 to 11 km) from the factory. At each site, the density of *Eriocrania* miners was estimated in 10 randomly selected birch trees (*Betula pubescens* Ehrh. or *B. pendula* Roth). The samples contained at least two *Eriocrania* species: *E. sangii* (Wood) and *E. semi-purpurella* (Steph.), and we analyzed the pooled data. Mines with penultimate and last instar larvae were collected from each site and kept in plastic vials. When full-grown larvae left the mines, the dried larval mass (LM) and the mass of frass within the mine (FM, produced during the whole larval stage) were determined. LM was an indicator of miner performance. The ratio between LM and FM was used as an index of plant quality (IPQ). High values of the ratio indicate high efficiency of food conversion to body mass and, consequently, a high leaf quality as food for miners. A sample of ten unmined leaves was taken from each tree for analysis of heavy metal levels.

Results

Despite the high variance of larval and frass masses within the sites, we found that the IPQ of the sites was significantly different ($F_{12,57} = 2.70$, $P < 0.01$). Highest larval masses, lowest frass masses and, consequently, the highest nutritional value of birch foliage, were found at sites 1-3 km from the pollution source. Larval mass showed a positive correlation with IPQ (r_s = 0.79, P < 0.01).

Levels of copper and nickel in leaves decreased exponentially with distance from the pollution source. LM, FM and IPQ did not correlate with the concentration of heavy metals in the foliage.

Eriocrania densities varied significantly among the sites (Kruskal-Wallis test: CHISQ = 44.56, P < 0.001), and tended to be low close to the factory, although the correlation between miner density and distance from the source of pollution was not significant (r_s = 0.48, P = 0.099).

The population density of *Eriocrania* did not correlate with IPQ (r_s = -0.12, P = 0.700), but had a significant negative correlation with levels of copper and nickel in the foliage of the host tree (r_s = -0.825, P < 0.01 and r_s = -0.836, P < 0.001, respectively).

Conclusions

1. The quality of birch foliage as food for *Eriocrania* miners can be conveniently measured under natural conditions, using the index describing the efficiency of leaf conversion to larval mass.
2. The quality of the foliage varied significantly among sites and was highest at the zone of moderate pollution.
3. There was a positive correlation between performance of individual miners and foliage quality.
4. The population density of miners did not reflect differences in the nutritional value of food.
5. Concentration of heavy metals in foliage did not correlate with larval mass, frass mass or nutritional quality of the leaves; instead, it correlated negatively with the population density of *Eriocrania* miners.

Proc. 8th Int. Symp. Insect-Plant Relationships, Dordrecht: Kluwer Acad. Publ.
S.B.J. Menken, J.H. Visser & P. Harrewijn (eds), 1992

Performance of *Neodiprion sertifer* on defoliated scots pine foliage

Päivi Lyytikäinen
Finnish Forest Research Institute, Dept of Forest Ecology, Vantaa, Finland

Key words: Induced resistance, needle quality, nutrients, *Pinus sylvestris*, sawfly

Insect feeding or simulated herbivory can induce short-term and long-term quality differences in plants, which affect the success of folivorous insects. Induced physico-chemical changes in á plant include the appearance of resinosis, *i.e.*, increased concentrations of fiber, and changes in secondary chemicals or nutrients. In woody plants, the differences in responses reported have been mainly from deciduous trees. Damage-induced defensive reactions in evergreen conifers have been found in *Pinus contorta* Douglas (Leather *et al.*, 1987) and *Pinus ponderosa* Dougl. ex Laws. (Wagner, 1986).

During the summers of 1985 and 1986, I carried out a study at the Archipelago Research Institute at Seili, SW Finland (60°15' N, 21°58' E). For the experiments, I used the larvae of *Neodiprion sertifer* Geoffroy (Hymenoptera, Diprionidae), which overwinter at the egg-stage; the gregarious larvae hatch in May-June. *N. sertifer* is the major damage agent among sawflies in Finland.

Materials and methods

The randomly selected even-aged (mean age 23.9 years) test trees (*Pinus sylvestris* L.) were divided into four defoliation levels according to removed needle biomass: 0% (control), 5-24%, 25-49% and 50-75%. The trees were defoliated at the beginning of the growing season preceding the tests by cutting mature needles from a certain proportion of branches with scissors to simulate the feeding of *N. sertifer*. For the nutrient analysis and water content measurements I collected one-year-old needles from the mid-third of the crown of a tree at every damage level. Concentrations of C and N were determined by using a CHN-analyser, Na, K, Mg and Ca by atomic absorption, and P after digestion by spectrophotometry.

Rearing was started when the larvae were at the first instar. The larvae grew in 1-litre plastic containers in groups of 30. For every three trees at each level there was one larval group, which I fed twice a week with a twig from a branch from defoliated and control trees. The temperature was kept at an outdoor temperature.

Results and discussion

Significant differences in larval performance took place only in the next growing season. The larvae that fed on the foliage from the highest defoliation level were the most delayed in their growth. The female cocoon weights were retarded (ANOVA, $F = 11.96$, $df = 3$, $P < 0.01$) and larval periods were prolonged by 2-6 days (ANOVA, $F = 4.61$, $df = 3$, $P < 0.05$). The needle water content decreased in the following summer at low and moderate defoliation levels, thus differing from the strongly increased content of the highest level (Student's paired comparison t-test, $t = -2.48$, $df = 8$, $P < 0.05$). The pooled values of N and Na increased and the values of P and Ca decreased in defoliated trees, but due to the small number of samples they just failed to be significant.

When the damage was over 50% defoliation affected the food quality of the sawflies mainly at the highest level. Low damage did not cause any response in *N. sertifer* larvae. The results were parallel to observations of *Neodiprion autumnalis* Smith and *Panolis flammea* D. & S. on defoliated *Pinus ponderosa* (Wagner, 1986) and *Pinus contorta* (Leather *et al.*, 1987). Significant changes in foliage quality have been detectable after a few days and have lasted for years (Wagner, 1986). In the present experiments the treatment effects in sawfly performance were seen after one years's defoliation.

The increment of the needle water content at the highest damage level can be interpreted as being compensatory. The decline at the lower levels corresponds to earlier observations. It has been suggested that P and K are important to folivores (Wagner, 1986; Leather *et al.*, 1987), whereas Ca and Na are not. The decreased amount of P may have affected the success of *N. sertifer*. Obviously, environmental conditions and tree species can also affect resistance (Tuomi *et al.*, 1988). In conclusion, Scots pine showed damage-induced changes in foliage quality after previous, extensive defoliation, which affected the sawflies using those trees.

References

Leather, S.R., A.D. Watt & G.I. Forrest (1987). Insect-induced changes in young lodgepole pine (*Pinus contorta*): the effect of previous defoliation on oviposition, growth and survival of the pine beauty moth, *Panolis flammea. Ecol. Entomol.* **12**: 275-281.
Tuomi, J., P. Niemelä, F.S. Chapin, J.P. Bryant & S. Sirén (1988). Defensive responses of trees in relation to their carbon/nutrient balance. In: W.J. Mattson, J. Levieux & C. Bernard-Dagan (eds), *Mechanism of Woody Plant Defenses Against Insects: Search for Pattern*, pp. 57-72. New York: Springer Verlag.
Wagner, M.R. (1986). Influence of moisture stress and induced resistance in ponderosa pine, *Pinus ponderosa* Dougl. ex Laws., on the pine sawfly, *Neodiprion autumnalis* Smith. *For. Ecol. Manage.* **15**: 43-53.

Proc. 8th Int. Symp. Insect-Plant Relationships, Dordrecht: Kluwer Acad. Publ.
S.B.J. Menken, J.H. Visser & P. Harrewijn (eds), 1992

Comparative studies of developmental biology, preference and feeding behavior of *Monellia caryella* on Juglandaceae native to North America

Michael T. Smith[1], Bruce W. Wood[2] and Ray F. Severson[3]
[1] *USDA-ARS, SIML, Stoneville (MS), USA*
[2] *USDA-ARS, SE. Fruit & Tree Nut Research Laboratory, Byron (GA), USA*
[3] *USDA-ARS, Phytochemical Research Unit, Athens (GA), USA*

Key words: Blackmargined aphid, host recognition, host selection, host suitability

Pecan, *Carya illinoensis* (Wang.) K. Koch, harbors a complex of three foliar-feeding aphid species, with *Monellia caryella* (Fitch) considered to be the most economically important among the three species. Development or improvement of non-chemical control methods for this aphid species requires a thorough understanding of the interaction between the

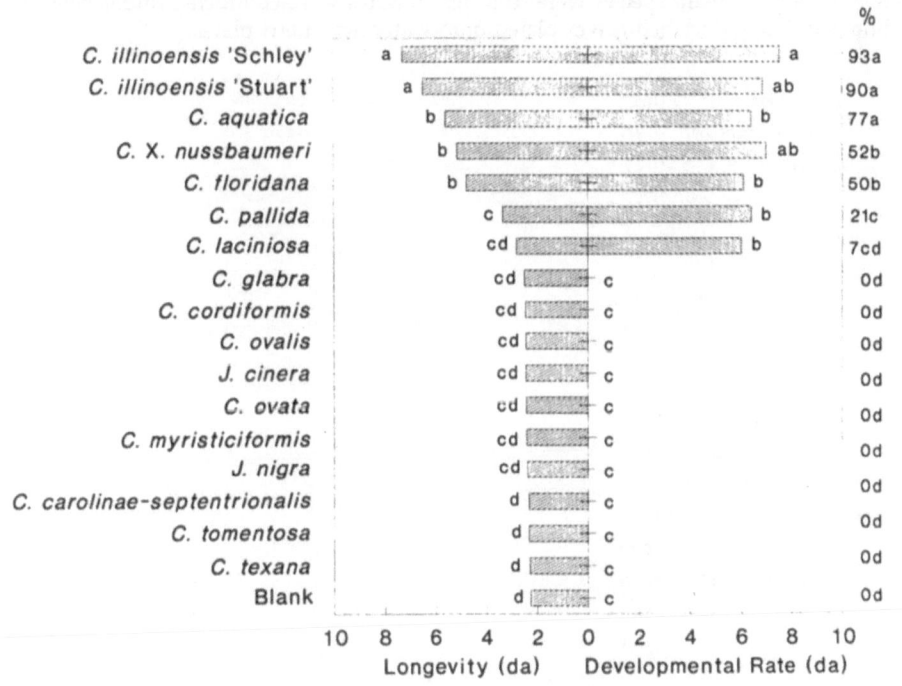

Figure 1. Nymph *Monellia caryella*: Longevity, developmental rates and percent nymphs developing to adult (%) among the Juglandaceae of North America.

aphid and its hosts. In a general survey of aphid species on Juglandaceae in North America, Bissell (1978) reported that *C. illinoensis* and *Carya cordiformis* (Wang.) K. Koch were the principle hosts of *M. caryella*, but that *M. caryella* had also been collected from *Carya aquatica* (Michaux f.) Nuttall, *Carya glabra* (Miller) Sweet, *Carya laciniosa* (Michaux f.) Loudon and *Carya ovata* (Miller) K. Koch. He did not believe that *M. caryella* inhabits any species of *Juglans*. Tedders (1978) reported that *M. caryella* was frequently collected from *C. aquatica* but never from other *Carya* spp., except *C. illinoensis*. These reports on the host specificity of *M. caryella* were based upon field observations. In the present paper we report results of investigations which more closely evaluate the biological and behavioral aspects of the relationship between *M. caryella* and the Juglandaceae native to North America. This information is prerequisite to elucidation and exploration of plant factors which control host plant specificity and the host selection processes of *M. caryella*.

Methods

Objectives of the research reported herein were: 1) to determine, under no-choice conditions, the longevity and developmental rates of nymphs and longevity and reproductive rates of adults of *M. caryella* among Juglandaceae native to North America (2 *Juglans* species, 13 *Carya* species and 1 interspecific pecan-hickory hybrid); and 2) to determine, under choice conditions, host plant preferences of nymph and adult *M. caryella* among species of Juglandaceae determined to be suitable host plants under the no-choice conditions. Observations of aphid behavioral activity (wandering as an indicator of searching, and settled as an indicator of probing and/or feeding) and aphid spatial position among the plant species were recorded. Studies were conducted under laboratory conditions utilizing detached leaves plated onto water agar petri plates.

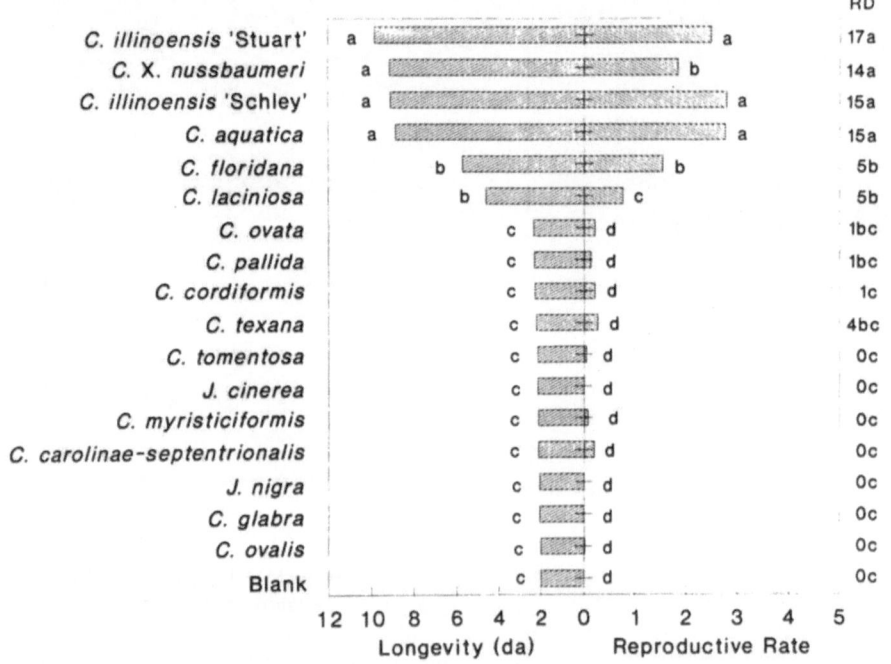

Figure 2. Adult *Monellia caryella*: Longevity, reproductive rates and reproductive days (RD = number days adults gave birth) among the Juglandaceae of North America.

Results and discussion

Results show that only 5 and 4 of the 15 Juglandaceae species were relatively suitable host plants for nymph and adult *M. caryella*, respectively (Figs 1 & 2). The interspecific hybrid cross and both *C. illinoensis* cultivars (Stuart and Schley) were also relatively suitable host plants for nymph and adult *M. caryella*. However, percentage of nymphs developing to adults, and adult reproductive days clearly show *C. illinoensis* (both cultivars), *C. aquatica* and *Carya X. nussbaumeri* Sarg. (interspecific hybrid) are the more suitable host species. Results from behavioral preference studies (Fig. 3) indicate that nymph and adult *M. caryella* prefer *C. aquatica*, *C. illinoensis* (both cultivars) and *C. X. nussbaumeri*, while adults also showed a preference for *C. laciniosa*.

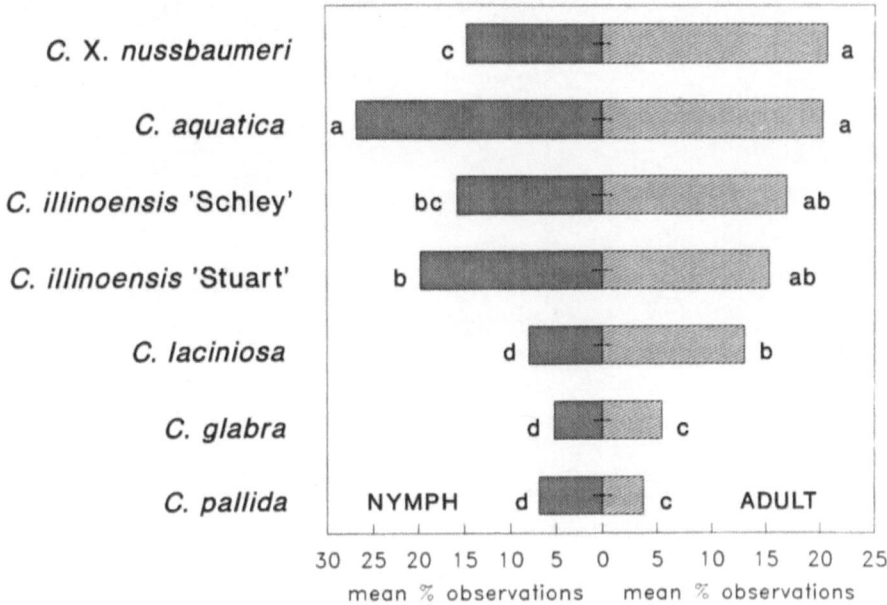

Figure 3. Preference of nymph and adult *Monellia caryella* among selected species of the Juglandaceae of North America.

These results correspond with the phylogenetic relatedness of the Juglandaceae species, with those species most closely related to *C. illinoensis* (*C. aquatica* and *C. X. nussbaumeri*) representing the most suitable and preferred host species of *M. caryella*. Both *C. illinoensis* and *C. aquatica* belong to the Apocarya section of the genus *Carya* ("pecan hickories"), while *C. X. nussbaumeri* is an interspecific hybrid cross of *C. illinoensis* and *C. laciniosa*. Therefore, *C. X. nussbaumeri* is more closely related to *C. illinoensis* than are the other two member species of the Apocarya (*C. cordiformis*, *Carya myristiciformis* (Michaux f.) Nuttall). Present investigations of leaf chemistry and electronic monitoring of *M. caryella* host selection and feeding behavior should identify mechanisms which control host specificity of *M. caryella*.

References

Bissell, T.L. (1978). Aphids on Juglandaceae in North America. *Maryland Agric. Exp. Station Misc. Pub. No. 911*, 78 pp.

Tedders, W.L. (1978). Important biological and morphological characteristics of the foliar-feeding aphids of pecan. *USDA Tech. Bull. No. 1579*, 29 pp.

Proc. 8th Int. Symp. Insect-Plant Relationships, Dordrecht: Kluwer Acad. Publ.
S.B.J. Menken, J.H. Visser & P. Harrewijn (eds), 1992

Host-plant selection by the tropical butterfly *Bicyclus anynana*

Rinny E. Kooi
Section of Evolutionary Biology, Dept of Biology, University of Leiden, Leiden, The Netherlands

Key words: Grasses, larval performance, oviposition choice, Satyridae

The African tropical butterfly *Bicyclus anynana* (Butler) (Satyridae) is present throughout the year in Malawi. In the wet season (November until April), there is a rich abundance of its grass food plants, which die away in the dry season.

B. *anynana* is oligophagous and its larvae feed on several tropical grasses. In this study, some aspects of its host-plant selection have been tested in experiments on oviposition choice and larval performance under dry and wet seasonal conditions.

Experiments

B. *anynana* was collected in Malawi (Brakefield & Reitsma, 1991). In an experiment on oviposition choice, three plants each of nine species of tropical grasses and one *Carex* species (Fig. 1) were offered simultaneously in a random block, to 20 females and 40 males in a climate room (5.9 x 2.25 x 1.90 m) (23°C, L12:D12, 90-100% RH) in two trials. After one week the number of eggs on each plant species was counted. Eggs had been laid on all of the species (Fig. 1). There was a significant difference in the number of eggs deposited on the ten species (ANOVA, $P < 0.001$), but no difference between the trials.

Forty-eight newly emerged first instar larvae were fed three grass species (Fig. 2) at 28°C (wet season) and at 17°C (dry season) (L12:D12, 90-100% RH) to investigate the suitability of food plants. The larvae developed well on all species tested. *Oplismenus compositus* appeared to be the most suitable food plant under both conditions as the

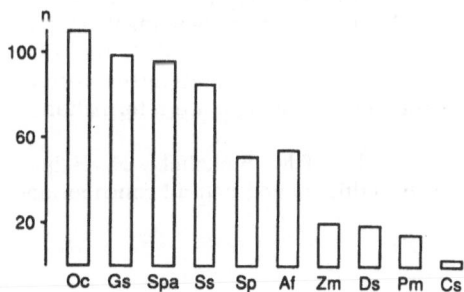

Figure 1. Number (n) of eggs deposited on ten grass species. Oc = *Oplismenus compositus* (L.) Beauv., Gs = *Ganotia stricta* Brongn., Spa= *Setaria palmifolia* (J.G. Koenig) Stapf, Ss = *Setaria* spec., Sp = *Setaria plicata* (Lamk.) Cooke, Af = *Axonopus flexuosus* (Peter) Troupin, Zm = *Zea mays* L., Ds = *Digitaria setifera* R. et S., Pm = *Panicum monticola* Stapf, Cs = *Carex* spec.

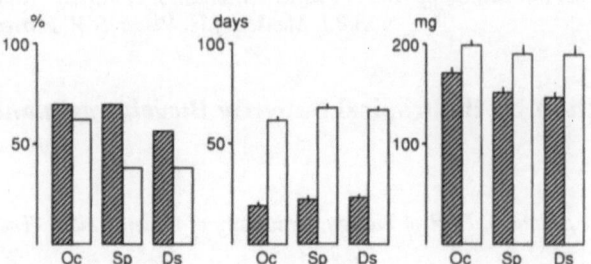

Figure 2. Larval performance on three grass species. Survival (left), larval developmental period (mid) and pupal weight (right), (mean ± SE). Hatched = 28°C; white = 17°C; see also Fig. 1.

survival was higher, the developmental period shorter and the pupal weight greater (ANOVA, P < 0.001). No interaction with temperature was found although the developmental period was much longer and the pupal weight was greater at 17°C.

Discussion

A more or less consistent pattern was discovered in the two experiments: the most preferred grass for oviposition appeared to be the most suitable food plant. This indicates that females are able to select suitable food plant species. As the insect accepts various plant species, presumably it can also shift her main food plant. This may be important for insects which are present in several generations per year.

These preliminary observations on the host-plant selection by *B. anynana* correspond with the general image that insects specialized for feeding on grasses are commonly oligophagous, feeding on grasses of more than one genus. They are mainly strictly graminivorous, sometimes including sedges (Cyperacea) in their diet (Bernays & Barbehenn, 1987). Remarkably, *B. anynana* larvae were able to complete development on the *Carex*-species used in the oviposition experiment.

B. anynana shows phenotypic plasticity. In the wet season their wings have conspicuous eyespots and a white band; in the dry season they are cryptic. Wet season butterflies are active and may use eyespots and white bands as active anti-predator devices, while dry season insects rest inactively on dead leaves.

As mentioned above, development time is dependent on food quality. Development time *per se* may be the fundamental factor controlling the wing pattern of *B. anynana* (Brakefield & Reitsma, 1991). Future work will investigate whether the effect of food plant on the developmental period also effects wing pattern formation.

Acknowledgements. I am grateful to P.M. Brakefield, H. Heijn, E. Schlatmann and J.F. Veldkamp for their constructive criticism and helpful contributions.

References

Bernays, E.A. & R. Barbehenn (1987). Nutritional ecology of grass foliage-chewing insects. In: F. Slansky, Jr. & J.G. Rodriguez (eds), *Nutritional Ecology of Insects, Mites, Spiders and Related Invertebrates*, pp. 147-175. New York: John Wiley & Sons.
Brakefield, P.M. & N. Reitsma (1991). Phenotypic plasticity, seasonal climate and the population biology of *Bicyclus* butterflies (Satyridae) in Malawi. *Ecol. Entomol.* **16**: 291-303.

Proc. 8th Int. Symp. Insect-Plant Relationships, Dordrecht: Kluwer Acad. Publ.
S.B.J. Menken, J.H. Visser & P. Harrewijn (eds), 1992

Interactions between host-plant information and climatic factors on diapause termination of two species of Bruchidae

J. Huignard, B. Tran, A. Lenga and N. Mandon
Institut de Biocénotique Expérimentale des Agrosystèmes, Tours, France

Key words: Coleoptera, imaginal alimentation, reproductive diapause

The Bruchidae (Coleoptera) are phytophagous insects developing in seeds of Leguminosae during the larval and nymphal stages. When the pods of their host plant are not available in the ecosystems, some species go into reproductive diapause. Diapausing enables insects to survive prolonged periods of unfavourable biocenotic and climatic conditions (Chippendale, 1982).

In this paper, we analyze the conditions of reproductive diapause termination in two species of specialist bruchids, namely, a temperate species *Bruchus rufimanus* (Boh.) developing in *Vicia faba* L. seeds and a tropical species *Bruchidius atrolineatus* Pic developing in cowpea seeds (*Vigna unguiculata* (Walp)). The main factor allowing diapause termination in both species, is the presence of host-plant inflorescences. Huignard *et al.* (1990) observed that when the host plant began to produce flowers *B. rufimanus* colonized the *V. faba* cultures and consumed much pollen. Experimental studies showed that the consumption of pollen was necessary for diapause termination of the females. When females consumed pollen of a male sterile variety (reduced to its exine) a limited number of females terminated their diapause (Tran & Huignard, 1992). The *B. atrolineatus* females did not penetrate the flowers and did not consume pollen. Previous experiments (Germain *et al.*, 1985) showed that the contact with host-plant inflorescences or with pods terminated diapause. Some chemical compounds of the *V. faba* pollen or of the *V. unguiculata* inflorescences probably influence the reproduction of the females by a trophic or a neurosensorial signal. In both species, a high proportion of males terminated their diapause when the climatic conditions were favourable, in the absence of host-plant stimuli.

The photoperiod was the main seasonal factor influencing the life cycle of both Bruchid species. Experimental studies were carried out to reproduce the photoperiodic conditions that prevailed when the host plant flowered in the field. The *B. rufimanus* males and females terminated their diapause only under conditions of long photoperiods (16:8 h LD or 18:6 h LD). Short photoperiods inhibited diapause termination, with or without host-plant inflorescences. The reproductive diapause of *B. atrolineatus* was terminated in the presence of host-plant inflorescences under climatic conditions prevailing at the end of the rainy season (the flowering period of *V. unguiculata*), *i.e.*, a short photoperiod (11:13 h LD) and a high air water content (14 g per kg). Longer photoperiods (11 h or 12 h L), low air water content (7 g per kg) and temperatures higher than 30°C, inhibited the influence of the host plant.

These interactions between climatic factors and host-plant stimuli are important for the regulation of insect reproduction. Females require more specific stimulations than males for diapause termination. Regulating reproduction by information from flowers is important in species that oviposit on pods; an egg laying substrate, available in nature for only a short period of time. It requires precise synchronization between the reproductive cycle of the plant and that of the beetles.

References

Chippendale, G.M. (1982). Insect diapause, the seasonal synchronization of life cycles and management strategies. *Entomol. exp. appl.* **31**: 24-35.

Germain, J.F., J. Huignard & J.P. Monge (1985). Influence des inflorescences de la plante hôte sur la levée de la diapause reproductive de *Bruchidius atrolineatus. Entomol. exp. appl.* **39**: 35-42.

Huignard, J., P. Dupont & B. Tran (1990). Coevolutionary relations between bruchids and their host plant. In: K. Fujii *et al.* (eds), *Bruchids and Legumes, Economics, Ecology and Coevolution,* pp. 171-179. Dordrecht: Kluwer Academic Publishers.

Tran, B. & J. Huignard (1992). Interactions between photoperiod and food affect the reproductive diapause in *Bruchus rufimanus* (Boh.). *J. Insect Physiol.* (in press).

Proc. 8th Int. Symp. Insect-Plant Relationships, Dordrecht: Kluwer Acad. Publ.
S.B.J. Menken, J.H. Visser & P. Harrewijn (eds), 1992

Assessing host-plant suitability in caterpillars: is the weight worth the wait?

E.D. van der Reijden and F.S. Chew
Biology Dept, Tufts University, Medford, Massachussets, USA

Key words: Fitness component, larval developmental time, *Pieris*, pupal weight

When we assess suitability of potential host plants for insects, we often measure components of fitness, *e.g.*, survivorship, larval developmental time, pupal weight, and adult fecundity. But intensity of selection on different fitness components might vary between species. The relative importance of two fitness components (larval developmental time and pupal weight) was studied, and found to differ between two species.

Two congeneric butterflies *Pieris napi oleracea* Harris (Nearctic) and *P. rapae* L. (naturalized to North America around 1860 from Palearctic sources (Scudder, 1889)) share similar habitats in New England, USA. However, their flight seasons differ greatly (Chew, 1981). *P. n. oleracea* is bivoltine with single-cohort broods, whereas *P. rapae* is multivoltine with 5 to 7 overlapping generations.

Larval developmental time was measured as date of hatching from the egg to date of pupation, inclusive. Pupal weight was measured 16-24 h after pupation.

We found that (a) *P. n. oleracea* varied widely in its larval developmental time, whereas *P. rapae*'s larvae developed faster within a narrower time range; (b) for *P. n. oleracea* males, heavier pupae had longer larval developmental times (r = 0.161, P < 0.05, n = 206; Fig. 1); and (c) for *P. rapae* males heavy pupae had shorter larval developmental times (r = -0.234, P < 0.05, n = 89; Fig. 1).

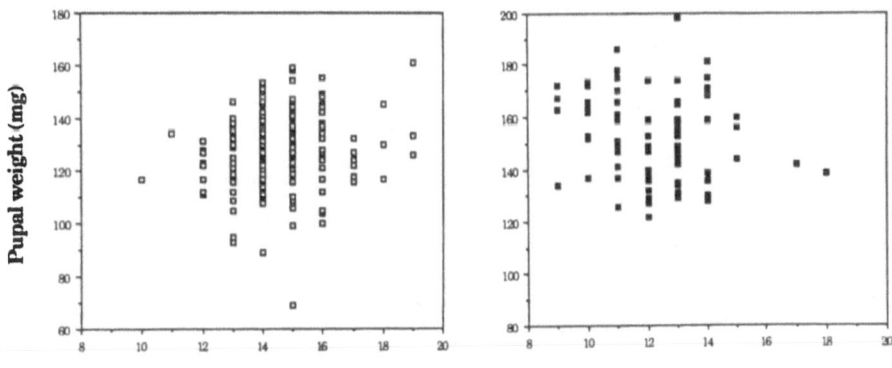

Figure 1. Pupal weight *vs* larval time of *Pieris* males. Left: *P. n. oleracea*: pupal weight increases with larval time. Right: *P. rapae*: pupal weight decreases with larval time. Trends for females were not significant.

These two species maximize different components of fitness. For *P. n. oleracea* males pupal weight gain is more important, whereas for *P. rapae* males, fast larval development is more critical.

When mating time is constrained by discrete flight periods (*P. n. oleracea*), there is little advantage for larvae to develop faster; and larval weight appears to be more critical. However, with continuous adult presence (*P. rapae*), faster-developing individuals could contribute an additional generation.

Host plants also influence larval time and pupal weight, and fitness components are used to study host plant effects. To accurately compare host-plant effects across insect species, we need to understand which fitness components are likely to be under intense or relaxed selection.

Since sex and temperature can also influence net growth results we need to account for these factors as well. Females in this study showed no significant trend; their growth responses, due to nutrition or behavior, could affect the net results we obtained. Temperature also affects growth responses and is currently under study.

Further research

To extend this work we plan to (a) determine whether pupal weight of a univoltine, monophagous North American congener (*P. virginiensis* Edwards) increases with larval developmental time; one might expect weight to be important because they only have one generation per year, but their host-plant season is extremely short. In addition we will examine (b) changes in larval time of *P. rapae* as voltinism varies from 2-3 generations per year in Alaska to continuous presences in Florida; and (c) study effects of food limitations (quantity and quality) on pupal weight and larval developmental time.

References

Chew, F.S. (1981). Coexistence and local extinction in two Pierid butterflies. *Am. Nat.* **118**: 655-672.

Scudder, S.H. (1889). *The Butterflies of New England*. Boston.

Proc. 8th Int. Symp. Insect-Plant Relationships, Dordrecht: Kluwer Acad. Publ.
S.B.J. Menken, J.H. Visser & P. Harrewijn (eds), 1992

Size, feeding ecology and feeding behaviour of newly hatched caterpillars

Duncan Reavey
Dept of Biology, University of York, UK

Key words: Larval weight, Lepidoptera, movement, silk, starvation

There is striking variation among Lepidoptera in the size of newly hatched larvae. Part of the explanation could be a link between size and feeding ecology.

Larval size could affect characteristics important in early larval life. For example, we might expect larger larvae to be more mobile, more tolerant of starvation and able to accept physically tougher food because they have larger mouthparts. We might therefore expect size to vary with the nature of the food plant and the spatial and temporal proximity of the eggs to the leaves on which the larvae will feed.

Here I examine patterns in the behaviour and survival of newly hatched larvae of 42 British species of Lepidoptera. I relate them to larval size and food plant characteristics. I find strong links between feeding ecology and larval behaviour. Surprisingly, however, there are few links between larval size and behaviour.

Results

Source of eggs. Eggs of 42 Lepidoptera species were laid in captivity by adults from the wild. Each egg was examined daily and on hatching each larva was weighed individually. Eggs were obtained from 13 families, mostly Noctuidae (17 spp.) and Geometridae (9 spp.). None of these overwinters as eggs. The mean size of newly hatched larvae spanned two orders of magnitude, from 0.022 mg to 2.91 mg.

Survival of starvation (40 spp.). Newly hatched larvae were placed in a wrapped petri dish at 15 °C. No food was provided. Each dish was examined daily until the death of the larva. Newly hatched larvae survived for a mean of 1.0 to 20.0 days without food. There was little variation within each species but great variation among species. There was no correlation with larval size (Fig. 1a). Grass feeders survived significantly longer than herb and woody plant feeders. The species surviving longest is the only lichen feeder studied.

Speed and direction of movement (35 spp.). Newly hatched larvae were placed midway up taut, 1.1 m vertical strings illuminated fom above. The speed and direction of movement (up or down) was recorded. Newly hatched larvae moved at a mean speed of 0.7 to 267.8 cm/h. There was no correlation with larval size (Fig. 1b) or with survival time (Fig. 1c). Grass feeders and woody plant feeders moved significantly faster than herb feeders. Grass feeders moved faster than woody plant feeders. Direction of movement varied widely among species. Woody plant feeders tended to move upwards, grass feeders downwards, and herb feeders both upwards and downwards (Fig. 1d).

Silking (12 spp.). Thirty newly hatched larvae were placed on a plastic grid suspended horizontally 20 cm above a tray. The proportion of larvae suspended by silk at any one time was recorded. There was a negative correlation between the proportion of larvae silking and larval weight (Fig. 1e); smaller larvae tended to silk more readily.

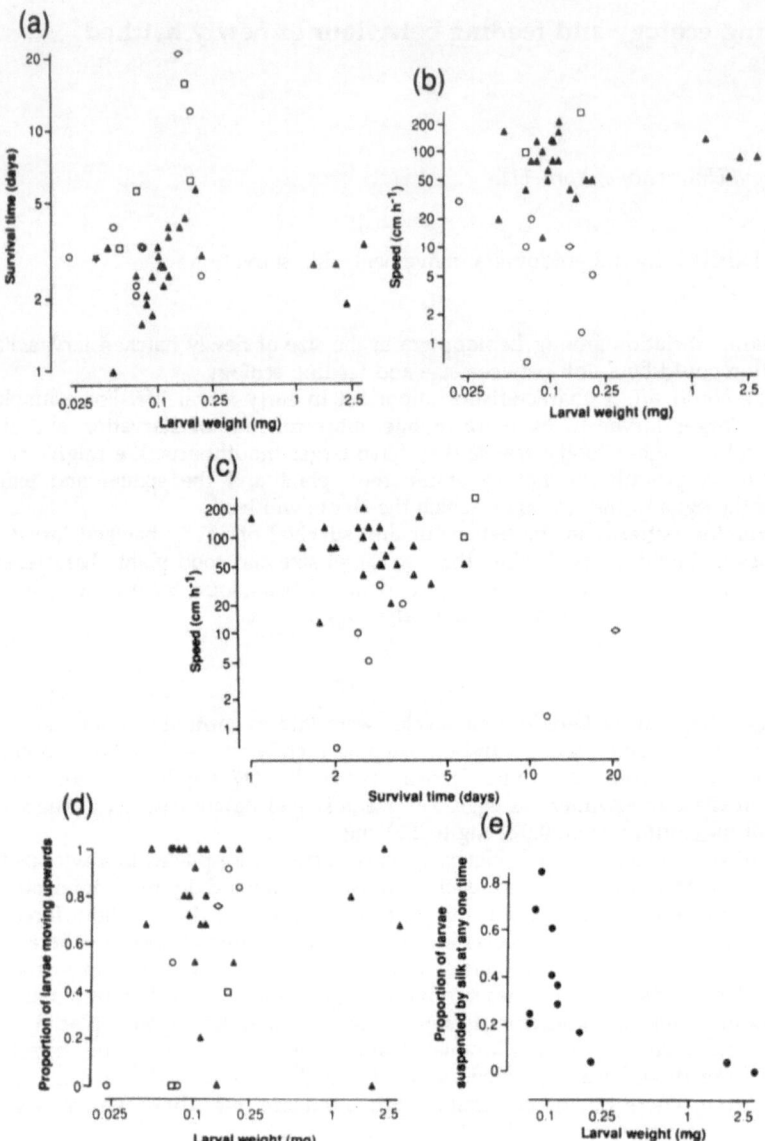

Figure 1. Relationships between (a) weight and survival time; (b) weight and speed of movement; (c) survival time and speed of movement; (d) weight and direction of movement; (e) weight and tendency to silk. For (a)-(d), food plant growth form for each species is indicated as follows: ▲ woody plant; o herb; □ grass; ◊ lichen; * multiple.

Discussion and conclusions

There are clear links between the feeding ecology and behaviour of newly hatched larvae:

Eggs and newly hatched larvae of woody plant feeders are larger than those of grass feeders which in turn are larger than those of herb feeders. Larger larvae could have mouthparts better suited to tougher food. Larger larvae also use foods more efficiently because they have a lower weight-specific basal metabolic rate. Adaptation of egg size to feeding requirements has been reported for predatory mosquitoes where egg size is related to the size of available prey.

While newly hatched larvae of woody plant feeders are bigger than those of herb feeders, survival time and speed and direction of movement are not correlated with body size. Rather, the patterns are characteristic of larvae feeding on plants of a particular growth form.

Surprisingly, there is no correlation between size and survival time. Larger size suggests greater energy reserves and a lower weight-specific metabolic rate. Survival time could be an adaptation to uncertainty or difficulty in obtaining food rather than a mere consequence of body size.

There is no correlation between size and speed. This is surprising as larval size determines stride length and absolute distance moved across an irregular surface such that larvae move more quickly as they grow. Speed could be an adaptation to the need to locate a distant or unpredictable food rather than a mere consequence of body size.

It is not possible to define particular combinations of starvation tolerance and behaviour that predominate. Other factors, like food proximity, food predictability and perhaps first instar dispersal complicate the picture.

Larval feeding is just one of many factors affected by size. Size could affect tolerance of desiccation and cold, vulnerability to predation, and time take to reach a critical size for diapause. Egg size could be kept to a minimum to maximise fecundity or it might simply reflect pressures for a particular size at another time in development if growth from egg to adult has little flexibility. Indeed, if use of a particular food plant requires newly hatched larvae to be a particular minimum size, it could require adults to be large enough to produce sufficient of these larger eggs.

Additionally, a larval size range of two orders of magnitude across different species has little effect on larval survival and behaviour. While particular features of larval size and behaviour might be more suited to feeding on woody plants or herbs or grasses, body size does not appear to constrain speed of movement and tolerance of starvation. Further features of these patterns are considered in Reavey (1992).

Reference

Reavey, D. (1992). Egg size, first instar behaviour and the ecology of Lepidoptera. *J. Zool.* (in press).

Proc. 8th Int. Symp. Insect-Plant Relationships, Dordrecht: Kluwer Acad. Publ.
S.B.J. Menken, J.H. Visser & P. Harrewijn (eds), 1992

Natural defence of pedunculate oak (*Quercus robur*) against the defoliating insect *Euproctis chrysorrhoea*

P. Scutareanu and R. Lingeman
University of Amsterdam, Dept of Pure and Applied Ecology, Amsterdam, The Netherlands

Key words: Minerals, nutrients, resistance, secondary compounds

After more than 30 years of research, several aspects of plant resistance to phytophagous insects still have to be proved. They include: appearance, predictibility and availability of chemical compounds, synchronisation and phenological protection, intraspecific variation in plant secondary chemistry and the pattern of allocation of defensive substances and nutrients for insect in tissues, and resource availability (Feeny, 1990).

We undertook studies carried out from 1986 - 1989 aimed at finding: (1) the level of secondary compounds that confer resistance to some pedunculate oak trees (*Q. robur* L.) against *Euproctis chrysorrhoea* L. (brown tail moth; Lepidoptera, Lymantryidae) attack; (2) the relationships among defensive substances, organic nutrients for caterpillars and mineral substances in leaves, as well as minerals in soil under the trees; (3) the quality of the leaves as food for caterpillars; (4) the transmission of the natural resistance from a parental tree to its vegetatively produced seedlings; (5) the frequency of resistant trees within an oak tree population and their specific phenotypic features, if any.

Material and methods

Location. Pure pedunculate oak stands in forest ecosystems located in northwest Rumania (forest Flora and Noroieni).

Insect pest. *E. chrysorrhoea* is an important defoliator which in Rumania has one generation per year, wintering as caterpillar in L2 instar.

Field work. This consisted of identifying and marking undefoliated resistant trees (N), and nearby heavily or completely defoliated unresistant trees (D), during outbreaks of *E. chrysorrhoea*. Leaf samples were collected from the two selected tree categories, 3-5 times a year: in April and May when the caterpillar were in their L2-L5 instars and at the end of August (L1-L2 instars). Furthermore we assessed phenotypic tree features, ingrafted seedlings using twigs detached from resistant trees, reared caterpillars from L2 to pupae on N and D tree leaves.

Laboratory work. This included chemical analysis using standard procedures, in order to characterize secondary compounds (flavonoids, polyphenols, organic acids, tannins, lignin), essential and unessential amino acids, organic nutrients for caterpillars (proteins and sugars) in leaves as well as mineral substances in leaves and soil under the tree (details will be provided elsewhere); biometric assessment of the caterpillars and leaves; biochemical analysis of the protein and lipid contents of caterpillars.

Results and conclusions

The natural defence of some resistant pedunculate oak trees was found to be constitutive and quantitative, due to a higher content of secondary compounds in their leaves. The concentrations of flavonoids and polyphenols in resistant (N) was always significantly higher than in unresistant (D) trees. This also holds for organic acid levels, which were often higher in young leaves, and tannins which were sometimes higher in mature leaves. Due to phenological differences in tree vegetation lignin presented some exceptions. At the time that the sampling was done in May, the content of lignin would have been normally higher in unresistant trees, which start budding earlier in spring. Essential amino acids (*e.g.*, valine) had higher values in D trees; unessential ones (*e.g.*, arginine) in N trees. Total protein and sugar, as well as protidic nitrogen was variable, but usually higher in N trees. Nitrogen and phosphorous was higher in N trees, while potassium, calcium and magnesium were lower, but we did not find the differences between the two categories to be statistically significant anywhere. In the soil underneath the resistant tree, the leaf litter is provided a high acidity.

The mean values of polyphenols, organic acids and tannin of the ingraft seedlings were equal to or even higher than those of the parental tree. Caterpillars fed with leaves picked

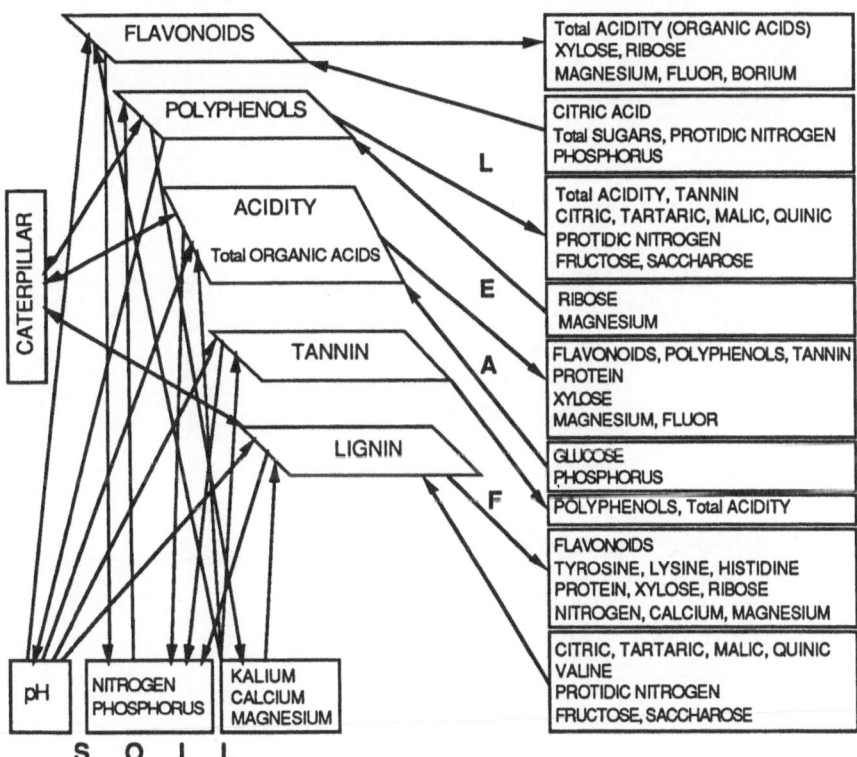

Figure 1. General pattern of interrelations among secundary defensive compounds in leaves of pedunculate oak resistent trees (N) and other compounds in leaves and soil influencing the development of *E. chrysorrhoea* caterpillars.

from resistant trees produced pupae that were lighter in weight than those fed with leaves from unresistant trees, whereas more leaves were consumed both in terms of number of leaves and percentage of damage on unresistant trees. We have drawn a general pattern of interrelations in a certain habitat (Fig. 1) based on (significant) regression lines and coefficients of correlation between the content in secondary compounds, nutrients for caterpillars and mineral substances in leaves which influence the growth of caterpillars, and between pH and mineral substances in the soil which influence the composition of leaves.

Finally, we can say that there is much intraspecific variability in plant secondary chemistry. The frequency of resistant trees within a pure pedunculate oak stand is over 20% of the total number of trees. Such resistant trees show specific phenotypic differences in branch insert angle and the depth of rhytidome cracks.

Reference

Feeny, P. (1990). Theory of plant chemical defense: a brief historical survey. *Symp. Biol. Hung.* **39**: 163-175.

Proc. 8th Int. Symp. Insect-Plant Relationships, Dordrecht: Kluwer Acad. Publ.
S.B.J. Menken, J.H. Visser & P. Harrewijn (eds), 1992

Estimating costs and benefits of the pyrrolizidine alkaloids of *Senecio jacobaea* under natural conditions

Klaas Vrieling and Catharina A.M. van Wijk
University of Leiden, Dept of Population Biology, Leiden, The Netherlands

Key words: Chemical defence, genetic variation, herbivory

Chemical defences in plants are subject to two opposing forces: benefits due to decreased herbivory and costs associated with the production and/or maintenance of the defence. Rhoades (1979), suggested that a varying herbivore pressure in time and space would maintain genetic variation in chemical defence. The aim of our experiment was to test this hypothesis under natural conditions over a three-year period. The biennial *Senecio jacobaea* L. and its herbivores was chosen as a model system. *S. jacobaea* is heavily attacked by several herbivores: pressure from most herbivores differs between populations and years. *S. jacobaea* contains pyrrolizidine alkaloids (PAs) which act as a chemical defence. The PA-concentration varies 10-fold between individuals and is under genetic control (Vrieling, 1991).

The following questions were derived from the hypothesis: 1) Do herbivores have a negative impact on plant fitness? 2) Does herbivore pressure vary in time and space? 3) Do herbivores discriminate between genotypes of *S. Jacobaea*? 4) Is herbivory negatively correlated with PA-concentration (benefits)? 5) Is plant fitness in herbivory-free treatments negatively correlated with PA-concentration (costs)?

Materials and methods

In April 1986, we cloned six genotypes of *S. jacobaea* from the dunes of Meijendel (near The Hague, the Netherlands) into 120 - 200 individuals per genotype. In April 1987, two artificial populations were established in the field. Each population was exposed to three treatments: a) a treatment in which we prevented herbivory (exclosure), b) a treatment where the larvae of *Tyria jacobaeae* L. were excluded (*Tyria* exclosure) and c) a control where all herbivores had free access. In May and September the rosette-diameter was measured to estimate biomass and number of flowerheads were counted in October. Experimental populations were visited frequently and visual damage by herbivores was recorded. In May, leaves were harvested for PA-analysis. PAs were extracted according to a procedure described by Vrieling *et al.* (1991).

Results

Herbivore pressure and plant fitness. Growth in rosette-diameter was less when herbivore pressure increased. The percentage of flowering plants and number of flowerheads produced was highest in the exclosure and lowest in the control.

Herbivore pressure in time and space. Herbivore pressure differed significantly between the two populations. Density of the different species of herbivores varied greatly between

77

years when corrected for rosette-diameter or number of flowerheads. Differences in density were even more pronounced when not corrected for available biomass.

Discrimination between genotypes by herbivores. In 8 out of 19 possible comparisons, herbivores discriminated between the six genotypes (among which the two most important herbivores of *S. jacobaea* in our study area, *T. jacobaeae* and the fleabeetle, *Longitarsus jacobaeae* Waterhouse), indicating genetic differences in acceptability.

Benefits of PAs. When herbivores significantly discriminated between genotypes the correlation between herbivory and PA-concentration was calculated. Herbivory of adults and larvae of *L. jacobaeae* significantly decreased with PA-concentration, indicating benefits of PAs.

Costs of PAs. Although differential growth between genotypes occurred in the greenhouse (before plants were transported to the field), PA-concentration was not significantly negatively correlated with the initial rosette-diameter. Moreover, growth during the summer season, percentage flowering and flowerhead production (corrected for rosette-diameter) were in none of the cases correlated with total PA-concentration in the exclosure.

Discussion

Fluctuations in herbivore pressure between populations and years are large enough to be important as a driving force in the maintenance of genetic variation in PA-concentration. If no interference with costs of other defence characteristics occurs, costs associated with PAs are rather small or absent, whereas benefits (reduced herbivory of *L. jacobaeae*), are present. Therefore, maintenance of genetic variation in PA concentration cannot be explained by a balance between cost and benefits. This suggests that another selective force acts on PA-concentration to counteract the benefits of reduced herbivory. Vrieling *et al.* (1991) put forward an alternative hypothesis involving a tritrophic interaction to explain genetic variation in PA-concentration.

Acknowledgements. This research was supported by the Foundation of Biological Research (BION), which is subsidized by the Netherlands Organisation for Scientific Research (NWO) and the NV. Duinwatermaatschappij Zuidholland.

References

Rhoades, D.F. (1979). Evolution of plant chemical defense against herbivores. In: G.A. Rosenthal & D.H. Janzen (eds), *Herbivores: Their Interaction with Secondary Plant Products*, pp. 3-54. New York: Academic Press.

Vrieling, K. (1991). Costs and benefits of alkaloids of *Senecio jacobaea* L. Ph.D. Dissertation. University of Leiden, The Netherlands.

Vrieling, K., E. van der Meijden & W. Smit (1991). Tritrophic interactions between aphids (*Aphis jacobaeae* Schrank), ant species, *Tyria jacobaeae* L. and *Senecio jacobaea* L. lead to maintenance of genetic variation in pyrrolizidine alkaloid concentration. *Oecologia* 86: 177-182.

Proc. 8th Int. Symp. Insect-Plant Relationships, Dordrecht: Kluwer Acad. Publ.
S.B.J. Menken, J.H. Visser & P. Harrewijn (eds), 1992

Induced chemical defence in *Cynoglossum officinale*

Nicole M. van Dam and Sheila K. Bhairo-Marhé
University of Leiden, Dept of Population Biology, Leiden, The Netherlands

Key words: *Ethmia bipunctella*, induced defence, induced response, pyrrolizidine alkaloids

The *Cynoglossum officinale* L. - *Ethmia bipunctella* F. relationship was chosen as a model to study induced chemical defence in plants. *C. officinale* (Boraginaceae), a biennial plant species, is commonly found in the sand dunes in Western Europe. Plants contain pyrrolizidine alkaloids (PAs), known to be a deterrent for generalist herbivores. Larvae of *Ethmia bipunctella* (Ethmiidae; Lepidoptera) are oligophagous and live on several species of Boraginaceae in the dunes. *Ethmia* caterpillars cause small holes, well dispersed over the leaves of *C. officinale* plants.

First, we assessed whether *C. officinale* shows an induced response for PAs. Second, we looked at the relation between the amount of damage inflicted and the strength of the response. Third, we compared the effect of artificial damage and natural damage caused by *Ethmia*.

Methods

Experiment 1. Seedlings were grown in a greenhouse (min. 20°C) for 8 weeks. 144 plants were randomly assigned to 6 groups. The scheme of damaging and harvesting is shown in Fig. 1. At T = 0 all the leaves of the plants of the first group (T3T0) were cut with a

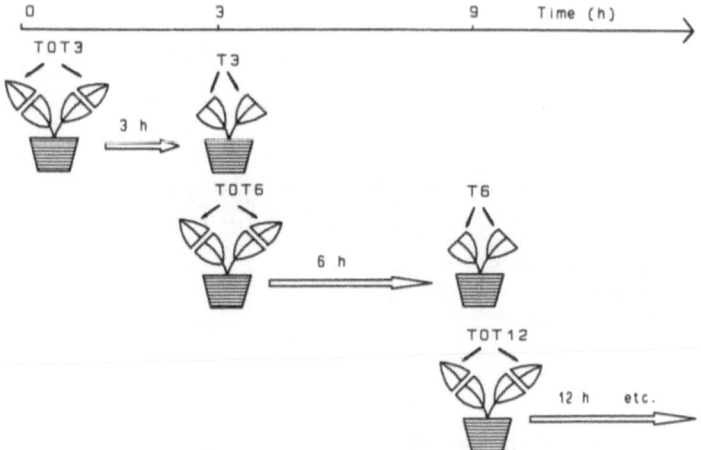

Figure 1. Time schedule of damaging and harvesting.

pair of scissors. The rest of the damaged plant was harvested 3 h later (T3) and simultaneously the plants of the next group (T6T0) were damaged and the leaftips subsequently sampled. In this way we were able to check any diurnal fluctuations. The samples were lyophyllized and extracted for PAs (Hartmann & Toppel, 1987). PA-concentrations were determined by a spectrophotometric method (Mattocks, 1967).

Experiment 2. Twenty-five plants were randomly divided into 5 groups and damaged by punching 5, 10, 15, 25 or 50 holes (3 mm) per plant. This type of damage mimics the damage of *Ethmia* larvae. The plants were harvested after 24 h and the PAs were measured. The undamaged leaves of the group with 5 holes (5 undam.) were collected separately from the damaged ones (5 dam.). We used two control groups: one at the beginning (con 0h) and one at the end (con 24h) of the experiment.

Experiment 3. *E. bipunctella* larvae were reared on *C. officinale* leaves in the laboratory. Fourth and 5th instar larvae were starved for 19 h and subsequently placed in leaf cages (one per plant) on the youngest, fully expanded leaf of a *C. officinale* plant. They were allowed to eat for 2 h. Then the larvae were removed and the place and amount of damage was copied on another group of plants. Naturally and artificially damaged leaves were collected at 0, 6, 12, 24 and 48 h after the larvae had been removed.

Results and discussion

C. officinale clearly shows an induced response after removing 50% of the leaf surface (Fig. 2). The ratio of $Tx/Tx,0$ (induced divided by control) increases according to the time between damage and harvest. *C. officinale* shows an induced response after severe mechanical damage and also as a reaction to moderate damage (Fig. 3). The PA-content reaches its maximum at 15 holes per plant. Plants with 10-15 holes are very commonly found in the field. *C. officinale* thus seems well adapted to the natural amount of damage.

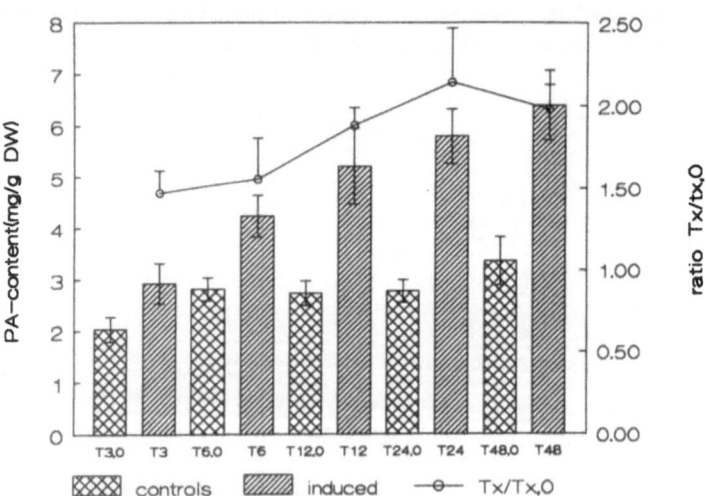

Figure 2. PA-contents (± SEM) in controls remain constant in time (ANOVA, P > 0.05) in induced plants. PAs increase with time after damage (ANOVA, P = 0.004). Ratio increases with time after damage.

Moreover, results indicate that induction of PAs is local. The undamaged leaves of the plants with five holes contain less PAs than the damaged ones of the same plants (Fig. 3). Such local induction can be the reason for the dispersed damage pattern of *Ethmia* larvae on *C. officinale*. Dispersion of damage might be one of the advantages of induced defences because the plant does not lose whole leaves (Edwards & Wratten, 1983). More experiments are needed to confirm this hypothesis.

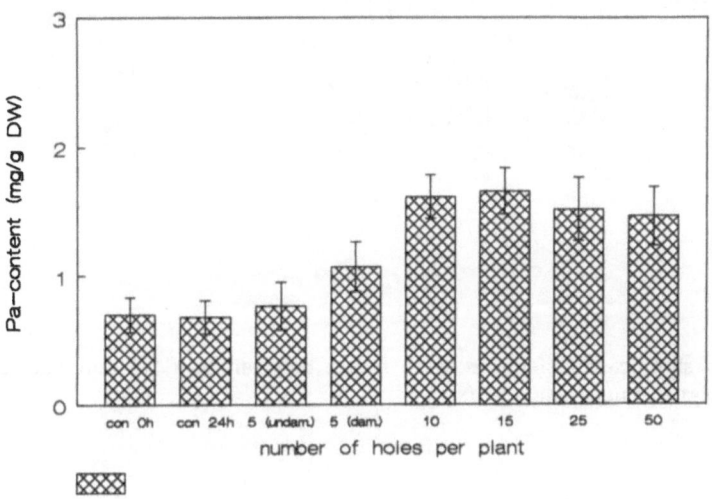

Figure 3. PA-content (± SEM) is significantly different between groups (ANOVA, P = 0.0005). Maximum response at 15 holes per plant.

Although no significant differences were found, the natural damage inflicted by *Ethmia* larvae seems to result in a stronger response (Fig. 4). The great variability of the response to natural damage (*e.g.*, 12 h after damage) might be caused by differences in the manner in which the larvae eat from the plant.

Future research should reveal whether the induced response affects the behaviour or fitness of *E. bipunctella* or other herbivores of *C. officinale*. Prins (1987) has already discovered that *Ethmia* larvae significantly preferred leaf discs from undamaged plants to discs cut from damaged plants.

References

Edwards, P.J. & S.D. Wratten (1983). Wound induced defences in plant and their concequences for patterns of insect grazing. *Oecologia* **59**: 88-93.
Hartmann, T. & G. Toppel (1987). Senecionine N-oxide; The primary product of Pa-biosynthesis in root cultures of *Senecio vulgaris*. *Phytochemistry* **26**: 1639-1643.
Mattocks, A.R. (1967). Spectrophotometric determination of unsaturated pyrrolizidine alkaloids. *Anal. Chem.* **39**: 443-447.
Prins, A.H. (1987). On the relationships between *Ethmia bipunctella* and its host plant *Cynoglossum officinale* L. *Med. Fac. Landbouw. Rijksuniv. Gent* **52**: 1335-1341.

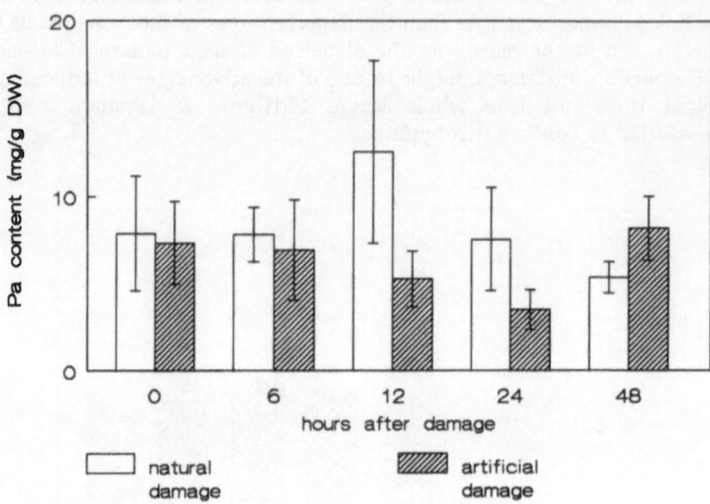

Figure 4. No significant differences were found between artificial and natural damage (Wilcoxon's signed ranks match pairs).

Proc. 8th Int. Symp. Insect-Plant Relationships, Dordrecht: Kluwer Acad. Publ.
S.B.J. Menken, J.H. Visser & P. Harrewijn (eds), 1992

Conversion of plant-derived pyrrolizidine alkaloids into insect alkaloids

Andreas Biller and Thomas Hartmann
Institut für Pharmazeutische Biologie der Technischen Universität, Braunschweig, Deutschland

Key words: Arctiidae, callimorphine N-oxide, *Tyria jacobaeae*

Larvae of the specialized arctiid *Tyria jacobaeae* L. sequester pyrrolizidine alkaloids (PAs) from their major host plant *Senecio jacobaea* L. (Asteraceae). The PAs are transferred via pupae to adults as described years ago by Rothschild *et al.* (1979). Plants and insects containing PAs are usually avoided by predators. Larvae and imagines are aposematically coloured to advertize their unpalatability to potential insectivores.

In the course of comparative studies of PAs as defensive chemicals in plants and insects we reinvestigated the sequestration of PAs by *T. jacobaeae*. We were particularly interested in elucidating the origin of callimorphine, first described as a "PA-metabolite" by Rothschild. Callimorphine is an insect-specific PA that has never been detected in plants.

Results

A comparison of the PA patterns of *T. jacobaeae* larvae and their host plant revealed that individual PAs are taken up without preference, except for O-acetylerucifoline, which is hydrolyzed in the gut and sequestered in its deacetylated form. Like plants, insects store PAs exclusively as N-oxides and they are able to N-oxidize any tertiary PA supplied in their food or injected into the haemolymph (Ehmke *et al.*, 1990).

The formation of callimorphine, present as N-oxide, was found to be restricted to the early stage of pupation. Tracer studies with ^{14}C-labelled retronecine and isoleucine revealed the formation of labelled callimorphine N-oxide. The label in the necine base moiety was specifically derived from [^{14}C]retronecine fed to larvae whereas feeding of [^{14}C]isoleucine led to a specific incorporation into the necic acid moiety (callimorphic acid). Thus, callimorphine N-oxide is formed in the insect by "partial biosynthesis", *i.e.*, esterification of a necine base of plant origin with a necic acid produced by the insect (Fig. 1). The recently described creatonotines synthesized by *Creatonotos transiens* (Walker) are another example of insect PA (Fig. 1; Hartmann *et al.*, 1990).

As the formation of callimorphine N-oxide in *T. jacobaeae* is restricted to the first stage of pupation one might argue that the process is related to the metabolic changes that take place during this critical metabolic transition. In particular, we have to expect changes in the storage behaviour of PAs. Detailed studies gave the following results:

a. [^{14}C]Retronecine fed to third instar larvae was incorporated into callimorphine N-oxide with the same efficiency as when fed to last instar larvae. This indicates the ability of larvae to store free retronecine until onset of pupation.

b. We compared the total amounts and concentrations of plant-derived PA-N-oxides, retronecine and callimorphine N-oxides in the haemolymph and body-tissue of last instar larvae and three early pupal stages (*i.e.*, 1) prepupae, 2) soft, yellow coloured,

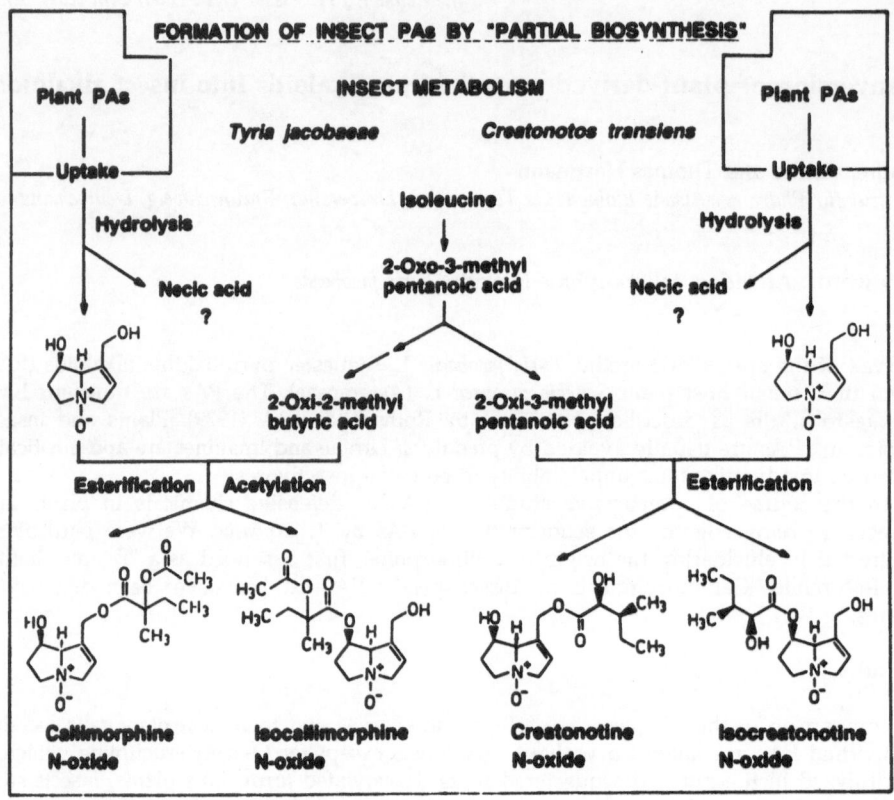

Figure 1. Formation of insect PAs by esterification of a plant-derived necine base with a insect produced necic acid.

age 0.5 day, 3) mature, 2 days). In both larvae and pupae most of total PAs were localized in the body-tissue (mainly integument). However, in the larvae, the PA concentrations in the haemolymph and body-tissue were found to be almost identical, whereas the PA concentration in the "haemolymph" (cell-free liquor) of the three pupal stages was almost twice as high as in the body-tissue. This clearly indicates the "storage-problems" in pupae.

c. In larvae free retronecine was almost exclusively found in the haemolymph; in the body-tissue the amount was negligible. A considerable proportion is rapidly transformed into callimorphine N-oxide during the prepupal stage. The newly formed callimorphine N-oxide is equally distributed between pupal haemolymph and body-tissue.

Conclusions

From the results described above we can offer two suggestions in answer to the question: why does *T. jacobaeae* produce callimorphine? Callimorphine is either produced as 1) a physiological need to transform free retronecine into an ester better suited for safe storage in pupae, or 2) an ecological need to recycle retronecine into a compound with better (?) properties for chemical defence.

References

Ehmke, A., L. Witte, A. Biller & T. Hartmann (1990). Sequestration, N-oxidation and transformation of plant pyrrolizidine alkaloids by the arctiid moth *Tyria jacobaeae*. *Z. Naturforsch.* **45**: 1185-1192.

Hartmann, T., A. Biller, L. Witte, L. Ernst & M. Boppré (1990). Transformation of plant pyrrolizidine alkaloids into novel insect alkaloids by arctiid moths (Lepidoptera). *Biochem. Sys. Ecol.* **18**: 549-554.

Rothschild, M., R.T. Aplin, P.A. Cockrum, J.A. Edgar, P. Fairweather & R. Lees (1979). Pyrrolizidine alkaloids in arctiid moths (Lep.) with a discussion on host plant relationships and the role of these secondary plant substances in the Arctiidae. *Biol. J. Linn. Soc.* **12**: 305-326.

Proc. 8th Int. Symp. Insect-Plant Relationships, Dordrecht: Kluwer Acad. Publ.
S.B.J. Menken, J.H. Visser & P. Harrewijn (eds), 1992

Phytoecdysteroids and insect-plant relationships in the Chenopodiaceae

Laurence Dinan
Dept of Biological Sciences, University of Exeter, Washington Singer Laboratories, Exeter, UK

Key words: Fat hen, flowering, goose-foot, pollination, polypodine B, 20-hydroxyecdysone

The functions of phytoecdysteroids remain enigmatic. The favoured hypothesis is that they reduce the extent of invertebrate predation on plants containing them, but definitive evidence for this is currently lacking (Lafont *et al.*, 1991). This is largely because there is presently no significantly detailed body of information about the identity and developmental titres of phytoecdysteroids for any one species of plant or any group of closely related species. One of the most effective ways of testing this hypothesis would be to establish from a normally phytoecdysteroid-containing plant species genetically stable lines with elevated or reduced phytoecdysteroid levels and to assess their relative susceptiblity to insect predation. However, prior to achieving this long-term goal, it would be necessary to identify an experimentally suitable species and ascertain as much as possible about the nature and distribution of phytoecdysteroids within the plants.

Current research. The Chenopodiaceae were chosen as the subject of the investigation because they are a large and important plant family, some members of which were known to contain ecdysteroids. Several chenopods are of significant agronomic importance being either crop species, weeds or the host plants for major insect pest species. We initially screened a number of chenopods and identified *Chenopodium album* L. as the major test system on the basis of its phytoecdysteroid content and growth potential (Dinan *et al.*, 1991). Phytoecdysteroids were isolated and identified as 20-hydroxyecdysone (69%), polypodine B (28%) and a mixture of at least eleven other unidentified ecdysteroids (3%). Only the two major phytoecdysteroids are biologically active in an insect (*Drosophila*) in vitro bioassay. In order to quantify phytoecdysteroids in small portions of *C. album* and hence be able to determine their precise distribution within individual plants, it was imperative to develop a sensitive micro-analytical method of extraction and quantification. This was achieved by coupling a simple solvent extraction procedure with an ecdysteroid-specific radioimmunoassay. Application of this procedure has revealed that the highest concentrations (ca. 0.1% of the dry wt) of phytoecdysteroids are present in the uppermost aerial portions of the plant, in the roots and in the flowers. Concentrations in root tissue fluctuate during development, indicating that this may be the site of phytoecdysteroid biosynthesis. The distribution in the vegetative parts of the plant is characteristic of a "qualitative" defence chemical. Within the flowers, the highest concentrations of phytecdysteroids are associated with the anther tissue (0.5% of the dry wt), yet the pollen contains negligible phytoecdysteroid levels. Analysis of several members of the genus *Chenopodium* has revealed that they either possess a similar association of phytoecdysteroids with flowering as *C. album* or contain no detectable phytoecdysteroid. This has led to the proposal that there may be a relationship between the presence of phytoecdysteroids and the mode of pollination of the species (Dinan, 1992). Thus, those species which are wind pollinated or pollinated by adapted insect species would contain phytoecdysteroids, while those species which are pollinated by unadapted insects would contain low levels of phytoecdysteroid. The absence of phytoecdysteroids from pollen from wind-pollinated species (such as *C. album*) may be explained by its need to be protected only during development, since pollen represents a considerable investment of resources by the plant, and hence the high levels in anther tissue. Once it is released there is only a low probablity of an individual pollen grain finding a flower to fertilise, so individual pollen grains are not worth protecting. Unfortunately, little conclusive data exists on the modes of pollination of most members of the Chenopodiaceae, so it is not currently possible to assess the validity of this hypothesis.

Future research. The microassay provides a means for the identification of individual plants of *C. album* which possess altered phytoecdysteroid levels. As a consequence of the simplicity, sensitivity and rapidity of the microassay procedure, it is feasible to remove single leaves from large numbers of individual plants in a population of *C. album* at, for example, the four-true-leaf stage and assess their ecdysteroid levels before the plants begin to flower. Genetic variability may already be present in natural populations or may be induced by mutagenesis. As a first step to exploiting natural variability, the repeatablity of phytoecdysteroid levels in the fourth and eighth leaves of 100 individual plants in a population was assessed (Fig. 1). This suggests that the variation of this character may be heritable. If the heritability is high enough, it should be possible to select lines with either high or low phytoecdysteroid levels within 3 to 4 generations.

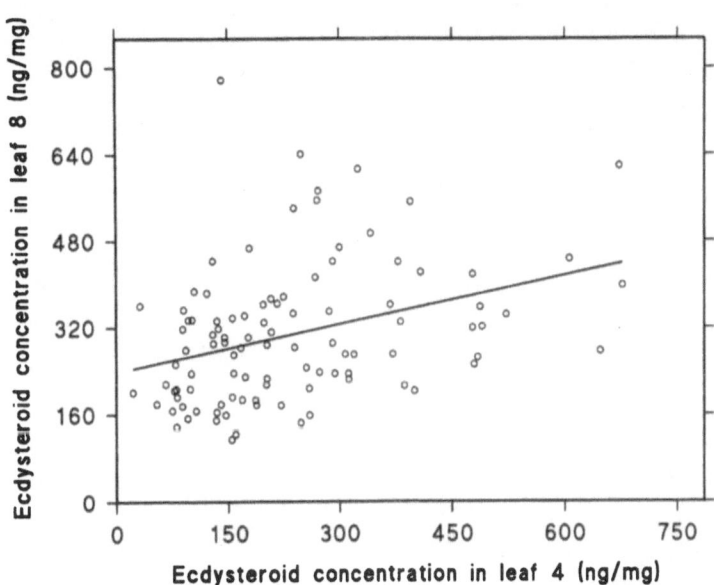

Figure 1. Repeatability of phytoecdysteroid concentration in leaves from individual plants of *Chenopodium album*. The uppermost leaf from 100 plants at the 4-true-leaf stage was removed, the plants were then allowed to develop to the 8-true-leaf stage. Again the uppermost leaf was removed. Phytoecdysteroids were extracted from the removed leaves and quantified by RIA. Equation of regression line: y = 0.299406x + 236.4285 (corr. coeff. = 0.342776).

Also, *C. album*'s widespread occurrence will be used to advantage to test populations from widely differing geographical locations. Should the natural variablity be inadequate, mutations will be induced chemically in germinating seeds with ethyl methanesulphonate. The M1 generation will be selfed and their progeny (M2 generation) assessed for ecdysteroid levels. This approach has previously been successfully used for the identification of mutants of lipid metabolism in *Arabidopsis thaliana* (L.) Heynh. (Browse *et al.*, 1985). Ultimately, it should be possible to determine the contribution of phytoecdysteroids to insect deterrence in the Chenopodiaceae.

References

Browse, J., P. McCourt & C.R. Sommerville (1985). A mutant of *Arabidopsis* lacking a chloroplast-specific lipid. *Science* **227**: 763-765.

Dinan, L. (1992). The association of phytoecdysteroids with flowering in fat hen, *Chenopodium album*, and other members of the Chenopodiaceae. *Experientia* (in press).

Dinan, L., S. Riseborough, M. Brading, C.Y. Clement, D.J. Witts, J. Smith, S. Colombe, V. Pettitt, D.A. Wheeler & D.R. Greenwood (1991). Phytoecdysteroids in the Chenopodiaceae (Goosefoots). In: I. Hrdý (ed.), *Insect Chemical Ecology*, pp. 215-220. Prague: Academia and The Hague: SPB Academic Publ.

Lafont, R., A. Bouthier & I.D. Wilson (1991). Phytoecdysteroids: structures occurrence, biosynthesis and possible ecological significance. In: I. Hrdý (ed.), *Insect Chemical Ecology*, pp.197-214. Prague: Academia and The Hague: SPB Academic Publ.

Proc. 8th Int. Symp. Insect-Plant Relationships, Dordrecht: Kluwer Acad. Publ.
S.B.J. Menken, J.H. Visser & P. Harrewijn (eds), 1992

The non-nutrional relationship of *Zonocerus* (Orthoptera) to *Chromolaena* (Asteraceae)

Michael Boppré[1], Andreas Biller[2], Ottmar W. Fischer[1] and Thomas Hartmann[2]
[1] *Forstzoologisches Institut der Universität Freiburg i.Br., Stegen-Wittental, Germany*
[2] *Institut für Pharmazeutische Biologie, Technische Universität Braunschweig, Germany*

Key words: Grasshoppers, pharmacophagy, pyrrolizidine alkaloids, sequestration, Siam weed

Zonocerus variegatus L. (Pyrgomorphidae) is a polyphagous, aposematic, West African grasshopper, the dry season populations of which are nowadays a pest in agriculture and forestry. The insects seem to gain protection from predation by storing pyrrolizidine alkaloids (PAs) from plant sources: (1) Bernays *et al.* (1977) reported storage of monocrotaline as the main PA in specimens reared on *Crotalaria retusa* (Fabaceae); (2) Boppré *et al.* (1984) found a close relative, Z. *elegans*, to be attracted to sources of dry PA-containing plants as well as to pure PAs, suggesting a pharmacophagous relationship of *Zonocerus* to PAs (*cf.* Boppré, 1986). In this paper, we report on sequestration of PAs by *Zonocerus* from the Siam weed, *Chromolaena odorata* King & Robinson (Asteraceae) and discuss a striking influence that secondary compounds of an introduced non-host plant can have on native insect populations.

Results and conclusions

Field-caught specimens of dry season populations of *Zonocerus variegatus*, collected in a teak plantation in Bénin, were found to contain four PAs in all stages and in both sexes as well as in eggs: rinderine (major PA), intermedine, lycopsamine and echinatine, all quantitatively as N-oxides. The PAs found in field-caught *Zonocerus* were not present in specimens raised indoors unless they had been given access to flowers of *Chromolaena*, which they readily consume.

Chemical analyses of plant material revealed rinderine (major PA), [7]O-angeloyl-retronecine, [9]O-angeloyl-retronecine, intermedine, and acetyl-rinderine in roots and flowers of *C. odorata*. The leaves, however, contained only traces of these PAs. Obviously, *Zonocerus* transforms intermedine into its isomer lycopsamine and rinderine into echinatine (3'R derivative into 3'S derivative form).

The discovery that *Zonocerus* stores PAs from *Chromolaena* is not just another case of utilization of secondary plant compounds by insects:

(1) Hatchlings and early instar hoppers partly feed on *Chromolaena*, and its thickets are preferred roosting sites. The leaves, however, are not heavily consumed and the plant turns out to be nutritionally inadequate for *Zonocerus*, not permitting normal development (*e.g.*, Chapman *et al.*, 1986). But, all stages of *Zonocerus* are strongly attracted to flowers of *Chromolaena*, which they consume in large numbers (Modder, 1984).

(2) *C. odorata* is not native to Africa; it is a South American species which has only recently been introduced into West Africa where it has become a serious pest (*e.g.*,

Ambika & Jayachandra, 1990).

Although Toye (1974) suggested the possibility of a correlation between the increase of *Zonocerus* populations and the spread of *Chromolaena*, he offered no clue regarding any possible causal link. Consequently the question remained: if this weed is nutritionally inadequate for *Zonocerus*, why do the insects become a pest in places where it occurs?

The apparent contradiction of the facts known becomes intelligible when considering the pharmacophagous trait of *Zonocerus*, supporting the suggestion forwarded by Boppré (1991): *Zonocerus* enjoys a non-nutritional association with *Chromolaena* which provides PAs; these secondary plant compounds are stored and chemically protect the grasshoppers and particularly their diapausing eggs from predators and parasitoids (*e.g.*, larvae of *Mylabris* beetles), contributing to the fitness and population density of dry season *Zonocerus*. Without *Chromolaena*, *i.e.*, either before its introduction, or in areas where it is lacking, or in the wet season when *Chromolaena* does not bloom, PAs seem to be a limited resource restricting the reproductive success of the grasshoppers.

Despite the coincidence of the spread of *Chromolaena* and the explainable pest status of dry season populations of *Zonocerus*, this is an example of the dramatic effect that the introduction of a plant may have on population dynamics of a native insect species although not used for nutrition. There is the danger that such ecologically harmful effects may occur with other introduced or exotic plants. By taking advantage of its pharmacophagous behaviour we could try to find an inexpensive means of controlling *Zonocerus*. We know, for instance, that Z. *variegatus* is attracted to various PA-containing plants (*Heliotropium, Crotalaria*) in the same way as to *Chromolaena* flowers, and pure PAs are effective lures, too.

Acknowledgement. Financial support by the Gesellschaft für Technische Zusammenarbeit (GTZ) is gratefully acknowledged.

References

Ambika, S.R., & Jayachandra (1990). The problem of *Chromolaena* weed. *Chromolaena odorata Newsletter* **3**: 1-6.

Bernays, E.A., Edgar, J.A. & M. Rothschild (1977). Pyrrolizidine alkaloids sequestered and stored by the aposematic grasshopper, *Zonocerus variegatus*. *J. Zool.* (Lond.) **182**: 85-87.

Boppré, M. (1986). Insects pharmacophagously utilizing secondary plant substances (pyrrolizidine alkaloids). *Naturwissenschaften* **73**: 17-26.

Boppré M. (1991). A non-nutritional relationship of *Zonocerus* (Orthoptera) to *Chromolaena* (Asteraceae) and general implications for weed management. In: R. Muniappan & R. Ferrar (eds) *Ecology and Management of Chromolaena odorata; Proc. 2nd Intern. Workshop on Biol. Control of Chromolaena odorata. (BIOTROP Special Publ No 44.) Bogor, Indonesia: ORSTOM and SEAMEO BIOTROP.*

Boppré, M., Seibt, U. & W. Wickler (1984). Pharmacophagy in grasshoppers? *Zonocerus* being attracted to and ingesting pure pyrrolizidine alkaloids. *Entomol. exp. appl.* **35**: 713-714.

Chapman, R.F., Page, W.W. & A.R. McCaffery (1986). Bionomics of the variegated grasshopper (*Zonocerus variegatus*) in West and Central Africa. *Annu. Rev. Entomol.* **31**: 479-505.

Modder W.W.D. (1984). The attraction of *Zonocerus variegatus* (L.) (Orthoptera: Pyrgomorphidae) to the weed *Chromolaena odorata* and associated feeding behaviour. *Bull. entomol. Res.* **74**: 239-247.

Toye, S.A. (1974). Feeding and locomotory activities of *Zonocerus variegatus* (L.) (Orthoptera, Acridoidea). *Rev. Zool. Afr.* **88**: 205-212.

Host-Plant Selection

Proc. 8th Int. Symp. Insect-Plant Relationships, Dordrecht: Kluwer Acad. Publ.
S.B.J. Menken, J.H. Visser & P. Harrewijn (eds), 1992

Search behaviour: strategies and outcomes

Rhondda E. Jones
Zoology Dept, James Cook University, Townsville, Queensland, Australia

Key words: Edge effect, *Epilachna varivestis*, *Eurema hecabe*, orientation, *Pieris rapae*,
resource concentration, simulation models

How an insect goes about finding its way to a food item or an oviposition site and starts
to use it, has traditionally been divided into a number of stages. The commonest
classification of the stages, following the parasitoid literature, is probably into habitat-fin-
ding, host-finding, host recognition, host acceptance, and host suitability. A much broader
operational subdivision which has proved very useful for flying insects is into "pre-
alighting" and "post-alighting" responses. The main advantage of this broader separation
is that it avoids the problem of trying to define functional transitions - for example, when
recognition stops and acceptance starts. Although the process is subdivided, an examination of the literature reveals that the greatest emphasis has been on the later stages; that
is, on host recognition and acceptance, and on post-alighting responses. This emphasis is
not surprising, for two reasons:
  Methodological difficulty. Study of pre-alighting behaviours requires analysis of how
insects move around. Except in large, dayflying insects like butterflies and insects which
walk rather than fly, analyses of flight patterns and alighting behaviour may require
formidable levels of ingenuity and environmental manipulation (see, *e.g.*, the methods
recently developed by Aluja *et al.* (1989) to monitor the behaviour of fruit flies foraging in
trees). More often, the best that can be done is some form of mark recapture study, where
all that is known is where the insect started and where it was finally caught, without any
real idea of the track it took to get there. Consequently, analyses of pre-alighting behaviour are heavily biased toward butterflies and occasionally beetles, and to insects which
walk rather than fly. The problems involved in obtaining the data are, moreover, not the
only methodological difficulties presented by such studies: the analytical tools available to
examine the properties of tracks are not particularly well developed, a problem discussed
in more detail later.
  Conceptual emphasis. Many of the most obvious and interesting questions relating to
host location and use by phytophagous insects were for many years based on developing
and testing ideas associated with plant-herbivore coevolution and plant chemistry:
questions of this kind often involve a focus on the mechanisms involved in insects'
response to different types of plant, and on plant attributes relevant to host selection and
use - that is, on host recognition and acceptance, and on the subsequent performance of
insects on the plants, rather than on how insects moved around to encounter objects to
which they might subsequently respond.
  Questions involving host recognition and acceptance and subsequent phytophage
performance are certainly of profound interest, but it is now also clear that an insect's
problems in encountering potential hosts may also have a considerable influence on its
relationships with alternative kinds and distributions of hosts, and it is this aspect of host

location that is considered here: that is, the tracks of insects as they move in search of host plants, what they do when they find them, and what the consequences of those processes may be. This paper will consider two case studies involving female pierid butterflies seeking oviposition sites, but will also compare the processes involved in these cases with those identified by Turchin (1986, 1987, 1988) for foraging by Mexican bean beetles. There are three questions which are useful in structuring an approach to analyses of search behaviour and movement patterns:

What are the rules? That is, what patterns can we identify in the decisions made by the insect in the process of generating a track and encountering potential host plants. I do not here mean the physiological and neurosensory mechanisms underlying those decisions, critical though these are, but the behavioural rules that result in a track. If they are to be useful, those rules will often be conditional ones: for example, a rule identified for Colorado potato beetles (Visser, 1988) might be "if you are hungry, it is light, and you smell a potato, then walk rapidly in a straight line upwind".

What are the consequences of those rules? Expressed in more operational terms, this question might translate into questions about how a change in any of the rules would affect the outcomes of the searching process. For example, will particular sets of movement rules make particular sorts of plant and particular spatial arrangements of plants more likely to be attacked? Will they increase dispersal distances? Will they make egg distributions clumped or random? Will they reduce or increase susceptibility to predators?

Where should we look for adaptation? That is, because any rule may have a multiplicity of consequences, what should we consider in trying to identify the selective pressures to which the behavioural processes are subjected.

Most of this paper will focus on the application of these three questions to a particular problem associated with the three case studies mentioned earlier. Before doing so, however, it is also useful to consider some methodological problems involved in answering the first two of them: that is, the problems involved in establishing the rules and evaluating their consequences.

The first problem was briefly alluded to earlier: that is, how to extract the rules from the tracks. The attributes of a track which need to be measured are more difficult to characterize than the binary choices involved in host recognition and acceptance. A track has attributes such as speed, altitude, direction, pauses and the frequency and pattern of turns, all of which have variances as well as means. Moreover, at various points along the track the insect may systematically change its behaviour or undertake actions such as oviposition: we may need to establish how tracks vary between individuals and populations, and how they are influenced by external features of the environment, including the distribution and attributes of host plants. There is, as yet, no real consensus about how best to do this in a consistent and rigorous way. The problem of characterization arises because tracks have directions and turning angles, and circular distributions - distributions of angles - are more difficult to manipulate than distributions of ordinary linear measurements. Some kinds of experimental situation are easily handled: for example, if the only question of interest is whether the insect is orienting with respect to a particular known stimulus, then the available tools are quite adequate to cope with identifying correlations between the direction of the stimulus and the net direction moved by the insect. But if we wish to ask any more difficult questions, then the available statistical tools are often inadequate. An example involving butterfly tracks will illustrate the problem. About 15 years ago, I constructed simulation models to describe the movement of Canadian and Australian female cabbage butterflies (Jones, 1977, 1987; Jones et al., 1980). One of the rules built into those models was that on any particular day, a butterfly had a preferred flight direction: that is, it was oriented with respect to some external cue, even if orientations differed between butterflies - which they appeared to do - and even though there was no information to suggest what this hypothetical cue might be. That decision could not then be adequately justified. It was made because the tracks of Australian butterflies were strongly directional, but this might have arisen simply because they were reluctant to change direction rather because they had a preferred direction. At the time, I chose the mechanism which seemed to me most biologically reasonable (on the grounds that it was not obvious how a butterfly in the absence of external cues could, after a landing and perhaps moving around on the plant and ovipositing, identify the direction of its previous flight in order to maintain it). Since then, several other workers have built butterfly movement simulations (see *e.g.*, Root & Kareiva, 1984; Zalucki & Kitching, 1982a, 1982b; Odendaal *et al.*, 1989). Those studies all made the opposite decision (on equally arbitrary grounds except in the last case). In other words, they used a rule which said that the butterflies had no orientation but did not like to make sharp turns between flights. Using one rule or the other makes no difference to the outcome of a movement episode on a small scale, but has very substantial effects indeed over long periods (Marsh & Jones, 1988). Since then, a test has been developed to distinguish between oriented and unoriented tracks (or more precisely, between biased and correlated random walks; Marsh & Jones, 1988). When applied to tracks collected by Jones (1977), the test suggests that Australian cabbage butterfly females do indeed have preferred directions (Canadian females were not sufficiently directional enough for the test to be able to discriminate between the alternative models). But tracks of the nymphalid butterfly *Euphydryas anicia* Doubleday & Hewitson to which the test has been applied (Odendaal *et al.*, 1989) suggest that this species is appropriately described by a correlated random walk, and thus is indeed able to remember the direction of a previous flight. The overall lesson from this is that in the absence of formal statistical test procedures, intuition is an unreliable guide to decisions about the appropriate models to describe observed tracks (which usually are relatively short), and that we are not at present well supplied with appropriate procedures.

Having extracted the rules from the tracks, the next problem is to evaluate the consequences of those rules, and how outcomes of search behaviour might change if the rules were different. The only procedure of sufficient flexibility to allow this involves the use of simulation modelling. Simulation models may also be used in part as analytical tools, since some of the parameter values which need to be extracted from the data may only be estimable by trial and error using the simulation itself. This practice increases the importance of requiring that a simulation model be subjected to appropriate experimental verification: that is, tested for its predictive power against data sets not used for parameter estimation. Used appropriately, simulation modelling is a powerful and very useful technique in this context. It allows simulated experiments of a range and variety which is not feasible in the field, and it allows the consequences of inferences from observations made at small spatial scales to be extrapolated to much larger scales. It has, however, two major limitations, outlined below.

A simulation model takes the rules governing behaviour which have been extracted from the data, and uses them to reconstruct hypothetical tracks according to those rules. The first limitation, therefore, is that the rules need to be correct, and as discussed earlier, they may not be easy to establish. The second limitation is that realistic movement simulations are themselves quite complex objects, and there is not yet any standard way of either putting them together or of reporting them in the literature. Consequently, making comparisons of results obtained by different people can be virtually impossible. This means that although simulations are at present one of the few tools we have to ask questions about the consequences of different sets of behavioural rules, it is to be hoped that they prove to be an interim solution, and that we will find better ways of achieving this goal.

The remainder of this paper discusses the three case studies mentioned previously in relation to a specific problem: that is, how does the spatial distribution of host plants, and specifically, whether they are isolated or in patches (and if in patches, their position in the patch) affect the likelihood that the plants are colonized by phytophagous insects.

Resource concentration and the edge effect

Some years ago, arising from a study of collard insects, Root (1973) proposed the "resource concentration hypothesis": that is, that herbivores (and especially specialized herbivores) are more likely to find and remain on hosts that are growing in dense or nearly pure stands. For some phytophages this is clearly the case, but for others it equally clearly is not: evidence bearing on the resource concentration hypothesis has been comprehensively reviewed by Kareiva (1982). Some species - including most of the butterflies which have been examined - show the opposite pattern. That is, isolated plants receive more eggs per plant than plants in groups; plants in sparse patches receive more eggs per plant than plants in dense groups, and plants on the edge of groups receive more eggs than plants in the centre. This reverse pattern has been called the "edge effect". The three questions initially discussed, applied to this phenomenon, then become (a) what are the behavioural rules which result in the existence of either resource concentration or the edge effect, (b) what additional consequences flow from those behavioural rules, and (c) where should we look for adaptation - from the insects point of view, is the reason for those behaviours that resource concentration or an edge effect has been historically advantageous and hence selected for?

Two of the cases to be discussed are pierid butterflies - the cabbage butterfly *Pieris rapae* L., and a tropical coliadine found in northern Australia called the grass yellow, *Eurema hecabe* Hübner. The third, as noted earlier, is the Mexican bean beetle *Epilachna varivestis* Mulsant. Of these three, the cabbage butterfly shows a clearcut edge effect, first documented by Cromartie (1975) - more eggs per plant are laid on the edge of patches than in the centre, more in sparse patches with the hosts widely spaced than in dense, closely-spaced patches, and more on isolated plants than on plants in groups. The Mexican bean beetle exhibits resource concentration - it is more common in large, dense patches of plants, and less common on isolated plants, or in sparse patches (Turchin, 1986, 1987, 1988). Within patches of plants, it is also more common towards the centre of the patch than it is around the edges. There is an additional difference between the distribution of Mexican bean beetles and cabbage butterfly eggs: Turchin showed that in sparse patches of hosts, the distribution of beetles per plant was close to random. In dense patches, by contrast, their distribution was highly aggregated. Cabbage butterflies reverse this pattern - although egg distributions are always aggregated, the degree of aggregation is greater in sparse patches than dense patches (Jones, 1977). There is less egg distribution data available for the grass yellow, but what there is shows neither strong resource concentration nor an edge effect: central and edge plants in large plots receive very similar numbers of eggs (Jones, unpubl.). If anything, more eggs were laid on central plants but the difference was very small. The first question therefore, is how the behavioural rules of these three species differ to produce these differences in distribution pattern.

What are the rules? The study of the Mexican bean beetle, as elucidated by Turchin (1986, 1987, 1988) used marked beetles whose positions and behaviour within plots of plants were regularly checked. The beetles made two kinds of moves: short moves, predominantly from plant to plant within the plot, which Turchin called "trivial moves", and longer "emigration flights" - higher straight flights out of the plot. As summarized in Table 1, the directions of movement were uniform, so there was no evidence of any directional tendencies within the plot. Even the "emigration flights" probably were not oriented, since the beetles quite often returned to the plots. Trivial moves were frequent when the host plants were close together, but very rare when host plants were widely spaced. More "emigration moves" occurred from widely spaced and edge plants. A critical feature of the movement patterns was that the beetles showed active aggregation: that is, the likelihood that they remained on a particular host increased with the number of beetles already on the plant. When these rules, with parameter values estimated from data collected in large host-plant patches, were built into a simulation, the model successfully generated realistic distribution patterns and departure frequencies, for both large and small host plant patches, and for dense and sparse host plant patches. It is clear that generation of the resource concentration effect in this system is critically dependent on both the rules governing trivial and emigration moves, and on the beetles' aggregation behaviour. That is, for this animal at least, interactions between individuals are important to generation of the resource concentration effect. As Turchin points out, interactions of this kind warrant much more attention in examining mechanisms which determine insect distribution patterns.

Table 1. Behaviour of mexican bean beetles *Epilachna varivestis* (from Turchin, 1986, 1987, 1988)

Directionality within host-plant patch	Nil
Movement outside patch	Long, higher straight flights "emigration moves"
Frequency of within-patch "trivial moves"	Frequent in dense patch Rare in sparse patch
Frequency of long "emigration moves"	Frequent from sparse patch Rare from dense patch
Probability of remaining on current host plant	Increases with number of beetles resident on plant

By contrast, the two butterfly species show no evidence of active aggregation: indeed, both are more likely to make longer flights after an encounter with a conspecific. To a casual observer, their oviposition behaviour appears very similar: both lay eggs singly, normally flying between each oviposition. Not every alighting results in an egg being laid, and not every flight results in a movement to a different plant: in fact "resettles" (that is, hops around on the same plant) are frequent in both species. Both species perceive possible hosts only from small distances away (that is, probably a metre or less) (Fahrig & Paloheimo, 1987). In the cabbage butterfly, detection at a distance is probably entirely visual, with final host identification from contact chemoreception (Traynier, 1979; Renwick & Radke, 1988). This sequence is common to many butterflies, and is probably also true for the grass yellow, as in other members of the genus *Eurema* (Mackay & Jones, 1989). Despite these similarities, the cabbage butterfly produces a strong edge effect and the grass yellow does not. Table 2 summarizes the results of analyzing observed tracks of each of these species. The results for Australian cabbage butterflies come from previous work (Jones, 1977, 1987; Jones *et al.*, 1980); those for the grass yellows are from unpublished data.

Table 2. Behaviour of cabbage butterfly (*Pieris rapae*) and grass yellow (*Eurema hecabe*)

	Pieris rapae	Eurema hecabe
Directionality within host-plant patch	High	Nil
Movement outside host-plant patch	Long straight flights	Long straight flights
Alighting probability within host-plant patch	High, varies with host plant	High, varies with host plant
Effect of oviposition on subsequent movement	Nil	Reduces frequency of resettles and increases average flight length
Eggs/alightings	Varies with host plant and between butterflies	Varies with host plant and between butterflies
Intensity of short-range attraction to host plants	Varies with egg load	Varies with egg load and between butterflies

There are a number of marked differences between the movement patterns of the two species. The cabbage butterflies, as noted earlier, are strongly directional, whereas the grass yellows, like the Mexican bean beetles, show no directional tendencies when moving within a patch of hosts. Flight lengths within a patch (or alternatively, landing probabilities) are comparable for the two species, varying in both cases with the type of host plant (for example, landing frequencies are lower when hosts are small, so that flight lengths then tend to be longer). After oviposition, the grass yellow shows a reduced tendency to resettle on the same plant, so that average flight lengths after oviposition tend to increase. This change decays gradually over several subsequent flights. No such effect of oviposition was detected in cabbage butterflies. In both species, the probability of oviposition during an alighting varied both with the quality of the host and between individual butterflies, but the range of values observed was similar for the two species. The strength of short-range attraction to hosts increased with egg-load in both species, and in the grass yellows at least, also varied between individual butterflies. At least some of the grass yellows appeared to show weaker host attraction than cabbage butterflies, though others appeared comparable.

When these rules are built into simulation models, the cabbage butterfly model generates an edge effect, and the grass yellow model slightly the reverse (i.e., a slight tendency for more eggs to be laid on plants in the centre of patches than on edge plants). But in this case, it is not so obvious which of the differences in behavioural rules is responsible for the differences in egg distribution!

One of the advantages of simulations, however, is that the investigator is not restricted to the organisms that nature has provided: model organisms can operate with any combination of rules. The next step, therefore, was to undertake a set of simulations in which the effects of each rule were tested in a standard analysis of variance factorial design in order to identify interactions as well as the main effects of each rule. Four factors were varied: (a) patch size - 16, 36, 64, or 100 plants, (b) directionality - either high or none, (c) effect of oviposition on subsequent flight present or absent, and (d) short-range attraction to host plants increases slowly or rapidly with egg load.

The resultant egg distributions for each combination of factors were then examined after the plot had been encountered by 50 model butterflies: here we examine only those main effects or interactions which influenced the edge effect. In order to standardize the results for differences in the total numbers of eggs laid during a run, the average numbers of eggs per plant laid on edge plants are divided by the average number of eggs per plant laid in the whole plot, so that values greater than 1.0 occur when edge plants receive more eggs than central plants, and values less than 1.0 occur when the central plants receive more eggs.

Significant effects were as follows:

a. Plot size: edge effects tended to be stronger in larger plots (Fig. 1), but no interactions involving plot size were significant.

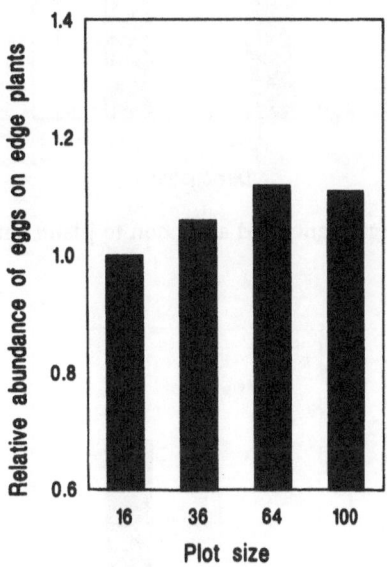

Figure 1. Effect of plot size on the edge effect.

b. There was a very strong interaction between directionality and attraction to plants (Fig. 2). A substantial excess of eggs on edge plants only occurred when butterflies were both strongly directional and showed strong short-range attraction to hosts. Non-directional butterflies with a strong attraction to hosts tended to concentrate eggs in the centre of plots. Weaker host attraction reduced both these effects (and supplementary runs in which host attraction was eliminated entirely, effectively eliminated both effects). The central concentration effect in the non-directional but strongly-attracted butterflies appears from supplementary runs also to be dependent on the switch to longer, straighter flights once the butterfly is outside the host patch.

c. There was a weaker, but still significant interaction between directionality and the existence of longer flights after oviposition (Fig. 3): this behaviour tended to weaken the edge effect slightly in directional butterflies, but had no effect on the edge effect in non-directional butterflies.

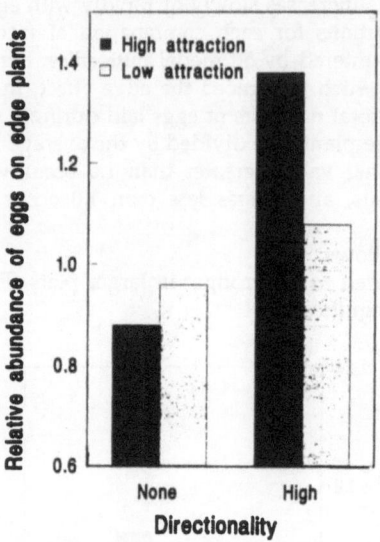

Figure 2. Effect of flight directionality and attraction to plants on the edge effect.

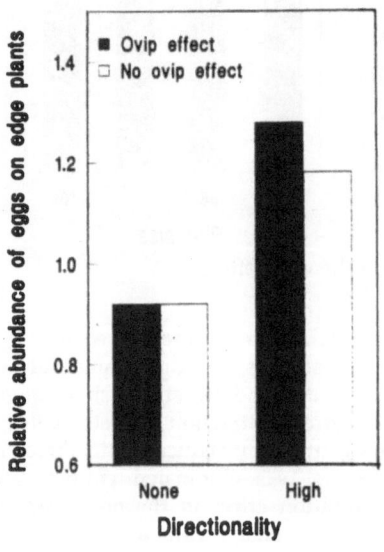

Figure 3.Effect of directionality and post-oviposition flight on the edge effect.

Overall, then, it appears that directional movement and short-range attraction to host plants are both necessary for the existence of an edge effect. When these two conditions are met, an edge effect appears, although its intensity may be modified by other behaviours (post-ovipositional flight lengthening) and by patch size. It also appears that if flight lengths and turning behaviour are very markedly different inside and outside patches of hosts, this may result in a concentration of eggs on central plants.

The necessity for short-range attraction to individual hosts suggests that edge effects should be less common when visual attraction to individual plants does not occur or is weakened - either because the insect species uses other perceptual cues, or because the host plants are particularly cryptic. We might also expect that resource concentration effects may be more common in species using olfactory cues to locate host patches from a distance, since the attraction will be to the patch as a whole rather than to individual plants within it (and, although this effect has not been examined here, because larger and denser patches may produce a stronger signal).

What are the consequences of the rules?

The analyses described above elucidate the effects of the behavioural rules on the existence or otherwise of edge effects. Each of the rules also has numerous other consequences, many of which are likely to be of much more significance to the butterfly's fitness than the existence or otherwise of edge effects or resource concentration. For example, more directional butterflies are also more dispersive, generate less aggregated egg distributions, invest a smaller fraction of their total egg load in any one patch of host plants, and require more flight time to lay a given number of eggs in patchy habitats (Jones, 1977).

Where should we look for adaptation?

Given the multiplicity of consequences arising from the operation of any one behavioural rule, it may well be quite misleading to consider the adaptive value of any one haphazardly chosen consequence in isolation: we should rather be considering the simultaneous effect of the whole set. That is, we should consider the adaptive value of the rule itself. There have been a number of studies comparing survivorship on isolated and edge plants *vs* clumped and central plants for several species showing an edge effect. The most common result is that either no difference is found or the difference is in the "wrong" direction. Such studies are certainly useful, but the above argument suggests that to ask "what is the adaptive value of the edge effect?" is to ask the wrong question. If edge effects are a consequence of directional flight and the use of visual cues, the right question is probably "when is it advantageous to use directional flight and visual cues?". We do not yet have good answers to this question, but it is unlikely that a common answer will be "when it is advantageous to generate edge effects"! For cabbage butterflies, a comparison of Australian and Canadian animals suggests that the answer may be "when flight time is not limiting and the costs of local crowding are high" (Jones, 1987). There may well be other answers in other circumstances.

References

Aluja, M., R.J. Prokopy, J.S. Elkinton & F. Laurence (1989). A novel approach for tracking and quantifying the movement patterns of insects in three dimensions under semi-natural conditions. *Environm. Entomol.* 18: 1-7.

Cromartie, W.J. (1975). The effect of stand size and vegetational background on the colonization of cruciferous food plants by herbivorous insects. *J. Appl. Ecol.* 12: 517-533.

Fahrig, L. & J.E. Paloheimo (1987). Inter-patch dispersal of the cabbage butterfly. *Can. J. Zool.* 65: 616-622.

Jones, R.E. (1977). Movement patterns and the egg distribution of cabbage butterflies. *J. Animal Ecol.* **46**: 195-212.

Jones, R.E. (1987). Behavioural evolution in the cabbage butterfly (*Pieris rapae*). *Oecologia* **72**: 69-76.

Jones, R.E., N.E. Gilbert, M. Guppy & V. Nealis (1980). Long-distance movement of *Pieris rapae*. *J. Animal Ecol.* **49**: 629-642.

Kareiva, P.M. (1982). Influence of vegetation texture on herbivore populations: resource concentration and herbivore movement. In: R.F. Denno & M.S. McClure (eds), *Variable Plants and Herbivores in Natural and Managed Systems*, pp. 259-289. New York: Academic Press.

Mackay, D.A. & R.E. Jones (1989). Leaf shape and the host-finding behaviour of two ovipositing monophagous butterfly species. *Ecol. Entomol.* **14**: 423-431.

Marsh, L.M. & R.E. Jones (1988). The form and consequences of random walk movement models. *J. Theor. Biol.* **133**: 113-131.

Odendaal, F.J., P. Turchin & F.R. Stermitz (1989). Influence of host-plant density and male harassment on the distribution of female *Euphydryas anicia* (Nymphalidae). *Oecologia* **78**: 283-288.

Renwick, J.A.A. & C.D. Radke (1988). Sensory cues in host selection for oviposition by the cabbage butterfly, *Pieris rapae*. *J. Insect Physiol.* **34**: 251-257.

Root, R.B. (1973). Organization of a plant-arthropod association in simple and diverse habitats: the fauna of collards (*Brassica oleraceae*). *Ecol. Monogr.* **43**: 95-12.

Root, R.B. & P.M. Kareiva (1984). The search for resources by cabbage butterflies (*Pieris rapae*): ecological consequences and adaptive significance of Markovian movements in a patchy environment. *Ecology* **65**: 147-165.

Traynier, R.M.M. (1979). Long-term changes in the oviposition behaviour of the cabbage butterfly, *Pieris rapae*, induced by contact with plants. *Physiol. Entomol.* **4**: 87-96.

Turchin, P. (1986). Modelling the effect of host patch size on Mexican bean beetle emigration. *Ecology* **67**: 124-132.

Turchin, P. (1987). The role of aggregation in the response of Mexican bean beetle to host plant density. *Oecologia* **71**: 577-582.

Turchin, P. (1988). The effect of host-plant density on the numbers of Mexican bean beetles. *Amer. Midl. Natur.* **119**: 15-20.

Visser, J.H. (1988). Host-plant finding by insects: orientation, sensory input and search patterns. *J. Insect Physiol.* **34**: 259-268.

Zalucki, M.P. & R.L. Kitching (1982a). The analysis of movement patterns in *Danaus plexippus* L. (Lepidoptera: Nymphalidae). *Behaviour* **80**: 174-198.

Zalucki, M.P. & R.L. Kitching (1982b). Component analysis and modelling of the movement process: the simulation of simple tracks. *Res. Popul. Ecol.* **24**: 239-249.

Proc. 8th Int. Symp. Insect-Plant Relationships, Dordrecht: Kluwer Acad. Publ.
S.B.J. Menken, J.H. Visser & P. Harrewijn (eds), 1992

Response of the oilseed rape pests, *Ceutorhynchus assimilis* and *Psylliodes chrysocephala*, to a mixture of isothiocyanates

E. Bartlet, I.H. Williams, M.M. Blight and A.J. Hick
AFRC, IACR Rothamsted Experimental Station, Harpenden, Herts, UK

Key words: Attraction, *Brassica napus*, cabbage seed weevil, cabbage stem flea beetle, traps, volatiles

Ceutorhynchus assimilis Payk., the cabbage seed weevil, and *Psylliodes chrysocephala* L., the cabbage stem flea beetle, are both important pests of oilseed rape (*Brassica napus* L.) in the UK. In electrophysiological experiments their antennae perceived 3-butenyl, 4-pentenyl and phenylethyl isothiocyanate (NCS), volatile metabolites of glucosinolates present in rape (Blight *et al.*, 1989). This study investigated the behavioural responses of the insects to a mixture of these three NCSs.

Responses of C. assimilis. The chemotactic responses of the seed weevil were tested in a linear track olfactometer. This olfactometer was designed by Sakumi and Fukami (1985) and adapted for testing the responses of the pea and bean weevil *Sitona lineatus* L. by Blight *et al.* (unpubl.). It has two chambers. Test material is put into one chamber and the other chamber acts as a control. Insects move out of a holding pot and along a wire to a T-junction. At the T-junction the insects can turn one way, into an airstream carrying odour from the test material or the other way, into the control airstream. The response of an insect to an odour is assessed by noting the number of insects turning towards it at the T-junction. For each replicate, the responsiveness of the weevils was first assessed by testing their reaction to 5 g of flowering rape (*cv.* Willi). The response to an equal (by weight) mixture of the three NCSs in pentane was then tested. The mixture was released into one of the chambers of the olfactometer by an automatic microapplicator at a rate of 1.7 µl/min. Twenty weevils were put into the holding pot each time. Turning responses were observed for ten minutes. Eight replicates of 20 weevils each were used. Statistical analyses used a generalized linear model (Blight *et al.*, unpubl.). Pentane alone was unattractive to the weevils (Table 1). Weevils were optimally attracted by a release rate of 15 µg total NCS over ten minutes. The odour of flowering rape was more attractive to the weevils than the NCS mixture.

Responses of P. chrysocephala. The responses of the cabbage stem flea beetle were investigated using field cages containing baited water traps. Nine field cages (2.74 x 2.74 x 1.83 m) each contained a seed tray (215 x 360 x 55 mm) filled with water. Water traps were either; unbaited, baited with 50 rape seedlings, *cv.* Topas, growth stage 1.2 (Sylvester-Bradley, 1985) or baited with an NCS mixture. An equal (by weight) mixture of the three NCSs in nonane was released from a glass vial with a polythene cap via a pipe cleaner wick. Initially 5 mg of each NCS in 3 ml of nonane was used giving a release rate of 1.6 mg total NCS/day. The three treatments were arranged in a latin square. On 18 September '91, 100 field collected cabbage stem flea beetles were released at each corner of every field cage. The number of beetles in each water trap was counted one week later.

Table 1. Movement of cabbage seed weevils towards NCS and towards flowering rape, tested in an olfactometer

Test material released over 10 minutes	% of weevils turning towards the odour	
	Test material	Flowering rape
pentane	56.9	85.6 ***
150 µg total NCS	62.4 *	86.0 ***
15 µg total NCS	73.8 **	74.4 ***
1.5 µg total NCS	65.9 *	82.3 **
0.15 µg total NCS	50.9	78.7 ***

Attraction significant at: * $P < 0.05$, ** $P < 0.005$, *** $P < 0.001$

The traps baited with seedlings caught more beetles than either the traps baited with NCS or the unbaited traps, although there was some evidence that the NCS mixture was attractive (Table 2). The baits were renewed on 27 September '91 and this time NCS was released at approximately 16 mg/day. One week later both the traps baited with seedlings and the traps baited with NCS were found to have caught more beetles than the control.

Table 2. Capture of cabbage stem flea beetles in baited water traps

Release rate (mg/day) total NCS	Mean no. of beetles caught (± SE)		
	Seedling traps	NCS traps	Control traps
1.6	22.0 ± 5.6	5.7 ± 1.8	2.3 ± 0.7
16.0	13.7 ± 4.6	13.0 ± 5.8	0.3 ± 0.3

Conclusions

These results show that both the seed weevil and the cabbage stem flea beetle are attracted by the NCS mixture. NCSs may assist the orientation of these insects to their host plant.

References

Blight, M.M., J.A. Pickett, L.J. Wadhams & C.M. Woodcock (1989). Antennal responses of *Ceutorhynchus assimilis* and *Psylliodes chrysocephala* to volatiles from oilseed rape. *Aspects of applied Biology* 23: 329-334.

Sakuma, M. & H. Fukami (1985). The linear track olfactometer: an assay device for taxes of the German cockroach, *Blattella germanica* L. (Dictyoptera: Blattellidae) towards their aggregation pheromone. *Appl. entomol. Zool.* 20: 387-402.

Sylvester-Bradley, R. (1985). Revision of a code for stages of development in oilseed rape (*Brassica napus* L.). *Aspects of applied Biology* 10: *Field trials methods and data handling.*

Proc. 8th Int. Symp. Insect-Plant Relationships, Dordrecht: Kluwer Acad. Publ.
S.B.J. Menken, J.H. Visser & P. Harrewijn (eds), 1992

Volatile plant metabolites involved in host-plant recognition by the cabbage seed weevil, *Ceutorhynchus assimilis*

M.M. Blight, A.J. Hick, J.A. Pickett, L.E. Smart, L.J. Wadhams and C.M. Woodcock
AFRC, IACR Rothamsted Experimental Station, Harpenden, Herts, UK

Key words: Behaviour, *Brassica napus*, gas chromatography, isothiocyanates, olfaction, traps, volatiles

The oilseed rape pest *Ceutorhynchus assimilis* Payk. feeds and develops only on Cruciferae and uses both olfactory and visual cues to orient to host plants. In this study, oilseed rape (*Brassica napus* L.) volatiles which interact with the seed weevil antenna were detected and identified. One group of compounds, the isothiocyanates, was then tested for behavioural activity in the field.

Volatiles emitted from oilseed rape were collected using standard air entrainment techniques (Blight, 1990), and electrophysiologically active compounds were located by gas chromatography (GC) coupled with either electroantennogram or single cell recordings (SCR; Wadhams, 1990). Active compounds were then identified using coupled GC-mass spectrometry.

Twenty-two of more than fifty compounds present in the air entrainment extract gave an antennal response (Table 1). Some volatiles, *e.g.*, the isothiocyanates, goitrin, guaiacol (2-methoxyphenol), methyl salicylate, *p*-anisaldehyde (4-methoxybenzaldehyde) and oct-1-en-3-ol were present only in trace quantities, although amounts emitted differed with the rape cultivar and the degree of tissue damage (Blight *et al.*, unpubl.). Most of the active compounds are ubiquitous plant volatiles, but the isothiocyanates and goitrin, which are glucosinolate metabolites, are characteristic of the Cruciferae.

Table 1. Oilseed rape volatiles perceived by the antenna of *C. assimilis*

Isothiocyanates	Fatty acid derivatives	Aromatic compounds
allyl	pentan-1-ol	*p*-anisaldehyde
3-butenyl	hexan-1-ol	2-phenylethanol
4-butenyl	hexan-1-al	phenylacetaldehyde
phenylethyl	(Z)-3-hexen-1-ol	methyl salicylate
	(Z)-3-hexenyl acetate	benzyl alcohol
	oct-1-en-3-ol	guaiacol

Terpenoids	Nitrogen-containing compounds
1,8-cineole	phenylacetonitrile
linalool	indole
(E,E,)-α-farnesene	goitrin

SCR studies showed that at least 30% of the olfactory cells found on the *C. assimilis* antenna responded specifically to the isothiocyanates. These cells were of three types (Blight *et al.*, 1989). The most abundant responded similarly to the 3-butenyl, 4-pentenyl and phenylethyl analogues, but the other types discriminated between the alkenyl and phenylethyl isothiocyanates. The allyl analogue elicited responses only at very high concentrations.

Field trapping. Attraction of *C. assimilis* by a mixture of these four isothiocyanates was demonstrated in the field (Table 2). Allyl isothiocyanate was released at 60 mg/day and the other analogues at 6 mg/day. Baited traps were significantly more attractive than controls (Factorial ANOVA, $P < 0.001$) but trap type did not have a significant effect on the numbers of weevils captured.

Table 2. Capture of *C. assimilis* in three different types of yellow trap baited with an isothiocyanate (NCS) mixture

Trap	Mean no. weevils caught per replicate	
	Control (nonane)	Baited (NCS in nonane)
Sticky box	6.7	28.5
Water trap (Petri dish)	6.5	42.5
Water trap (bowl)	11.0	30.5

A total of 707 weevils were caught in 6 replicates (SE = 12.1). Means within columns are not significantly different, (LSD test, $P = 0.05$).

Conclusions

The data suggest that certain isothiocyanates play an important role in host plant recognition by *C. assimilis*. However, modern oilseed rape cultivars emit very little of the 3-butenyl, 4-pentenyl and phenylethyl isothiocyanates and the allyl analogue is absent or present only in trace quantities (Blight *et al.*, unpubl.). The role of the other physiologically active compounds in cueing orientation is therefore being investigated.

References

Blight, M.M. (1990). Techniques for isolation and characterization of volatile semiochemicals of phytophagous insects. In: A.R. McCaffery & I.D. Wilson (eds), *Chromatography and Isolation of Insect Hormones and Pheromones*, pp. 281-288. London: Plenum Press.

Blight, M.M., J.A. Pickett, L.J. Wadhams & C.M. Woodcock (1989). Antennal responses of *Ceutorhynchus assimilis* and *Psylliodes chrysocephala* to volatiles from oilseed rape. *Aspects of Applied Biology* 23: 329-334.

Wadhams, L.J. (1990). The use of coupled gas chromatography: electrophysiological techniques in the identification of insect pheromones. In: A.R. McCaffery & I.D. Wilson (eds), *Chromatography and Isolation of Insect Hormones and Pheromones*, pp. 289-298. London: Plenum Press.

Proc. 8th Int. Symp. Insect-Plant Relationships, Dordrecht: Kluwer Acad. Publ.
S.B.J. Menken, J.H. Visser & P. Harrewijn (eds), 1992

The olfactory and behavioural response of seed weevils, *Ceutorhynchus assimilis*, to oilseed rape volatiles

K.A. Evans
Dept of Crop Science and Technology, The Scottish Agricultural College, Edinburgh, UK

Key words: Anemotaxis, attraction, *Brassica napus*, electroantennograms, gas chromatography, traps

The behavioural and sensory responses of cabbage seed weevils *Ceutorhynchus assimilis* Payk. to oilseed rape *Brassica napus* L. odour were studied. At seed weevil emergence sites, significantly more weevils were caught in yellow water traps baited with Industrial Methylated Spirit (IMS) extracts of oilseed rape leaf and flower odour (970 and 722 weevils respectively) than the IMS control baits (seven weevils). The attraction of seed weevils to oilseed rape leaf and flower odour was also demonstrated in the laboratory using a four-choice olfactometer (Vet *et al.*, 1983).

Analysis of components. The volatile compounds present in the odour of oilseed rape were adsorbed on Tenax GC and identified using gas chromatographic and mass spectrometric techniques. Electroantennograms (EAGs) were obtained from seed weevils in response to the compounds identified from oilseed rape odour using the methodology described by Visser (1979). The strongest EAG responses were recorded to the relatively minor rape leaf volatiles, 3-butenyl isothiocyanate, 4-pentenyl isothiocyanate (both < 0.5% of the composition of rape leaf odour) and the major rape flower volatile (E,E)-α-farnesene (> 60% of the composition of rape flower odour).

Artificial rape leaf and flower odour was prepared in the laboratory based on the proportions of the volatiles present in Tenax GC sampled rape odour. EAGs from seed weevils exposed to the artificial equivalents of rape leaf and flower odour were comparable to those obtained from Tenax GC sampled odour. When 3-butenyl isothiocyanate and 4-pentenyl isothiocyanate were omitted from the artificial rape leaf odour, significantly reduced EAGs were obtained from male weevil antennae; female antennae also exhibited a reduction in sensory response. Similarly, when the two green leaf volatiles, Z-3-hexen-1-ol and Z-3-hexenyl acetate were omitted from the artificial leaf odour, reduced EAGs were obtained, significantly so in the case of female weevils. These two volatiles accounted for over 90% of the composition of rape leaf odour. Omission of α-farnesene from the artificial rape flower odour resulted in significantly reduced EAGs in both male and female weevils.

Behavioural tests. In the olfactometer, the behaviourally attractive response of seed weevils to artificial rape flower odour was absent when α-farnesene was omitted from the odour. Seed weevils were behaviourally unresponsive to α-farnesene in the olfactometer.

In an inflatable wind tunnel (Jones *et al.*, 1981), male and female seed weevils moved upwind via odour mediated anemotaxis in the presence of rape leaf or rape flower odour. There was a significant reduction in the proportion of female weevils moving upwind when 3-butenyl and 4-pentenyl isothiocyanate were absent from artificial rape flower odour. There was also a reduction in the proportion of male and female weevils moving upwind when α-farnesene was omitted from artificial flower odour.

At a seed weevil emergence site, nearly three times as many weevils were captured in yellow water traps baited with artificial rape flower odour (175 weevils) compared to artificial flower odour with no α-farnesene (60 weevils).

Odour-mediated anemotaxis. Upwind anemotaxis mediated by oilseed rape odour was demonstrated in the field by the recapture of marked seed weevils released from the centre of a circular array (40 m in diameter) of eight baited yellow water traps. Significantly fewer weevils were recaptured when IMS was the bait (17%), compared to the rape leaf (27%) and rape flower (36%) baits. The flight direction of weevils was random in the case of females, but tended to be downwind for males when IMS was used as the bait. In mark and recapture experiments with rape leaf or rape flower odour as baits, both sexes of weevils were significantly recaptured upwind from their release point. These results demonstrate that seed weevils are capable of detecting and responding via odour-mediated anemotaxis to oilseed rape odour from a distance of at least 20 m from the source of the odour.

Conclusions

Seed weevils can locate a source of oilseed rape odour by odour-mediated anemotaxis from a distance of at least 20 m. Omission of key constituents of oilseed rape odour can diminish or nullify the anemotactic response of weevils to rape odour. The decline in behavioural response is accompanied by a reduction in the sensory response of olfactory receptors on weevil antennae.

References

Jones, O.T., R.A. Lomer & P.E. Howse (1981). Responses of male mediterranean fruit flies, *Ceratitis capitata*, to trimedlure in a wind tunnel of novel design. *Physiol. Entomol.* **4**: 353-360.

Vet, L.E.M., J.C. van Lenteren, M. Heymans & E. Meelis (1983). An airflow olfactometer for measuring olfactory responses of hymenopterous parasitoids and other small insects. *Physiol. Entomol.* **8**: 97-106.

Visser, J.H. (1979). Electroantennogram responses of the Colorado beetle, *Leptinotarsa decemlineata*, to plant volatiles. *Entomol. exp. appl.* **25**: 86-97.

Proc. 8th Int. Symp. Insect-Plant Relationships, Dordrecht: Kluwer Acad. Publ.
S.B.J. Menken, J.H. Visser & P. Harrewijn (eds), 1992

Dispersive flight of the cabbage stem weevil

Christian Kjær-Pedersen
National Environmental Research Institute, Dept of Terrestrial Ecology, Silkeborg, Denmark

Key words: Anemotaxis, *Ceutorhynchus quadridens*, flight orientation, olfaction, phototaxis

The cabbage stem weevil *Ceutorhynchus quadridens* Panz is an oligophagous herbivore that feeds on Brassicaceae. This study describes the stimuli used as directing cues during host finding, and explores whether host specific chemicals have any influence on the flight behaviour.

Material and methods

Field captured cabbage stem weevils were starved for 12 h and then released into a cage with two walls of terylene net and two walls of glass. A trial consisted of counting the number of alightments on each wall in one minute. To establish whether cabbage stem weevils behave differently inside and outside a host odour plume, I placed the cage alternately within a winter rape field, and in areas without host plants. The cage was frequently rotated so that first a net wall and second a glass wall faced upwind.

Results and discussion

Flight of the cabbage stem weevil was directed by the sun and wind. At low windspeeds (0-1 m/s) outside the host crop odour plume, significantly more weevils flew up against the wind than in any other direction (Fig. 1A). At higher windspeeds (1-2.5 m/s) significantly more weevils alighted on wall 3 (downwind) than any other wall (Fig. 1B). I was not able to determine whether the sun exposure of wall 3 is a dominating factor in the profound preference for wall 3. The observed up- and downwind flight is optimal in detecting odour plumes, according to the geometrical considerations of Sabelis & Schippers (1984). Their conclusion is based upon the assumption that wind directions fluctuate at least 30°.

The weevils showed a profound phototaxis when a glass wall was turned upwind (Fig. 1A, B and D). Moreover, the observed up- and downwind flight was not directed by visual cues as the flight direction was seen to change as the wind stimuli ceased. Orientation in relation to both the sun and the wind is a way of avoiding flying over the same area more than once.

Situated within the odour plume of a host, a significant part of the weevils flew up against the wind. As volatile chemicals are transported by the wind, upon meeting host-specific chemicals, flying upwind in search of the odour source would seem a good strategy.

When the cage was turned with a glass wall up against the wind the weevils were devoid of the wind stimulus. Under these circumstances, significantly more weevils alighted on the walls exposed to the sun. This behaviour often seems to be unfavourable

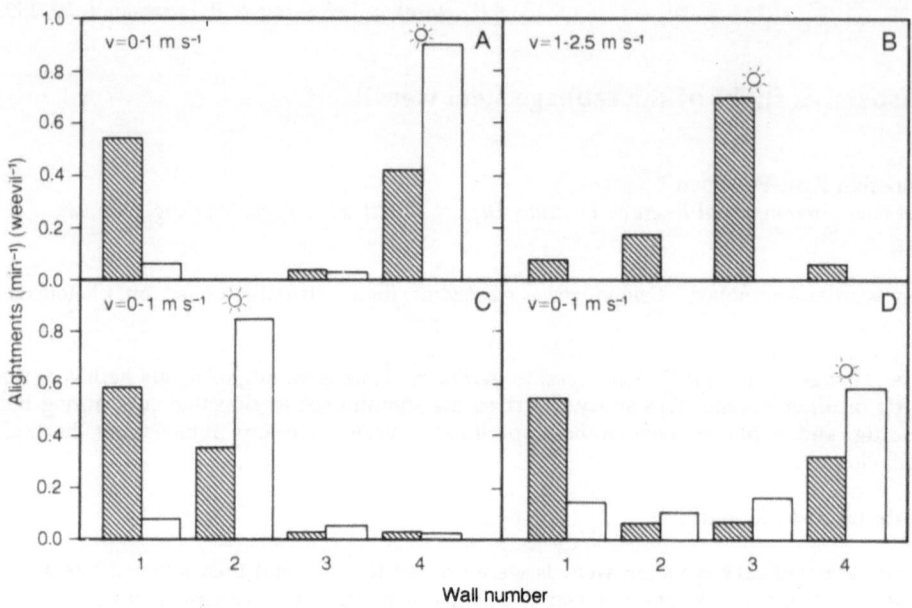

Figure 1. Relative flight orientation of *C. quadridens*, at different sun positions. Wall number 1 refers to the wall turned against the wind, clockwise, followed by numbers 2, 3 & 4. Wall number 1 was either made of net (hatched bars) or glass (clear bars). A-B: The cage was outside the host odour plume. C-D: The cage was inside the host odour plume. Flight orientation was measured as the number of alightments on one wall in proportion to all alightments. The sun position is indicated by ☼. v is the range of windspeeds encounted during a trial.

in the process of finding hosts, but natural conditions in which the weevil will experience this situation must be rare, because the weevils can sense and orientate to the wind at windspeeds of between 0 and 1 m/s. Furthermore, the plume will be small so that the weevils can see the plant when it encounters the odour plume.

Observations with the same weevils showed a significant decrease in flight activity from 0.285 alightments/min/weevil under direct sun to 0.068 under cloudy conditions (paired t-test). This observation indicates that the weevils apparently use none other than wind and sunlight as guiding stimuli during flight. When a net wall was turned upwind, a large proportion of the weevils still flew towards the sun exposed walls. I believe that this, however, is an artefact of the experimental design. Normally, wind directions vary over time due to turbulence. Such changes in wind direction will, in combination with the presence of glass walls, create sheltered areas in the cage. Weevils departing from such sheltered areas will, according to the observed behaviour of weevils devoid of wind, fly toward the sun exposed wall.

Conclusions

The cabbage stem weevil uses wind and sun as directing stimuli during distant host search. Wind stimuli are preferred to the sun. Without these stimuli flight activity decreases markedly. Apparently, no other directing stimuli are used during flight.

Reference

Sabelis, M.W. & P. Schippers (1984). Variable wind directions and anemotactic stategies of searching for an odour plume. *Oecologia* **63**: 225-228.

Proc. 8th Int. Symp. Insect-Plant Relationships, Dordrecht: Kluwer Acad. Publ.
S.B.J. Menken, J.H. Visser & P. Harrewijn (eds), 1992

Responses of the black bean aphid, *Aphis fabae*, to a non-host plant volatile in laboratory and field

Rufus Isaacs[1], Jim Hardie[1], Alastair J. Hick[2], Lesley E. Smart[2] and Lester J. Wadhams[2]
[1] Dept of Biology, Imperial College, Silwood Park, Ascot, Berkshire, UK
[2] AFRC, IARC, Rothamsted Experimental Station, Harpenden, Herts, UK

Key words: Aphids, behaviour, isothiocyanate, predators, repellency, slow release formulation

Plant-produced volatile chemicals are known to influence host location and selection behaviour by aphids (Pickett *et al.*, 1992). Coupled gas chromatography and single cell recording techniques have shown that sensory cells in the proximal primary rhinarium of winged virginoparous *Aphis fabae* Scopoli respond to alkenyl isothiocyanates (Nottingham *et al.*, 1991), although these compounds are not found in the aphid's host plants. 4-Pentenyl isothiocyanate was electrophysiologically the most potent. In the same study, behavioural bioassays showed that this chemical repels *A. fabae*. The present report explores the behavioural response of the aphid to specific amounts of 4-pentenyl isothiocyanate in olfactometer bioassays and tests the effectiveness of a chemical slow release formulation under field conditions.

Methods

4-Pentenyl isothiocyanate was released from a 10 µl glass microcapillary (Drummond Ltd.) held at 45° to the vertical, in the centre of a side arm of a perspex linear track olfactometer (Nottingham *et al.*, 1991) and the amount diffusing into the 500 ml/min airstream was measured. Twenty five alate *A. fabae* were starved for 24 h, then placed in the bottom of the olfactometer. Aphids which climbed a vertical bar towards an overhead light were scored for their initial turning choice on reaching a horizontal bar. This was either towards treated or clean air. Each replicate ran for 10 min, and the olfactometer was rotated 180° between runs to allow for directional bias. Release rates were varied by controlling the amount of chemical in the microcapillary. Control runs were made without chemical.

A preliminary field evaluation of 4-pentenyl isothiocyanate was made in late sown field beans, *Vicia faba* (Moench) with the precursor [3,5-bis-(4-pentenyl)-1,3,5-thiadiazine-2-thione] which decomposes to the isothiocyanate on exposure to water vapour. Five treatments were applied electrostatically to the 3 x 3 m plots on six occasions between June 4th and July 31st 1991, the number of applications being limited by poor weather conditions in June. Treatments were as follows: (a) no spray, (b) precursor 28 g/ha, (c) precursor 280 g/ha, (d) tetrahydrofurfuryl alcohol (THFFA) solvent 10.4 l/ha, (e) pirimicarb 280 g/ha. Laboratory studies showed that at 20°C and 100% RH the precursor releases the isothiocyanate at a rate of approximately 25 µg/h and g. Numbers of winged *A. fabae* landing were recorded on thirty plants per plot on eight occasions between 12 June and 12 August along with colony size, assessed on a logarithmic scale.

Two pitfall traps in each plot were emptied weekly between 4 June and 13 August and the numbers of predators and *A. fabae* recorded.

At bean maturity, yield samples were taken from a six row x 1 m area in the centre of each plot.

Results

Olfactometer. No directional bias was found in control runs. Aphids were significantly repelled when release of chemical exceeded 210 ng/h, *i.e.*, 3.3×10^{10} molecules/ml of air (Table 1).

Table 1. Results from olfactometer assays of 4-pentenyl isothiocyanate with winged virginoparous *A. fabae*

Release rate (ng/h)	n	Mean no. aphids ± SE		t	P
		Laden air	Control		
0	10	11.2 ± 1.18	10.5 ± 0.89	1.413	NS
25	8	5.6 ± 1.30	5.3 ± 1.76	0.552	NS
60	8	9.6 ± 0.91	12.3 ± 1.25	2.072	NS
90	8	9.3 ± 1.07	11.5 ± 0.89	1.528	NS
120	8	7.8 ± 1.56	10.9 ± 1.38	1.630	NS
150	8	6.4 ± 1.13	9.6 ± 1.58	2.260	NS
210	8	6.4 ± 1.79	11.0 ± 1.66	3.360	*
430	10	9.0 ± 0.92	15.0 ± 0.94	3.540	**

* $P < 0.02$,   ** $P < 0.01$ (Related sample t-test)

Field. Only five winged *A. fabae* were recorded during June, when temperature was below average and rain was recorded on 28 days. Analysis of variance showed that the two treatments of the 4-pentenyl isothiocyanate precursor did not significantly reduce aphid colony size compared to the solvent-treated control plots on any of the assessment dates.

Analysis of the pitfall trap catches showed that the experimental treatments had no significant effect on the predator fauna of Carabidae, Staphylinidae, and Coccinellidae throughout the period of colony growth. Although catches of apterous *A. fabae* showed no treatment effect, they increased dramatically in mid July, when bean plants ceased vegetative growth. No more *A. fabae* were trapped after July 30. Analysis of bean weights from each plot showed that they were significantly increased only by the pirimicarb treatment ($P < 0.05$).

Discussion

The initiation of repellency when the isothiocyanate was released at and above 210 ng/h suggests that there is a threshold for the behavioural response of the aphid.

In the field, winged aphid numbers were low throughout the season making it impossible to assess treatment effects on colonising insects. Wet weather may also have washed the precursor formulation from the plants prematurely. The experimental treatments had no significant effect on aphid colony size.

Movement of apterous *A. fabae* detected by pitfall trapping was probably in response to decreased nutritional value of the host plant since plant nutrients would then be directed towards pod production. This is in agreement with laboratory studies by Hodgson (1991) on other aphid species.

References

Hodgson, C.J. (1991). Dispersal of apterous aphids (Homoptera: Aphididae) from their host plant and it's significance. *Bull. Entomol. Res.* **81**: 417-427.

Nottingham, S.F., J. Hardie, G.W. Dawson, A.J. Hick, L.J. Wadhams & C.M. Woodcock (1991). Behavioural and electrophysiological responses of aphids to host and non-host plant volatiles. *J. Chem. Ecol.* **17**: 1231-1242.

Pickett, J.A., L.J. Wadhams, C.M. Woodcock & J. Hardie (1992). The chemical ecology of aphids. *Annu. Rev. Entomol.* **37**: 67-90.

Proc. 8th Int. Symp. Insect-Plant Relationships, Dordrecht: Kluwer Acad. Publ.
S.B.J. Menken, J.H. Visser & P. Harrewijn (eds), 1992

Aggregation in a flower bud-feeding weevil

W.L. Mechaber and F.S. Chew
Dept of Biology, Tufts University, Medford, Massachussets, USA

Key words: *Anthonomus musculus*, host plant finding, *Vaccinium corymbosum*,
V. macrocarpon

We studied a Nearctic flower bud-feeding insect pest, *Anthonomus musculus* Say
(Coleoptera: Curculionidae) cranberry weevil, to determine whether a combination of
adult host plant finding ability and feeding preferences for host plants account for weevil
distribution on host plant species. *A. musculus* feeds on as many as 12 native, ericaceous
plant species. In addition, it is an economic pest on two native agricultural crops in two
different regions of the eastern United States: *Vaccinium corymbosum* L. (Ericaceae), high
bush blueberry in New Jersey, and *Vaccinium macrocarpon* Ait., large cranberry in
Massachusetts. However, at any one time during the growing season, adult weevils are
distributed on several host plant species.

This research reports our preliminary investigations with Y-tube olfactometer and
terrarium bioassays into the mechanisms of host plant finding for *A. musculus*. Since
adult *A. musculus* life spans, in days, exceed the number of days that flower buds of
individual host plant species are available (adult weevils live in laboratory culture from a
minimum of 32 days to a maximum of 13 months), we hypothesized that there would be
evidence of attraction to host plant volatiles, as adult weevils need to move between host
plant species as the growing season progresses. Further, since females require flower
buds for oviposition sites, while males and females can feed on leaves of certain host
plants when flower buds are not available, we wondered if females would differ from
males in their responses to host plant volatiles.

Results and discussion

We were unable to demonstrate response by the majority of the weevils we tested in
Y-tube olfactometers to host plant odors alone. Since field observations of weevils on both
native and commercial host plants revealed that distribution of the insect is often patchy,
we considered that weevil-produced pheromone(s) might be a source of volatile
attractants for this insect, as has been demonstrated in other anthonomine weevils.

Results from Y-tube olfactometer bioassays demonstrated that the majority of the
weevils, 69 to 81% depending upon the assay, selected volatiles from feeding-damaged
host plant (host plant with conspecific feeding-damage, frass and any altered plant
volatiles as a result of herbivory) when offered these volatiles against either purified air or
intact host plant (host plant with no feeding damage). There were no significant
differences in preference by sex of weevil. The pattern of response was similar for weevils
tested during 1990 on vegetative cranberry vines and weevils tested during 1991 on
cranberry vines with flower buds.

Weevils in terrarium microhabitat bioassays responded similarly when offered intact host plant versus feeding-damaged host plant with or without the presence of conspecific weevils. This pattern of preference was observed on three different host plant species: *V. pallidum*, an early season native host plant, *Gaylussacia baccata*, a native host plant that flowers between early and mid-season host plants, and *V. macrocarpon*, the commercial mid-season host plant.

The results from both bioassays are based on one time responses of weevils to test conditions and volatiles. Since the experimental weevils had been recently field-collected, we suggest that this variation in attraction to volatiles from intact host plant versus feeding-damaged host plant may mimic actual field responses.

Field observations of *A. musculus* during the early season in Massachusetts have revealed that weevils are found on commercial cranberry bogs before any new growth is present. At that time of the season, there are no flower buds for feeding or oviposition. In laboratory feeding preference assays, weevils collected on cranberry and offered cranberry versus another early season host plant, fed preferentially on the alternative host plant. One interpretation of these data is that *A. musculus* may remain on a less-preferred host plant due to an inability to find the preferred host plant.

These findings raise many questions for further study. In particular, we hope to identify the source of volatile attractant(s) and determine over what distance the volatile(s) are effective. In addition, we hope to determine what additional cues may be used in host plant finding by *A. musculus*.

Proc. 8th Int. Symp. Insect-Plant Relationships, Dordrecht: Kluwer Acad. Publ.
S.B.J. Menken, J.H. Visser & P. Harrewijn (eds), 1992

Electroantennogram responses of aphids to plant volatiles and alarm pheromone

W.A. van Giessen[1,2], J.K. Peterson[1] and O.W. Barnett[2]
[1] *USDA, ARS US Vegetable Laboratory, Charleston, South Carolina, USA*
[2] *Dept of Plant Pathology and Physiology, Clemson University, South Carolina, USA*

Key words: *Acyrthosiphon pisum*, (E)-ß-farnesene, *Macrosiphum euphorbiae*, *Myzus persicae*, kairomones, olfaction, primary rhinaria

Following a migratory or trivial flight, aphids are attracted to yellow and green colors when looking for a new host plant for feeding and reproduction. Accumulating evidence indicates, however, that not only vision, but also olfaction plays a role during host-plant location (*e.g.*, Visser & Taanman, 1987). The proximal and distal primary rhinaria (PPR and DPR, respectively), which are located on the fifth (PPR) and sixth (DPR) antennal segments, are thought to be general plant odor receptors.

Electroantennograms. In the present study the primary rhinaria were examined for their sensitivity to plant odors and aphid alarm pheromone. Electroantennograms (EAGs) were recorded from excised antennae of *Acyrthosiphon pisum* (Harris), *Macrosiphum euphorbiae* Thomas, and *Myzus persicae* Sulzer. It was possible to obtain electro-antennogram recordings from the PPR and DPR separately. The sum of the relative EAG responses from the DPR and PPR was equal to the relative EAG response from the total antenna. Stimulus-response curves were made for a number of plant volatile components, plant extracts, and aphid alarm pheromone.

All three species showed a high selectivity for the green leaf volatiles. The PPR of *A. pisum* was more sensitive to lower stimulus concentrations of hexanal than the DPR, while the reverse was true for (Z)-3-hexen-1-ol (Fig. 1). The DPR exhibited higher saturation responses than the PPR for both hexanal and (Z)-3-hexen-1-ol, while no PPR responses were found for the terpenoid (+)-carvone. For paraffin-oil extracts of mustard and celery, the DPR and PPR showed virtually equal saturation responses. The PPR, however, was much more sensitive to these extracts than the DPR.

The PPR and DPR were stimulated with a series of mono-saturated alcohols with increasing numbers (3-9) of carbon atoms. While the DPR exhibited a characteristic optimum around C6 and C7 (Yan & Visser, 1982), the PPR did not show such selectivity.

The primary rhinaria have also been shown to be the receptors of alarm pheromone (Wohlers & Tjallingii, 1983). EAG responses of the PPR and DPR were measured when stimulated by the main component of aphid alarm pheromone, E-(ß)-farnesene (EBF). In all three species, the DPR was more sensitive to EBF than the PPR. Compared to *A. pisum* and *M. euphorbiae*, the DPR of *M. persicae* was particularly sensitive to EBF, while the PPR of this species showed a relatively lower sensitivity to EBF.

These results strongly suggest that there is a basic difference in olfactory (EAG) response between the primary rhinaria. We hypothesize that specialization exists between receptors in the two primary rhinaria.

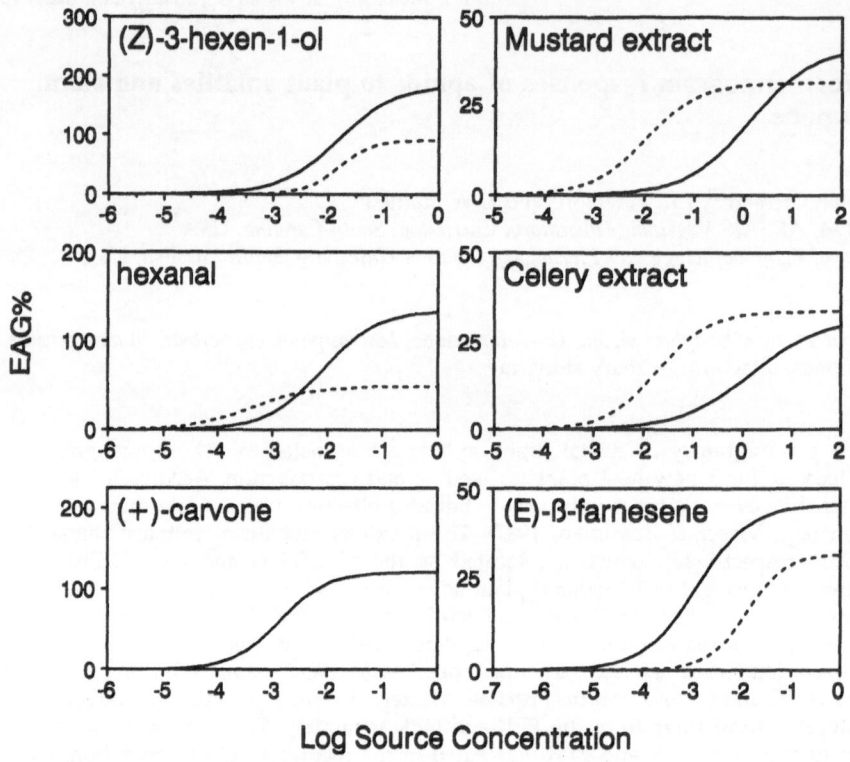

Log Source Concentration

Figure 1. Dose-response curves of virginoparous *A. pisum* (n = 6) to serial dilutions (in paraffin oil) of the host volatile components (Z)-3-hexen-1-ol, hexanal, (+)-carvone, paraffin oil extracts of mustard and celery, and (E)-ß-farnesene. Relative EAGs are expressed as percentage of the response to (Z)-3-hexen-1-ol (10^{-2} v/v). EAGs were recorded from the DPR (solid lines) and the PPR (dashed lines) separately. Curves were fitted by non-linear regression using the logistic function $R = \alpha(1+\text{ß}e^{-\tau(\log C)})^{-1}$.

References

Visser, J.H. & J.W. Taanman (1987). Odour-conditioned anemotaxis of apterous aphids (*Cryptomyzus korschelti*) in response to host plants. *Physiol. Entomol.* **12**: 473-479.
Wohlers, P. & W.F. Tjallingii (1983). Electroantennogram responses of aphids to the alarm pheromone (E)-ß-farnesene. *Entomol. exp. appl.* **33**: 79-82.
Yan, F.-S. & J.H. Visser (1982). Electroantennogram responses of the cereal aphid *Sitobion avenae* to plant volatile components. In: J.H. Visser & A.K. Minks (eds), *Proc. 5th Int. Symp. Insect-Plant Relationships*, pp. 387-388. Wageningen: Pudoc.

Proc. 8th Int. Symp. Insect-Plant Relationships, Dordrecht: Kluwer Acad. Publ.
S.B.J. Menken, J.H. Visser & P. Harrewijn (eds), 1992

The role of host-plant odour and sex pheromones in mate recognition in the aphid *Cryptomyzus*

J. Adriaan Guldemond[1], A.F.G. Dixon[2], J.A. Pickett[3], L.J. Wadhams[3] and C.M. Woodcock[3]
[1] *DLO-Research Institute for Plant Protection, Wageningen, The Netherlands*
[2] *School of Biological Sciences, University of East Anglia, Norwich, UK*
[3] *AFRC Institute of Arable Crops Research, Rothamsted Experimental Station, Harpenden, Herts, UK*

Key words: Aphididae, olfactometer, reproductive isolation

The sex pheromones of a number of aphids have been shown to consist of two terpenoids, nepetalactol and nepetalactone (Dawson *et al.*, 1987). The ratios of these two components differ between the various species and some species specificity has been demonstrated, although males are attracted to a range of ratios (Hardie *et al.*, 1990). However, unless aphid males are able to disciminate between small differences in the ratios, it is uncertain whether this ratio alone can account for species-specific mate recognition. One factor that might increase the specificity of the signal is the odour of the host plant on which mating takes place.

This hypothesis was tested by using aphids of the genus *Cryptomyzus*, which has two closely related (sibling) species, *C. galeopsidis* (Kalt.) and *C. maudamanti* Guldemond, and a more distantly related species, *C. ribis* (L.), all living on *Ribes rubrum*, red currant. Furthermore, an intraspecific form (host race) of *C. galeopsidis* that lives on *R. nigrum*, black currant, was used (Guldemond, 1991). No response of the males to pheromone-releasing females of different species and forms, and to synthetic pheromones and host plant odour, was measured in an olfactometer (Pettersson, 1970). The first choice was the field of the olfactometer that a male initially entered and was analysed using a X^2-test; the time spent in each of the four fields in the first three minutes was used as a measure of prolonged attraction/arrestment of a male and was analysed using Friedman's method for randomized blocks.

Results

Host plant odour did not attract males of the black currant host race of *C. galeopsidis*: 28 and 23% of the time was spent in the fields containing leaf odour from black and red currant, respectively ($X^2 = 0.01$, df = 3, NS). Also, host plant odour did not increase the attractiveness of natural or synthetic aphid sex pheromones. However, males were attracted, both in terms of their first choice and time spent in the odour fields, by sex pheromones. In no-choice tests, males of *C. galeopsidis* were attracted to females of the other species and host races, but, in a choice test, they were preferentially attracted to conspecific females (Guldemond & Dixon, unpubl.). This indicates that specific mate recognition is well developed, which could reduce the chance of interspecific hybridization. Because *C. galeopsidis* males were attracted to a leaf which had previously hosted pheromone-emitting females it would appear that sex pheromone is absorbed and retained by leaves for at least several hours (Fig. 1). Since females only release pheromone for a part of the day (Guldemond & Dixon, unpubl.), this response possibly enables males to find females over a longer period.

Males responded to different ratios of synthetic sex pheromone components only in their first choice. Analysis of entrainment samples from pheromone-emitting females using coupled gas chromatography - mass spectrometry (GC-MS), revealed that the sex pheromones of the various species and forms of *Cryptomyzus* had similar lactone : lactol ratios (1:30). This was surprising, because males of *C. galeopsidis* distinguished between the sex pheromone emitting females of the different *Cryptomyzus* species, which strongly suggests that the sex pheromone consists of more than two components.

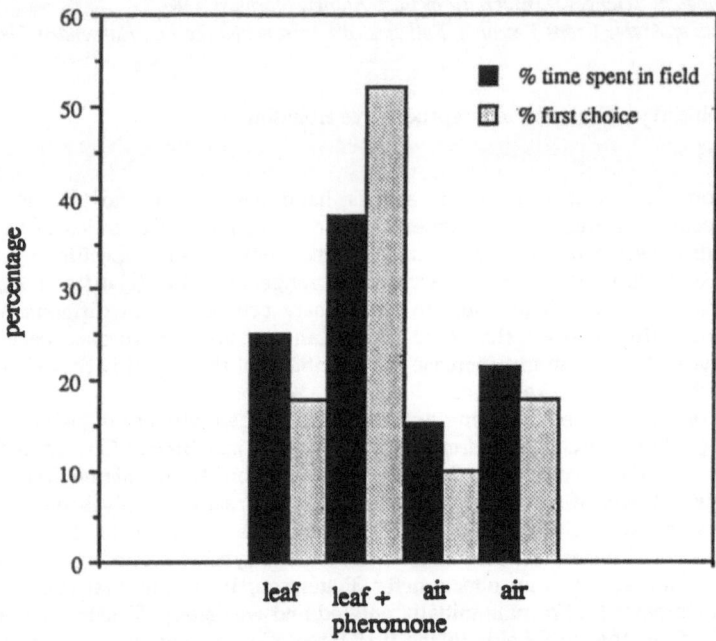

Figure 1. Percentage of time spent in each of the odour fields and the first choice of male *C. galeopsidis* (red currant host race) in response to the odour of a red currant leaf, a red currant leaf which had previously hosted sex pheromome releasing females, and two control (air) odour fields. (n = 50; time spent (leaf versus leaf+odour): X^2 = 4.79, df = 1, P < 0.05; first choice: X^2 = 23.28, df = 3, P < 0.001)

Conclusion

We conclude that, in *Cryptomyzus*, 1) host plant odour does not act as a synergist to increase the attractiveness of sex pheromone, 2) males can distinguish between pheromones of closely related species, and 3) sex pheromones are likely to consist of more than two components. Therefore, sex pheromone communication in aphids may resemble, the sophisticated system in, for instance, moths more than previously acknowledged.

Acknowledgements. JAG was funded by a European Science Exchange Program Fellowship and travel grants from the Netherlands Organization for Scientific Research (NWO) and the Royal Society.

References

Dawson, G.W., D.C. Griffiths, N.F. Janes, A. Mudd, J.A. Pickett, L.J. Wadhams & C.M. Woodcock (1987). Identification of an aphid sex pheromone. *Nature* **325**: 614-616.

Guldemond, J.A. (1991). Biosystematic and morphometric study of the *Cryptomyzus galeopsidis/alboapicalis* complex (Homoptera, Aphididae), with a key to and notes on the *Cryptomyzus* species of Europe. *Neth. J. Zool.* **41**: 1-31.

Hardie, J., M. Holyoak, J. Nicolas, S.F. Nottingham, J.A. Pickett, L.J. Wadhams & C.M. Woodcock (1990). Aphid sex pheromone components: age-dependent release by females and species-specific male response. *Chemoecol.* **1**: 63-68.

Pettersson, J. (1970). An aphid sex attractant. I. Biological studies. *Entomol. scand.* **1**: 63-73.

Proc. 8th Int. Symp. Insect-Plant Relationships, Dordrecht: Kluwer Acad. Publ.
S.B.J. Menken, J.H. Visser & P. Harrewijn (eds), 1992

Comparison of electroantennogram responses by females of the black swallowtail butterfly, *Papilio polyxenes*, to volatiles from two host-plant species

Robert Baur and Paul Feeny
Sect. Ecology and Systematics, Cornell University, Ithaca, New York, USA

Key words: *Daucus carota*, gas chromatography, *Pastinaca sativa*

Females of many butterfly species make use of visual cues and nonvolatile chemicals on the plant surface to recognize their host plants for oviposition. For the black swallowtail butterfly, *Papilio polyxenes* F. (Lepidoptera, Papilionidae), whose host range in North America includes various species of Apiaceae (Umbelliferae), plant odor provides additional cues for host plant selection. Volatiles from carrot (*Daucus carota* L.), a typical host species, increased both the landing frequency of females and the number of eggs laid in laboratory experiments with surrogate plants (Feeny et al., 1989).

In subsequent work, fractionation and bioassay of headspace volatiles from carrot foliage revealed that the stimulatory activity could be attributed entirely to the polar fraction. Gas chromatographic separation of this fraction, coupled with EAG (electroantennogram) recording, revealed that females perceive a profile of compounds, several of which were identified by GC-MS and comparison with reference compounds (Baur et al., unpubl.).

GC-EAG recordings. Here, using the same techniques, we report that black swallowtail females also perceive a profile of compounds in the polar headspace fraction of another host plant, parsnip (*Pastinaca sativa* L.). In GC-EAG chromatograms of carrot and parsnip volatiles, both recorded from the same female under identical conditions, some peaks in both chromatograms had identical retention times, while others were unique to one of the traces (Fig. 1). Peak identifications confirmed that some compounds were responsible for EAG activity in both plant species, though the important peaks at 14.6 min were evoked by (Z)-sabinene hydrate in carrot and by n-nonanal in parsnip. Of the compounds that were most active electrophysiologically (Fig. 2), benzaldehyde and n-nonanal were found only in parsnip. The two sabinene hydrate isomers and 4-terpineol, all hydroxygenated monoterpenoids present in relatively high concentrations in carrot, were not detected in parsnip. Of particular interest was the finding that the compounds with the highest specific EAG activities of all, namely n-decanal and two unknown compounds A and B, are apparently present in minute concentrations in both plant species (Fig. 2).

Conclusions

Though the identities of several of the perceived compounds have yet to be established, and though behavioral assays are needed to determine the stimulant activity of individual compounds and mixtures, it now seems likely that the effect of plant odor on oviposition by *P. polyxenes* is based on responses to partially overlapping profiles of volatile compounds in different host plants. However, since some of the most active compounds

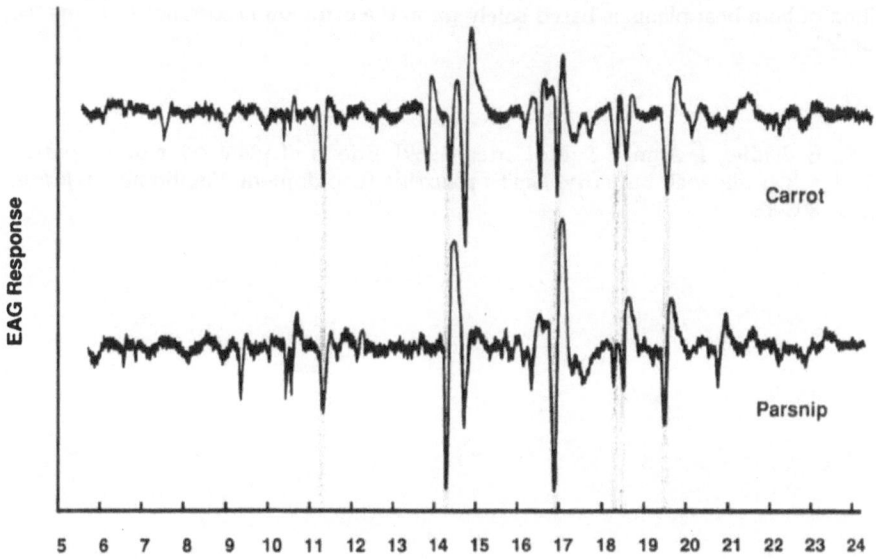

Figure 1. GC-EAG responses to the polar fractions of carrot and parsnip volatiles. Peaks connected with shaded bars were evoked by the same compounds in both plants.

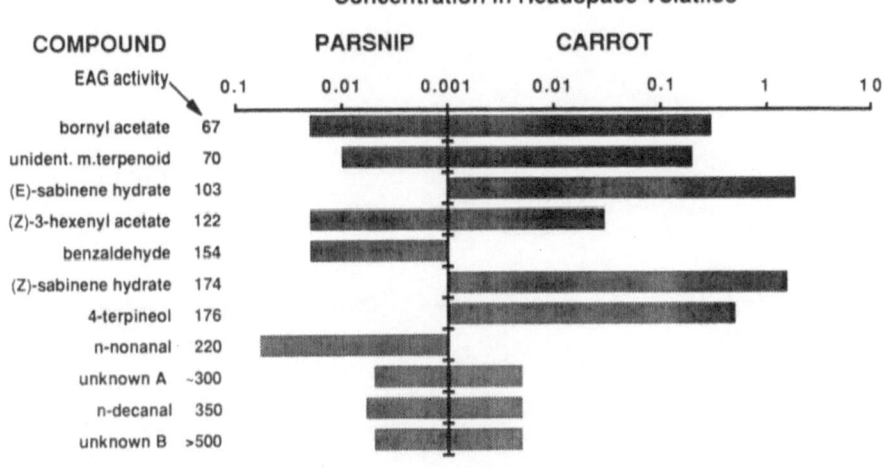

Figure 2. Concentrations of compounds with major EAG activity in parsnip and carrot volatiles, expressed as per cent of concentration of most abundant volatile compound in respective plant, *i.e.*, (Z)-ß-ocimene in parsnip and sabinene in carrot, both with low EAG activity. EAG activity of 100 ppm of each compound is indicated as per cent of response to 100 ppm of (E)-2-hexanal.

were found in both carrot and parsnip, we cannot yet eliminate the hypothesis that recognition of both host plants is based solely upon the common occurrence of a few "key compounds".

Reference

Feeney, P., E. Städler, I. Åhman & M. Carter (1989). Effects of plant odor on oviposition by the black swallowtail butterfly, *Papilio polyxenes* (Lepidoptera: Papilionidae). *J. Insect Behav.* **2**: 803-827.

Proc. 8th Int. Symp. Insect-Plant Relationships, Dordrecht: Kluwer Acad. Publ.
S.B.J. Menken, J.H. Visser & P. Harrewijn (eds), 1992

Olfactory and visual cues in host-finding in the Burnet moth, *Zygaena trifolii*

Peter Ockenfels[1] and Frank Schmidt[2]
[1] *Zoologisches Forschungsinstitut und Museum A. Koenig, Bonn, Germany*
[2] *Institut für Organische Chemie und Biochemie der Universität Hamburg, Hamburg, Germany*

Key words: Antennography, GC-MS, *Knautia*, Lepidoptera, olfaction, orientation, plant-insect relation, Zygaenidae, (-)-verbenone

Zygaena trifolii (Esper) moths are strongly attracted to the flowers of Dipsacaceae, such as *Knautia arvensis* (L.) Coulter and other scabious-like flowers. This attraction is particularly obvious in habitats where other nectar sources are readily available. It may thus be assumed that during the life cycle of Burnet moths, scabious flowers offer resources that are of special significance. We report on the physiological significance of volatile compounds of *K. arvensis* flower-scents to *Z. trifolii* moths, as tested by their antennal response and their attractiveness in the field.

Methods

Investigations were made by testing the EAG-activity of five fractions of the crude extract of *K. arvensis*, of synthetic compounds and of the synthetic female sex pheromone. Volatile compounds of *K. arvensis* were collected using the CLSA-technique (Grob, 1973). The five fractions were obtained by preparative gas chromatography. Pure synthetic samples formed the reference stimuli. EAG experiments were carried out with freshly excised antennae by following the procedure described by Van der Pers (1981). The chemical composition of volatile compounds in the scent of *K. arvensis* was identified by capillary gas chromatography and mass spectrometry. Field tests were carried out using coloured cardboard discs of 35 mm ø, which differed in the spectral composition of the colour, providing a nectar source and rubber caps, which bore the test compounds. We recorded only the complete feeding reactions (proboscis reaction).

Results and discussion

Fractions 1, 2 and 3 were clearly dominated by α-pinene, 1,8-cineole and verbenone, while the rest of the odour contained only small amounts of multicomponent mixtures. The verbenone of fraction 3 was shown to be the (-)-enantiomer of a very high optical purity. The average quantity of (-)-verbenone in crude extracts is about 200 ng/50 flower heads. The tests with the crude extracts of *K. arvensis* and synthetic compounds clearly showed that an olfactoric stimulus is provided in the flower scent, resulting in great biological activity at the electrophysiological level, and in field tests. This high electrophysiological activity led to the hypothesis, that *K. arvensis* flowers contain olfactoric stimuli for the feeding behaviour of Burnet moths.

Our field data demonstrate that the attraction of wild flying *Z. trifolii* moths to artificial flower heads is dependent on both a visual stimulus and the scent stimulus provided by from the crude extract. In the absence of either, settling behaviour and the feeding reaction were not observed. Changes in the spectral composition of the colour immediately led to a break in the biological reaction. These findings agree with the concept of Schneider (1986), that insect orientation in the field does not depend on a single stimulus, but on a number of different stimuli that are registered and evaluated at the same time.

References

Grob, K. (1973). Organic substances in potable water and its precursor, Part I. *J. Chrom.* **84**: 255.

Schneider, D. (1986). Plant recognition by insects: A challenge for neuro-ethological research. In: V. Labeyrie, G. Fabres & D. Lachaise (eds), *Insects - Plants*, pp. 117-123. Dordrecht: Junk Publishers.

Van der Pers, J.N.C. (1981). Comparison of electroantennogram response spectra to plant volatiles in seven species of *Yponomeuta* and in the tortricid *Adoxyphyes orana*. *Entomol. exp. appl.* **30**: 181-192.

Proc. 8th Int. Symp. Insect-Plant Relationships, Dordrecht: Kluwer Acad. Publ.
S.B.J. Menken, J.H. Visser & P. Harrewijn (eds), 1992

Specialization of receptor neurons to host odours in the pine weevil, *Hylobius abietis*

Atle Wibe and Hanna Mustaparta
Dept of Zoology, Univ. of Trondheim-AVH, Dragvoll, Norway

Key words: Electrophysiology, gas chromatography, headspace, olfaction, volatiles

In previous studies of host odours, the sensitivity of single sensory cells to various compounds known to be present in host plant materials has been tested (Masson & Mustaparta, 1990). The results suggest that individual plant odour receptor neurons respond to several compounds and are therefore less specialized than pheromone neurons. In screening tests of various synthetic compounds, it is highly probable that many constituents of the natural plant volatile complex are lacking. In our present study of the pine weevil, we linked gas chromatography to electrophysiological recordings to test single receptor neurons for all compounds present in plant volatile bouquets.

The pine weevil *Hylobius abietis* L. is strongly attracted by volatiles from their host trees, pine and spruce (Escherich, 1923). Injured trees, *e.g.*, by beetles feeding on them, release a higher concentration of volatiles and attract more beetles than uninjured trees (Tilles *et al.*, 1986a). Previous electrophysiological studies, testing synthetic compounds known to be present in conifers, have shown that α-pinene and ß-pinene stimulate a large number of receptor neurons (Mustaparta, 1975a). Behavioural studies in the laboratory (Mustaparta, 1975b) and in the field (Tilles *et al.*, 1986b) have revealed that α-pinene attracts the pine weevil. When combined with ethanol, the attraction is synergized (Tilles *et al.*, 1986b), whereas addition of limonene eliminates the attraction (Norlander, 1990). Electrophysiological studies also indicated that the natural signals for numerous receptor cells were lacking in the test repertoire. The present study was undertaken to identify the fractions of the volatiles in host and non-host plants received by the receptor neurons of the pine weevil.

Coupled gas chromatography - sensillum recording. Headspace volatiles from host (pine and spruce) and a non-host plant (juniper) were trapped in porapaque Q by sucking purified air over the plant material, through the absorbant. The plant substances were then washed out by hexane. Recordings from single receptor neurons on the antennal club of the weevil, were tested with a sample of each plant wash (evaporated on filter paper). Samples which elicited responses in a receptor neuron were then injected into the column of the gaschromatograph (GC). After separation, one half of the effluent was lead to the GC-detector and the other half into an air stream blowing over the antennae. Thus all fractions were tested in sequence on the receptor neuron.

Receptor cells which were strongly activated by one particular sample of the whole plant mixture, responded to only one or a few fractions of it. When two or more fractions were effective, the response to one fraction was always stronger. Some cells were strongly activated by one or two of the major fractions, α-pinene, ß-pinene, limonene and myrcene, while others by minor fractions. In some receptor neurons being tested with several GC-separated plant samples (host as well as non-host), we detected responses to fractions with corresponding retention times. For example, one receptor neuron was strongly activated by the non-host sample (juniper) as well as by the host (pine and spruce) samples. When tested by GC, each strongest response recorded was to a fraction with a retention time corresponding to that of myrcene. Several other fractions in the juniper sample elicited weaker responses. This might simply be due to the fact that the juniper samples were more concentrated than the pine and spruce samples. Alternatively, the fractions present in juniper may not be present in pine and spruce.

Conclusions

The results indicate that receptor neurons of the pine weevil, responding to plant volatiles rather are specialized than broadly tuned to many compounds. They seem to be respond to one, or perhaps two compounds, but may also respond to other analogues in the plant material. The quality of the host odour appeared to be perceived mainly via "labelled-lines", each carrying information about one particular compound, However, the presense of various analogue compounds, the amount of which may vary from one tree to another, may modify some labelled lines. Thus, when responding to two compounds a receptor cell may mainly convey information about the compound present in highest concentration. When the various fractions have been chemically identified, interactions of compounds on a receptor cell has to be tested.

Acknowledgements. The study was financially supported by Borregaard Research Fond and Norwegian Research Council for Sciences and Humanities NAVF (Project no. 452.92/002).

References

Escherich, K. (1923). *Die Forstinsekten Mitteleuropas* 2. Berlin: Paul Parey.
Masson, C. & H. Mustaparta (1990). Chemical information processing in the olfactory system of insects. *Physiol. Rev.* 70: 199-245.
Mustaparta, H. (1975a). Responses of single olfactory cells in the pine weevil *Hylobius abietis* L. (Col.: Curculionidae). *J. Comp. Physiol.* 97: 271-290.
Mustaparta, H. (1975b). Behavioural responses of the pine weevil *Hylobius abietis* L. (Col.: Curculionidae) to odours activating different groups of receptor cells. *J. Comp. Physiol.* 102: 57-63.
Norlander, G. (1990). Limonene inhibits attraction to α-pinene in the pine weevil *Hylobius abietis* and *H. pinastri*. *J. Chem. Ecol.* 16: 1307-1320.
Tilles, D.A., G. Norlander, H. Nordenhem, H.H. Eidmann, A.-B. Wassgren & G. Bergström (1986a). Increased release of host volatiles from feeding scars: a major cause of field aggregation in the pine weevil *Hylobius abietis* L. (Coleoptera: Curculionidae). *Environ. Entomol.* 15: 1050-1054.
Tilles, D.A., K. Sjödin, G. Norlander & H.H. Eidmann (1986b). Synergism between ethanol and conifer host volatiles as attractants for the pine weevil *Hylobius abietis* L. (Coleoptera: Curculionidae). *J. econ. Entomol.* 79: 970-973.

Proc. 8th Int. Symp. Insect-Plant Relationships, Dordrecht: Kluwer Acad. Publ.
S.B.J. Menken, J.H. Visser & P. Harrewijn (eds), 1992

Plant chemicals involved in honeybee-rapeseed relationships: behavioural, electrophysiological and chemical studies

M.H. Pham-Delègue[1], M.M. Blight[2], M. Le Métayer[1], F. Marion-Poll[1], A.L. Picard[1], J.A. Pickett[2], L.J. Wadhams[2] and C.M. Woodcock[2]
[1] *INRA-CNRS, Laboratoire de Neurobiologie Comparée des Invertébrés, Bures sur Yvette, France*
[2] *AFRC, IACR, Rothamsted Experimental Station, Harpenden, Herts, UK*

Key words: *Apis mellifera ligustica, Brassica napus*, electroantennograms, gas chromatography, olfaction, pollination, proboscis extension, volatiles

Oilseed rape (*Brassica napus* L.) is a major arable crop within the EC. Although the crop is primarily self-fertile, pollination by honeybees is beneficial, increasing both the number of seeds per pod and the proportion of early flowers that set pods. Besides, pollination by honeybees is essential for the production of hybrid rapeseed; readily available and efficient insect pollinators are required to transfer pollen from male fertile to male sterile lines. Recent genetic engineering of oilseed rape has led to the development of plants with new qualities, such as disease and insect resistance. Therefore, an evaluation of the possible effects of these new plants on insect pollinators is necessary prior to their being placed on the market. This work seeks to investigate the odour cues produced by oilseed rape that are important to bees for recognizing the flowers and for stimulating their foraging. The incorporation of appropriate cues into rape cultivars by plant breeders should insure more effective pollination.

Conditioned proboscis extension. Behavioural studies were conducted using the conditioned proboscis extension assay (CPE) (De Jong & Pham-Delègue, 1991). After a two-hour starvation period, foragers were conditioned by exposure for six seconds. After the first three seconds the bee was induced to extend her proboscis by application of sugar solution to the antenna and was then rewarded with the sugar solution during the final three seconds of odour stimulation. This procedure was repeated three times for each insect. Bees were conditioned to three concentrations (1000 ng, 100 ng, 10 ng) of a synthetic mixture of six rapeseed flower volatiles, *i.e.*, linalool, 2-phenylethanol, methylsalicylate, benzyl alcohol, (*E*)-2-hexenal, 1-octen-3-ol (Tollsten & Bergström, 1988; Blight *et al.*, 1992). The ability of these compounds to elicit the CPE response was then examined by testing the components individually. The threshold concentration for activity was 10-100 ng. Linalool, 2-phenylethanol and methylsalicylate were the most active, producing responses at the 1000 ng level from more than 50% of the conditioned bees.

Electroantennograms. Electroantennograms (EAGs) for benzyl alcohol, 2-phenylethanol, methylsalicylate and linalool were recorded at the 100 ng level. The two last mentioned compounds, which elicited the highest proportion of CPE responses, produced the highest amplitudes (> 1.5 mV).

Coupled gas chromatography. When bees were conditioned to oilseed rape flowers, 85% responded to an air entrainment extract of flower volatiles, and 60% of those conditioned to the rape extract responded to the flowers. This clearly demonstrated that the extract accurately reflected the spectrum of volatiles produced by the flowers. Bees were then conditioned to the extract and their responses to the individual components were examined by coupled gas chromatography (GC-CPE). Automatic acquisition and treatment of the GC and biological data was done using a software program. Thirty-two peaks elicited a response from at least one individual, but only five, including (E,E)-α-farnesene - a major component of the extract - induced a response in more than 35% of the bees.

A coupled GC-EAG-CPE recording system, now being developed, will provide a rapid and reliable method for investigating the perception of floral compounds by the insect's peripheral olfactory system and the associated central nervous system integration of the chemical signals.

References

Blight, M.M., A.J. Hick, J.A. Pickett, L.E. Smart, L.J. Wadhams & C.M. Woodcock (1992). Volatile plant metabolites involved in host plant recognition by the cabbage seed weevil, *Ceutorhynchus assimilis* Payk. In : S.B.J. Menken, J.H. Visser & P. Harrewijn (eds), *Proc. 8th Int. Symp. on Insect-Plant Relationships*, pp. 105-106. Dordrecht: Kluwer Acad. Publ.

De Jong, R. & M.H. Pham-Delègue (1991). Electroantennogram responses related to olfactory conditioning in the honey bee (*Apis mellifera ligustica*). *J. Insect Physiol.* **37**: 319-324.

Tollsten, L. & G. Bergström (1988). Headspace volatiles of whole plants and macerated plants parts of *Brassica* and *Sinapis*. *Phytochemistry* **27**: 4013-4018.

Proc. 8th Int. Symp. Insect-Plant Relationships, Dordrecht: Kluwer Acad. Publ.
S.B.J. Menken, J.H. Visser & P. Harrewijn (eds), 1992

Volatiles from soybean foliage detected by means of a TCT-HRGC system: their possible role in insect-plant relationships

G. Giangiuliani[1], M. Castellini[2], P. Damiani[2] and F. Bin[1]
[1] *Entomology Institute, Faculty of Agriculture, Perugia, Italy*
[2] *Bromatologic Chemistry Institute, Faculty of Pharmacy, Perugia, Italy*

Key words: Headspace, *Nezara viridula*, olfaction, *Trichopoda pennipes*, *Trissolcus basalis*

How phytophagous insects locate and recognize their host plants and whether the associated parasitoids might use the same chemical cues, has been studied by several researchers. Host acceptance by an insect involves biochemical plant parameters and morphological plant characteristics, and its selection patterns usually consists of a sequence of behavioural responses to an array of stimuli associated with non-host and host plants (Visser, 1986).

Soybean is one of the most widely cultivated crops in the world that has been increasingly grown in Italy over the past few years. *Nezara viridula* L. is one of the most important pests of soybean and in Italy is considered to be the "key-pest" of this crop. For these reasons we have attempted to clarify the role of this plant-insect interaction. This paper describes the extraction method and a new method of analysis (HRGC-TCT) to detect soybean leaf volatiles.

Extraction and analysis method

In the past, several methods to isolate plant volatiles have been applied, *e.g.*, steam distillation, vacuum steam distillation, solvent extraction, cold condensation, absorbent trap etc. (see *e.g.*, Visser *et al.*, 1979; Buttery & Ling, 1984). The most important aspect when designing such a procedure is to maximize the recovery of plant volatiles while minimizing contaminants.

We used a gentle method (Tenax GC trapping) for isolating volatiles from plants. We took the leaves and stems of soybean *cv.* Canton, at vegetative stages R3-R5 (about 40 g for each analysis). For each analysis, the time taken to collect the volatiles was 30 min and the absorbent was Tenax 60-80 mesh in test tubes adapted for the analysis method and ready to be fixed in the TCT injection system. The purging gases were hyperpure Nitrogen and Helium at a flow of 100 ml/min. GC analysis was done by a gas chromatograph Chrompack CP 9000 equipped with a TCT (Thermal Desorption Cold Trap) injection system under the following conditions: desorption for 6 min at 230°C, followed by criofocusing at -120°C. The hydrogen flow was about 25 ml/min, the make up Nitrogen flow about 20 ml/min and the air flow 250 ml/min. The used column was a Supelcowax 10 30 m long, 0.32 mm i.d. The oven conditions were programmed from 50°C to 210°C (10°C/min).

Results and conclusions

The first aim of this research was to find the best approach for volatile isolation and HRGC separation. The TCT system producing a very "clean" detection of the substances seemed to be the best. The first result showed by the gas chromatograms obtained with this method is the presence of two different groups of substances. These different "detection areas" differ in complexity and the quantitative and correlative differences are probably dependent on the vegetative stages of the plant.

Future goals of this research are: (a) identification of the substances by mass spectroscopy, (b) analysis of the plant volatiles at different vegetative and reproductive stages, (c) bioasssays of the identified compounds with *N. viridula*, and (d) bioassays of the same compounds with its egg parasitoid *Trissolcus basalis* Woll. and adult parasitoid *Trichopoda pennipes* F. to identify their possible role as synomones.

References

Buttery, R.G. & L.C. Ling (1984). Corn leaf volatiles: identification using Tenax trapping for possible insect attractants. *J. Agric. Food Chem.* **32**: 1104-1106.

Visser, J.H. (1986). Host odor perception in phytophagous insects. *Annu. Rev. Entomol.* **31**: 121-144.

Visser, J.H., S. van Straten & H. Maarse (1979). Isolation and identification of volatiles in the foliage of potato, *Solanum tuberosum*, a host plant of the Colorado beetle, *Leptinotarsa decemlineata. J. Chem. Ecol.* **5**: 13-25.

Proc. 8th Int. Symp. Insect-Plant Relationships, Dordrecht: Kluwer Acad. Publ.
S.B.J. Menken, J.H. Visser & P. Harrewijn (eds), 1992

Host-finding by *Phoracantha semipunctata*: host volatiles, electroantennogram recordings and baited field traps

E.N. Barata[1], P.J. Fonseca[2], E. Mateus[3] and J. Araujo[1]
[1] *Dept de Biologia, Univ. de Evora, Evora, Portugal*
[2] *Dept Zool., Fac. Ciências, Univ. de Lisboa, Lisboa, Portugal*
[3] *Dept Ciências e Engenharia do Ambiente, F.C.T., Univ. Nova de Lisboa, Portugal*

Key words: Cerambycidae, Coleoptera, *Eucalyptus globulus*, headspace, monoterpenes, olfaction

Phoracantha semipunctata F., a specific bark borer of *Eucalyptus* trees, causes significant damage in the stands in southern Portugal. It readily attacks trees that are under physiological stress, and felled trees (Chararas, 1969). The aim of this study was to test the hypothesis that the odour released by the host tree plays an important role in the host-finding behaviour of this nocturnal insect.

GC analysis. The gas chromatographic analysis of the volatiles released from the bark, leaves and fruits of *Eucalyptus globulus* Labill using the headspace technique showed the presence of α-pinene, ß-pinene, 1,8-cineole, myrcene, α-terpinene, γ-terpinene, p-cymene, d-limonene, camphene, terpinolene, and ethanol. These compounds were identified by co-injection of standards in apolar (OV-101) and polar (SUOX-0.1) capillary columns. Identification of α-pinene and 1,8-cineole was confirmed by mass spectrometry.

EAG recordings. Electroantennogram recordings (EAG) of responses to the odour released by bark, node, leaf, and fruits of *E. globulus*, as well as to some volatile compounds present in its odour blend, were made on living females and males of *P. semipunctata*. This was done by inserting a tungsten recording electrode into the tip of the antenna and placing the tungsten reference electrode into the ipsilateral eye. Stimulus consisted of a 3 ml odour puff injected in a 80 ml/s air stream 1.5 m away from the animal. This was accomplished using a syringe attached to a pasteur pipette containing the odour source: 1.5 g of *E. globulus* cut material or a piece of filter paper impregnated with 25 µl of a dilution of the volatile compound in paraffin oil (v/v). Significant differences were not detected in male and female responses. Throughout the experiment the amplitude of the antennal responses to a standard stimulus (1,8-cineole 1:10) did not decrease. Stimulation with air and paraffin oil did not elicit any measurable response. Bark, node, leaf and fruit elicited the following EAG amplitudes (mean ± 95% confidence interval): 0.16 ± 0.12 (n = 6), 0.24 ± 0.23 (n = 6), 0.36 ± 0.11 (n = 10) and 0.83 ± 0.26 (n = 10) mV, respectively. In Table 1 we present EAG amplitudes of *P. semipunctata* in response to several volatile compounds. The insects showed a high olfactory sensitivity to α-pinene, and a moderate sensitivity but intense response to camphene at the highest concentration tested (1:1); 1,8-cineole, α-terpinene, and ethanol elicited moderate responses but only to high stimulus concentration; stimulation with p-cymene and terpinolene did not elicit any measurable response.

Table 1. EAG amplitudes (mV) of *P. semipunctata* in response to volatile compounds of *E. globulus* (mean ± 95% confidence interval, n = 9)

Compound	Dilution of the compound in paraffin oil (v/v)				
	1:10000	1:1000	1:100	1:10	1:1
α-Pinene	0.13 ± 0.06	0.16 ± 0.09	0.21 ± 0.08	0.38 ± 0.11	0.49 ± 0.14
Camphene	0.05 ± 0.05	0.06 ± 0.05	0.14 ± 0.06	0.38 ± 0.12	0.71 ± 0.27
α-Terpinene	0.06 ± 0.06	0.07 ± 0.06	0.08 ± 0.08	0.17 ± 0.08	0.24 ± 0.10
1,8-Cineole	0.02 ± 0.04	0.03 ± 0.07	0.03 ± 0.08	0.15 ± 0.07	0.25 ± 0.12
Ethanol	0.01 ± 0.04	0.02 ± 0.05	0.03 ± 0.03	0.06 ± 0.04	0.18 ± 0.07

Field traps. During the first three weeks of July 1991, empty traps, and traps baited with either a dry log, a freshly cut log, or leaves of *E. globulus* were placed in a stand according to a six randomized block design. The logs were approximately 1 m in length, diameter 14 cm. The leaves were put into a wire net cylinder with the same dimensions as the logs. The baits were suspended vertically inside a 1 m long wire net cylinder 25 cm in diameter covered with tanglefoot. The baits were renewed weekly and their place within each block randomized. Significantly more adults were caught in the traps baited either with a freshly cut log or with leaves, than in empty traps or traps baited with a dry log (Fig. 1). The proportion of males to females caught in all of the traps was constant (3:1). However,

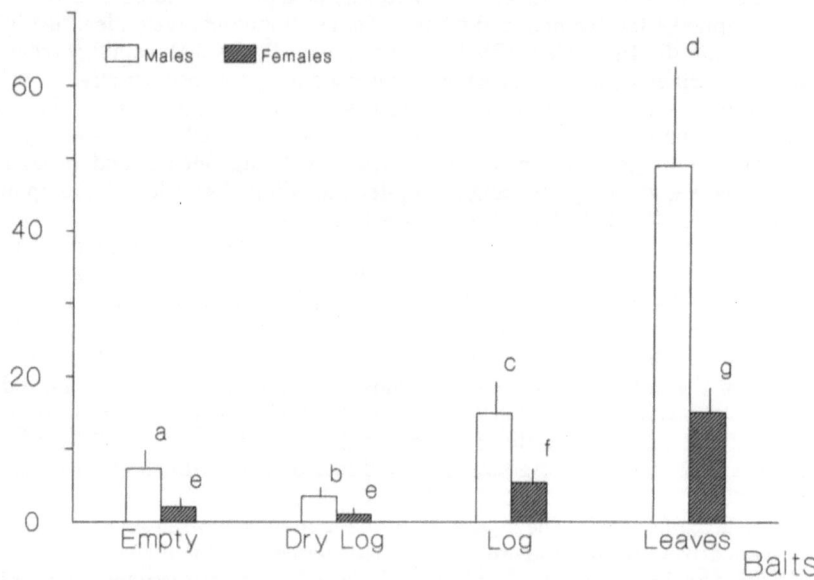

Figure 1. Average number of adults caught in empty traps and in traps baited with one dry log, one freshly cut log or with leaves of *E. globulus*. Mean values with different letters differ significantly (Duncan's New Multiple Range Test, P < 0.01). Vertical bars indicate 95% confidence intervals.

no differences were found between male and female emergences (surveyed from May to September). These results confirm the preliminary data obtained in the previous year.

Conclusions

Powell (1978), suggested that *P. semipunctata* locates its host tree by an olfactory stimulus. The field results presented here show that both males and females can locate freshly cut *E. globulus* material from a distance, probably in response to the odour released, perceptible by the insect as revealed in the EAG recordings. Although the EAG results obtained should be regarded as preliminary, it is clear that this antennal olfactory system is selective for the compounds present in the *E. globulus* odour. Further research on identifying the volatile compounds present in the odour blend of host trees and screening their EAG activity will tell us more about the olfactory stimulus involved in *P. semipunctata* host-finding behaviour. By understanding host selection in this insect better measures for *Eucalyptus* protection could be devised.

Acknowledgement

The work was partially funded by Junta Nacional de Investigação Cientifica e Tecnológica (JNICT) and by Associação de Produtores de Pasta de Papel, ACEL.

References

Chararas, C. (1969). Biologie et ecologie de *Phoracantha semipunctata* F. (Coléoptère Cerambycidae xylophague) ravageur des eucalyptus en Tunisie, et methodes de protection des peuplements. *Ann. Inst. Nat. Rech. Forest. de Tunisie* 2: 1-37.
Powell, W. (1978). Colonization of twelve species of *Eucalyptus* by *Phoracantha semipunctata* F. (Coleoptera: Cerambycidae) in Malawi. *Bull. ent. Res.* 68: 621-626.

Proc. 8th Int. Symp. Insect-Plant Relationships, Dordrecht: Kluwer Acad. Publ.
S.B.J. Menken, J.H. Visser & P. Harrewijn (eds), 1992

Chemical recognition of diverse hosts by *Pieris rapae* butterflies

K. Sachdev-Gupta, C.D. Radke and J.A.A. Renwick
Boyce Thompson Institute, Ithaca, New York, USA

Key words: *Brassica juncea*, glucosinolate, oviposition, Pieridae

The major oviposition stimulant responsible for recognition of cabbage, *Brassica oleracea* L. (Cruciferae) by the cabbage butterfly, *Pieris rapae* L. (Pieridae), has been identified as glucobrassicin (Renwick *et al.*, 1991). However, many host plants lack detectable quantities of glucobrassicin. We therefore examined the source of stimulatory activity in a range of plants that are acceptable for oviposition. These include *Tropaeolum majus* L. (Tropaeolaceae), *Carica papaya* L. (Caricaceae), and *Brassica juncea* L. (Cruciferae), along with *Isatis tinctoria* L. (Cruciferae), which is known to be rich in indole glucosinolates, including glucobrassicin and the structurally related neoglucobrassicin and gluco-brassicin-1-sulfonate (Elliot & Stowe, 1971).

Chemical fractionation

Alcoholic extracts of *T. majus*, *C. papaya*, *B. juncea*, and *I. tinctoria* were defatted with hexane and then partitioned between n-butanol and water. In each case the aqueous fraction was chromatographed using a reverse-phase C_{18} flash chromatography column. The chromatographic fractions were pooled on the basis of their TLC pattern on silica gel plates in solvent system 1: ethyl acetate-methanol-acetic acid-water (4:1:1:0.5). All the fractions were desulfated with aryl sulfatase enzyme (Minchinton *et al.*, 1982). Each desulfated fraction was analyzed by HPLC on a C_{18} column and by TLC on silica gel in solvent system 2: chloroform-methanol-water (14:10:1). Details of the analytical methods, conditions for growing all the plants and for rearing the insects were as previously described (Renwick *et al.*, 1991). Bioassays were performed by spraying the test solutions (1 gram leaf equivalent, fresh weight/plant) on Sieva bean (*Phaseolus vulgaris* L., *cv.* Sieva) foliage (Renwick *et al.*, 1991).

The active fractions 5-8 and 9-13 (Fig. 1) of *T. majus* showed a single spot (hRf = 48.0) on TLC in solvent system 1. Desulfation of these fractions yielded a single product (hRf in solvent system 2 = 69.2 and HPLC Rt = 30.5 min). On the basis of co-TLC and co-HPLC with the desulfated product of an authentic sample as well as ^{13}C NMR data of the desulfated product (Table 1), it was identified as glucotropaeolin. An active compound present in fractions 6-7 and 8-14 (Fig. 1) of *C. papaya*, exhibiting similar properties, was also identified as glucotropaeolin.

Column chromatographic fractions 1-3 and 4-9 of *B. juncea* were stimulatory to *P. rapae*. These fractions were found to contain sinigrin (hRf = 41.9; desulfated product: hRf = 60.2, HPLC Rt = 14.99 min and ^{13}C NMR data (Table 1)). Some activity in fractions 12-15 is attributed to the presence of a small amount of glucobrassicin (hRf = 53.0; desulfated product: hRf = 58.9, HPLC Rt = 36.78 min and ^{13}C NMR data (Table 1)).

Similar chromatographic analysis of the post-butanol aqueous extract of *I. tinctoria*

136

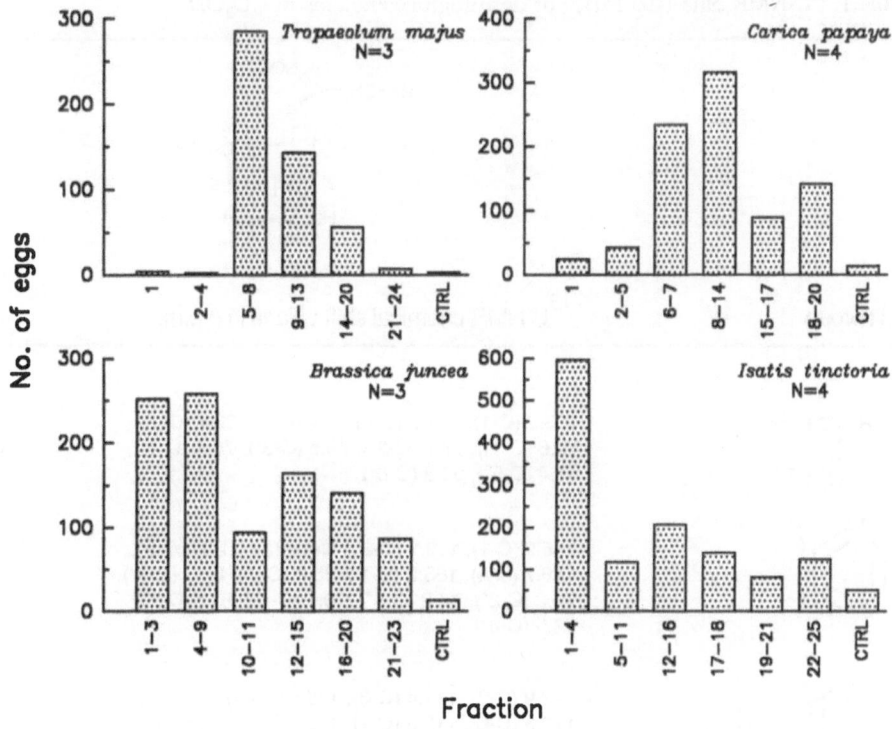

Figure 1. Oviposition by *P. rapae* in response to a choice of column fractions from each plant species.

yielded two active components present in fractions 1-4 and 12-16. Fractions 12-16 were found to contain glucobrassicin. Fractions 1-4 showed a major spot at hRf = 28 on TLC in solvent system 1. Desulfation of this material generated only desulfoglucobrassicin. Since glucobrassicin itself was not detected in fractions 1-4 prior to desulfation, it is assumed that the original compound in this fraction was glucobrassicin-1-sulfonate.

Conclusions

P. rapae strongly prefers glucobrassicin over sinigrin (Renwick *et al.*, 1991; Traynier & Truscott, 1991). Glucoiberin, a glucosinolate containing sulfur in the alkyl side chain, is completely inactive (Renwick *et al.*, 1991). The present study indicates that *P. rapae* is stimulated to oviposit by two other glucosinolates *i.e.*, glucotropaeolin and glucobrassicin-1-sulfonate. On the basis of these findings, it appears that *P. rapae* responds preferably to the presence of aromatic glucosinolates. Preliminary evidence suggests that glucosinolates containing an alkenyl side chain may also stimulate oviposition, but those with a sulfur atom in the side chain are inactive. Further studies on this structure-activity relationship are in progress.

Table 1. ^{13}C NMR data (400 MHz) of desulfoglucosinolates in CD$_3$OD

R group	^{13}C NMR chemical shift values in p.p.m.
$\underset{2}{CH_2}\!=\!\underset{1}{CH}$ —	133.2 (C-1), 119.5 (C-2), 154 (C-1'), 38.1 (C-2'), 82.6 (C-1"), 73.7 (C-2"), 78.8 (C-3"), 70.1 (C-4"), 81.4 (C-5"), 62.3 (C-6")
[phenyl ring, positions 1–6]	135.9 (C-1), 130.1 (C-2 & C-6), 129 (C-3 & C-5), 128.7 (C-4), 153.8 (C-1'), 38.8 (C-2'), 82.5 (C-1"), 74.2 (C-2"), 79.2 (C-3"), 70.1 (C-4"), 81.6(C-5"), 62.5 (C-6")
[indole ring, positions 2–9, N-H]	123.9 (C-2), 111.4 (C-3), 119.5* (C-4), 119.8* (C-5), 122.5 (C-6), 112.3 (C-7), 138.2 (C-8), 128.2 (C-9), 154 (C-1'), 30.3 (C-2'), 82.7 (C-1"), 74.4 (C-2"), 79.3 (C-3"), 71.1 (C-4"), 82 (C-5"), 62.7 (C-6")

* Values may be interchanged.

References

Elliott, M.C. & B.B. Stowe (1971). Distribution and variation of indole glucosinolates in woad (*Isatis tinctoria* L.). *Plant Physiol.* **48**: 498-503.

Minchinton, I., J. Sang, D. Burke & R.J.W. Truscott (1982). Separation of desulpho-glucosinolates by reversed-phase high performance liquid chromatography. *J. Chromatogr.* **247**: 141-148.

Renwick, J.A.A., C.D. Radke, K. Sachdev-Gupta & E. Städler (1991). Leaf surface chemicals stimulating oviposition by *Pieris rapae* on cabbage. *Chemoecology* (in press).

Traynier, R.M.M. & R.J.W. Truscott (1991). Potent natural egg-laying stimulant for cabbage butterfly, *Pieris rapae*. *J. Chem. Ecol.* **17**: 1371-1380.

Proc. 8th Int. Symp. Insect-Plant Relationships, Dordrecht: Kluwer Acad. Publ.
S.B.J. Menken, J.H. Visser & P. Harrewijn (eds), 1992

Role of nutrients found in the phylloplane, in the insect host-plant selection for oviposition

S. Derridj[1], V. Fiala[2], P. Barry[1], P. Robert[1], P. Roessingh[3] and E. Städler[3]
[1] INRA station de zoologie, Versailles Cedex, France
[2] INRA station du métabolisme et de la nutrition des plantes, Versailles Cedex, France
[3] Eidgenösische Forschungsanstalt CH, Wädenswil, Switzerland

Key words: Carbohydrates, contact chemoreceptors, *Ostrinia nubilalis*

Visual, olfactory and gustatory cues are involved in the ovipositon site selection by phytophagous insects. The majority of them alight on the phylloplane (leaf surface) in the course of their host selection and before ovipositing. There they are in contact with waxes, leaf cuticle, and with substances originating from the leaf tissues and the environment. Under natural conditions different communities of epiphytic microorganisms are also present on the maize phylloplane. To explain the host oviposition preference of the generalist insect *Ostrinia nubilalis* Hübner (Lepidoptera) we collected the water soluble substances from the phylloplanes of maize, sunflower and leek. The metabolites we found such as low molecular carbohydrates, free amino acids, and organic acids occur in very small quantities (10^{-5} to 10^{-6} moles m^{-2}). A high correlation was found between the possibility to trace host plants suitable for oviposition and the difference between carbohydrate quantities found in the plant phylloplanes. Fructose was particularly important. There was no real significant effect of amino acids observed, except for proline which seemed to have a synergistic effect with sugars. We designed other experiments to confirm these previous findings which were insufficient proof of a causative relationship. First, we had to verify the presence of carbohydrates in the corn phylloplane and to determine their location; second, to precision their action on the insect oviposition with artificial substrates; and third, to explain how they are perceived by the insect (behavioural events which lead to the detection, sensorial receptors concerned).

Methods

Procedures used to visualize and localize reducing sugars (glucose, fructose, etc.) in the phylloplane necessitated avoiding their leaching and solubilization by water. Sugars were precipitated in alcohol solutions and were visualized in deposits of monossacharides silver complex. They were observed by scanning electron microscopy associated with microanalysis with X ray spectrometry on dry corn leaf samples.

The oviposition responses of *O. nubilalis* to carbohydrates were studied on dried filter paper strips, into which the quantities of sugars collected in the maize phylloplane had been incorporated. These were natural and derived from analytical data of substances, collected on the maize ear leaf at twilight. The observations were pursued for six days. The ovipositon preference behaviour on filter paper and maize leaves was directly observed for several hours in the laboratory using an infra-red video system. According to these observations, electrophysiological recordings were made from sensilla of contact receptors present on the ovipositor and the tarsae. This was done by using tip recording techniques.

Results

Monosaccharides were detected basically at two types of locations. The first one, in which they were situated in great numbers, consisted of scattered areas where they were localized in depressions along the anticlinal walls of the epidermal cells. The second one was found more scantily in the shape of "craters" of 0.2 to 0.4 μm diameter. The distribution seemed more uniform when drops of water had been deposited on the phylloplane before the observations. Artificial substrates simulated this distribution more closely.

Oviposition data showed that when sugars were applied singly onto artificial substrates, fructose stimulated oviposition, whereas sucrose and glucose were either neutral or even inhibitory. Experiments offering a choice of the three sugars on the same substrate in different multiples of concentration against the reference pattern were likely to be closer to field conditions. Ratios were important and the modification of one or two sugar quantities was likely to destroy the stimulatory effect of the active pattern or to induce a non-preference effect. Other preliminary experiments on entire plants show that substances like organic acids were correlated with the insect oviposition.

The females alight several times on the phylloplane before ovipositing. The number of alightings are less numerous on plants than on artificial substrates with sugars. To test the phylloplane the insect curves its abdomen and surveys the surface by scanning with its ovipositor for several seconds, which is desicive for egg-laying. The appendages in contact with the leaf surface are the six legs and the ovipositor.

Electrophysiological recordings from contact chemoreceptors on the sensilla chaetica present on the rim of the ovipositor valves indicated a fructose threshold of 10^{-4} M while sucrose appears to be ten times less active. Glucose was not stimulatory. Mixture experiments of fructose and sucrose did not show any evidence of the existence of separate cells for each sugar. Preliminary recordings from contact chemoreceptor sensilla placed distal on the tarsi showed a sensitivity to the three sugars.

Conclusions

These results emphasize the role of the phylloplane as the interface between plants and the environment. The insect is able to get information concerning plant metabolites without any feeding. For this polyphagous insect *O. nubilalis*, sugars are already shown to be involved in the acceptability and suitability of maize by the progeny. Now we have demonstrated that they also act as kairomones in its host-plant selection for oviposition. This opens the way to new investigations. The phylloplane contains the major part of the metabolites and phytophagous insects are in contact with them when they are on the leaf surface, whether they are generalist or specialist, and whether it is to oviposit or to feed.

Proc. 8th Int. Symp. Insect-Plant Relationships, Dordrecht: Kluwer Acad. Publ.
S.B.J. Menken, J.H. Visser & P. Harrewijn (eds), 1992

Oviposition stimulant for the cabbage root fly: important new cabbage leaf surface compound and specific tarsal receptors

Peter Roessingh, Erich Städler, Jakob Hurter and Thomas Ramp
Eidg. Forschungsanstalt, Wädenswil, Switzerland

Key words: *Brassica oleracea, Delia radicum*, gustation

Oviposition by the cabbage root fly, *Delia radicum* (L.) (Dipt., Anthomyiidae) is governed to a large extent by chemicals present on the leaf surface of the host plant *Brassica oleracea* L. Although this fact has been recognized for some time, knowledge about the nature of the phytochemicals is still far from complete. Glucosinolates (mustard oil glucosides) and their breakdown products, the isothiocyanates (mustard oils), have been known to be involved. The glucosinolates stimulate oviposition and the tarsal D-sensilla contain chemoreceptors sensitive to this group of compounds (Städler, 1978; Roessingh *et al.*, unpubl.).

Pure glucosinolates, however, were shown to be far less active than a crude methanolic leaf surface extract, indicating a possible role for other compounds present in the extract (Roessingh *et al.*, unpubl.). Since no a priory knowledge of the compound(s) involved is available, a "brute force" search strategy has to be used: starting from the crude surface extract, and assaying successive fractions produced in the purification process for biological activity (in this case oviposition stimulation, see also Roessingh & Städler, 1990).

Purification process. Fractions were purified using ion exchange and reverse phase chromatography (MPLC and HPLC) and one of the major oviposition stimulants was isolated. Structural analysis is still in progress but the available evidence (retention times, UV spectrum, NMR data) shows that this compound is probably not a glucosinolate.

One major obstacle for a search strategy like this is the large number of bioassays needed, since after each purification step (often many), the active fraction(s) have to be located again. One way to increase the efficiency of the purification cycle is the application of a rapid bioassay using only a little of the test material. This proved to be possible with electrophysiological recordings from relevant receptors. Although behavioural tests are always necessary, much information can be obtained from sensory data. Unfortunately, identifying the right receptors is often difficult when no pure compounds are available. In the case of the cabbage root fly the receptor was indeed only found after a small amount of pure compound became available, from purification runs using behavioural assays. In later runs, active fractions were largely identified on the basis of spike counts, significantly improving the speed in the purification cycle.

Electrophysiological observations. Receptor cells for the new compound were only found in the ventro-medial pairs of C-sensilla (terminology of Grabowski & Dethier, 1954) on the fifth tarsomere. The chemoreceptors in the D-sensilla do not respond to it. Unexpectedly, however, it turned out that both aliphatic and aromatic glucosinolates stimulated receptors in the C-sensilla, in concentrations comparable to those found effective for the D-sensilla. Mixture experiments indicate that the sensitivity is located on

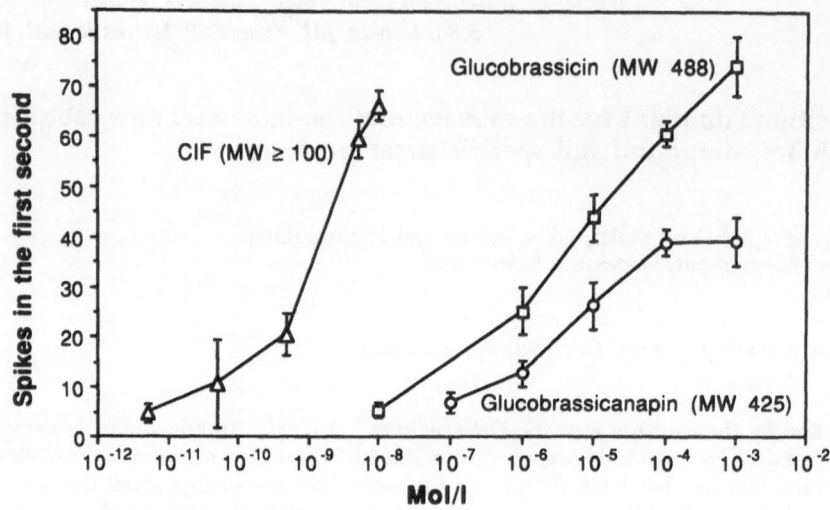

Figure 1. Dose-response curves of ventro-medial C-sensilla on the fifth tarsomere of *Delia radicum* for three *Brassica* chemicals. CIF: Newly isolated purified non-glucosinolate oviposition stimulant with still unknown structure (molecular weight assumed to be ≥ 100, concentration estimated from spectroscopic and analytical data).

the same receptor cells. On a molar basis, however, the cells are several orders of magnitude more sensitive to the new compound (Fig. 1).

Conclusions

The presence of the new compound contributes significantly to the recognition of cruciferous plants by *Delia radicum*, but more work investigating its presence and distribution is clearly needed.

Acknowledgement. This research was supported by grant 31-30059.90 of the Swiss National Science Foundation to E.S.

References

Grabowski, C.T. & V.G. Dethier (1954). The structure of the tarsal chemoreceptors of the blowfly, *Phormia regina* Meigen. *J. Morphol.* **94**: 1-19.
Roessingh, P. & E. Städler (1990). Foliar form, colour and surface characteristics influence oviposition behaviour in the cabbage root fly, *Delia radicum. Entomol. exp. appl.* **57**: 93-100.
Städler, E. (1978). Chemoreception of host plant chemicals by ovipositing females of *Delia (Hylemya) brassicae. Entomol. exp. appl.* **24**: 511-520.

Proc. 8th Int. Symp. Insect-Plant Relationships, Dordrecht: Kluwer Acad. Publ.
S.B.J. Menken, J.H. Visser & P. Harrewijn (eds), 1992

Tarsal contact chemoreceptors of the cherry fruit fly, *Rhagoletis cerasi*: specificity, correlation with oviposition behaviour, and response to the synthetic pheromone

Erich Städler[1], Beat Ernst[2], Jakob Hurter[1], Ernst Boller[1] and Marek Kozlowski[3]
[1] *Eidg. Forschungsanstalt, Wädenswil, Switzerland*
[2] *Ciba Geigy AG, Basel, Switzerland*
[3] *Dept of Entomology, SGGW-AR, Warszawa, Poland*

Key words: Gustation, host marking, oviposition-deterring pheromone

Females of the European cherry fruit fly (*Rhagoletis cerasi* L.) (Dipt., Tephritidae) mark the host fruits (cherry, *Prunus avium* (L.) L.) after oviposition by dragging their ovipositor over the fruit surface, leaving behind a trace of droplets containing a pheromone. This host-marking pheromone trail deters subsequent ovipositions by the same or other flies, and has been called oviposition-deterring pheromone (ODP).

Experiments and discussion. Contact-chemoreceptor sensilla were identified (D-sensilla) on the ventral surface of the tarsi of both sexes which contain a receptor cell sensitive to the ODP. The specificity of the ODP receptor cell was studied using the following stimuli occurring in the fly's natural environment: (1) Fly faeces, components of insect faeces: allantoin, urea, uric acid, (2) Glucosides of host and non-host plants: phloridzin, salicin, phenyl-glucopyranoside, (3) Methanolic leaf surface extracts of yew, cabbage and cherries. With the exception of the raw extract of fly faeces (Boller & Hurter, 1985) all substances (selected examples in Table 1) did not stimulate the ODP cell, thus documenting its selectivity.

To study the origin of the ODP we collected the faeces of eight females daily until the 7th day of life. The faeces of five males were collected on the 4th and 5th day. The individual faecal droplets were dissolved in 2 µl of water and used to stimulate the D-sensilla. The results of these recordings (Fig. 1) clearly show that the tested male faeces was not, or hardly active, whereas female faeces strongly stimulated the ODP cell from the 3rd day of life. To verify the relationship between the oviposition deterrency and ODP spike activity, we used the same raw extract of faeces and flies from the same batch of pupae for oviposition and electrophysiological experiments. The oviposition deterrency indices and the spike counts produced a similar dose response curve. The correlation between the oviposition discrimination index (Boller & Hurter, 1985) and the mean spike frequency of the ODP receptor cells was highly significant (P < 0.001, Spearman correlation coefficient: 0.903).

Hurter *et al.* (1987) isolated and identified the ODP from fly faeces and Ernst & Wagner (1989) first synthesized the 8R15R isomer which we compared with the natural raw extract at different concentrations. The raw extract and the synthetic ODP molecule stimulated the ODP cell in a concentration dependent manner (Fig. 2). The slopes of the two regression lines fitted to the means of each concentration were not different ($F_{1, 208}$ = 1.131, P >> 0.05). This indicated that the same cell was sensitive to both compounds and this was further supported by the observation that the raw extract, the synthetic ODP, as

Table 1. Stimulation of the ODP receptor cell by different substances which occur in the environment of the cherry fruit fly

Stimulus	Concentration	Repetitions sensilla/flies	Number of spikes mean ± SE[1]
NaCl (solvent)	30 mM	109/34	3.1 ± 0.2
ODP raw extract	0.1 mg/ml	138/36	37.8 ± 1.8
Food:			
sucrose/yeast hydrolysate 4:1	1 mg/ml	12/3	0.1 ± 0.04
honeydew	natural	8/2	0
pollen (Lonicera)	saturated	4/1	0
Host plant (Prunus avium):			
fruit juice	100%	12/3	0.2 ± 0.04
fruit surface	10% MeOH extract[2]	5/2	0.3 ± 0.3
leaf surface	10% MeOH extract[2]	5/2	0.6 ± 0.4

[1] Spikes in the first second of stimulation (Städler et al., 1987).
[2] Concentrated extracts of 50 undamaged cherry fruits, preferred for oviposition or 5 undamaged cherry leaves, respectively.

well as the mixture, produced only one type of spike and that no additions or 'doubles' did occur. The horizontal distance between the regression lines (log concentration/log spikes) was calculated to be 1.935 logarithmic concentration steps, or about a factor of 100. This means that the natural raw extract contained about 1% of the ODP. The perception threshold for the raw extract was 10^{-5} mg/ml, and the synthetic ODP 10^{-7} mg/ml which is equal to $1.73 \cdot 10^{-10}$ M.

References

Boller, E.F. & J. Hurter (1985). Oviposition-deterring pheromone in *Rhagoletis cerasi*: behavioral laboratory test to measure pheromone activity. *Entomol. exp. appl.* **39**: 163-169.

Ernst, B. & B. Wagner (1989). Synthesis of the oviposition-deterring pheromone (ODP) in *Rhagoletis cerasi*. *Helv. Chim. Acta* **72**: 165-171.

Hurter, J., E.F. Boller, E. Städler, B. Blattmann, N.U. Bosshard, H.-R. Buser, L. Damm, M.W. Kozlowski, R. Schöni, F. Raschdorf, R. Dahinden, E. Schlumpf, H. Fritz, W. Richter & J. Schreiber (1987). Oviposition-deterring pheromone in *Rhagoletis cerasi* L.: Purification and determination of the chemical constitution. *Experientia* **43**: 157-164.

Städler, E., R. Schöni & M.W. Kozlowski (1987). Relative air humidity influences the function of the tarsal chemoreceptor cells of the cherry fruit fly (*Rhagoletis cerasi*). *Physiol. Entomol.* **12**: 339-346.

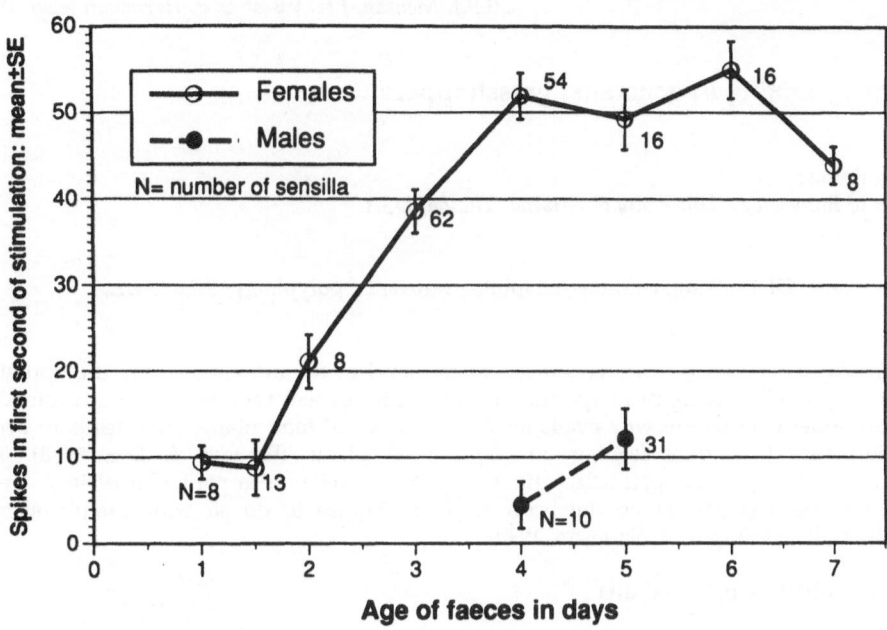

Figure 1. ODP activity of the raw extract of individual faecal droplets from males and females of different ages.

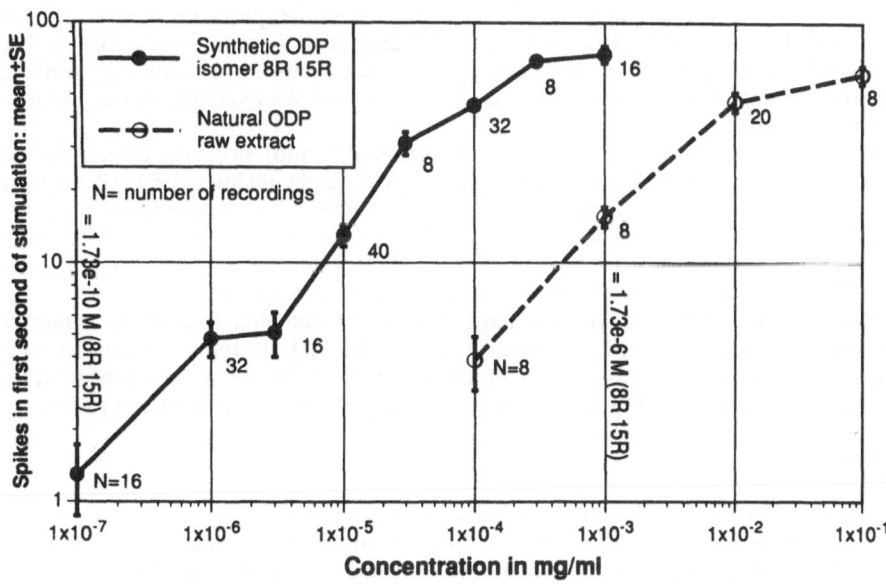

Figure 2. Recordings from the same ODP receptor cells to the raw extract of faeces and the synthetic N[15(ß-glucopyranosyl)-oxy-8-hydroxypalmitoyl]-taurine.

Proc. 8th Int. Symp. Insect-Plant Relationships, Dordrecht: Kluwer Acad. Publ.
S.B.J. Menken, J.H. Visser & P. Harrewijn (eds), 1992

Dietary mixing in generalist grasshoppers

E.A. Bernays
Dept of Entomology, University of Arizona, Tucson, USA

Key words: Diet mixing, learning, neophilia, nutrients, polyphagy, *Schistocerca*

Polyphagous species that are generalists at the level of the individual have the option to forage optimally among plant species with different nutrient profiles. This is presumed to be an underlying reason why availability of mixtures of food plants often leads to better performance than does feeding on single food plants (Bernays & Lee, 1988). The possibility that grasshoppers select different plants in relation to particular dietary needs, has also been suggested on the basis of their abilities to do so with complementary artificial diets (Simpson & Simpson, 1990).

How to obtain a balanced diet

Mechanisms that might allow grasshoppers to achieve a balanced diet from a selection of less suitable foods include:

1. Direct nutrient feedbacks. Shortage of carbohydrate or protein in the diet results in lowered levels of sugar and amino acids in the haemolymph, which in turn causes changes in the relative sensitivity of mouthpart chemoreceptors to sugars and amino acids. This has been postulated to directly influence food selection, such that feeding is stimulated most by what is most related to need (Simpson & Simpson, 1990). This requires that sugars and amino acids accurately represent the available carbohydrates and proteins in foods.

2. Learned associations. Among food with unsuitable nutrient levels, insects may learn to avoid them and to prefer novel tastes (Bernays & Raubenheimer, 1991; Champagne & Bernays, 1991). If this were to occur on a limited number of successive foods there would be a tendency to obtain a diet approaching the optimal (Bernays & Lee, 1988). Learned positive associations are also possible (Simpson & White, 1991), and may provide a suitable mechanism in combination with some kind of nutrient feedback.

However, in a situation where a large number of potential food items occur in an unpredictable or fine grained arrangement, it may be that an approximation to nutrient selection may occur to some extent by random switching between a variety of food items. An example may be found in *Taeniopoda eques* (Burmeister) (Bernays *et al.*, in press).

Experimental

Taeniopoda eques is an individual generalist, switching frequently between foods in the field, and ingesting up to eight different items even within one meal (Raubenheimer & Bernays, in press). The habitat in which it lives in the southwest United States contains a great diversity of small annuals including seedlings, and diverse perennials, many of which have flowers that are eaten. Regions and seasons also vary and in one region plant profiles also change during the life of an insect. Patches of any one food item tend to be small and individuals wandering throughout the day may encounter many plant species only once. Evidence has been presented that this species is apparently a "compulsive" switcher (Bernays et al., in press): all food items become less acceptable over time while novel ones are accepted. What is the physiological basis for such switching behavior?

We carried out an experiment to test the hypothesis that in this species odor novelty provides a stimulus for feeding. We exposed young adult female *T. eques* to a single odor for one hour and then provided individuals with an artificial diet cake (Bernays et al., in press) containing very low concentrations of the same or a different odour. Two chemically contrasting odors were used - coumarin and citral. Insects were observed singly at 30°C and the behavior recorded for two hours. The latency to feeding was significantly greater when the odor was the same than when it was different. Further, the meal lengths were greater (at least during hour 1) when the odor was novel than when the odor had been experienced (Fig. 1).

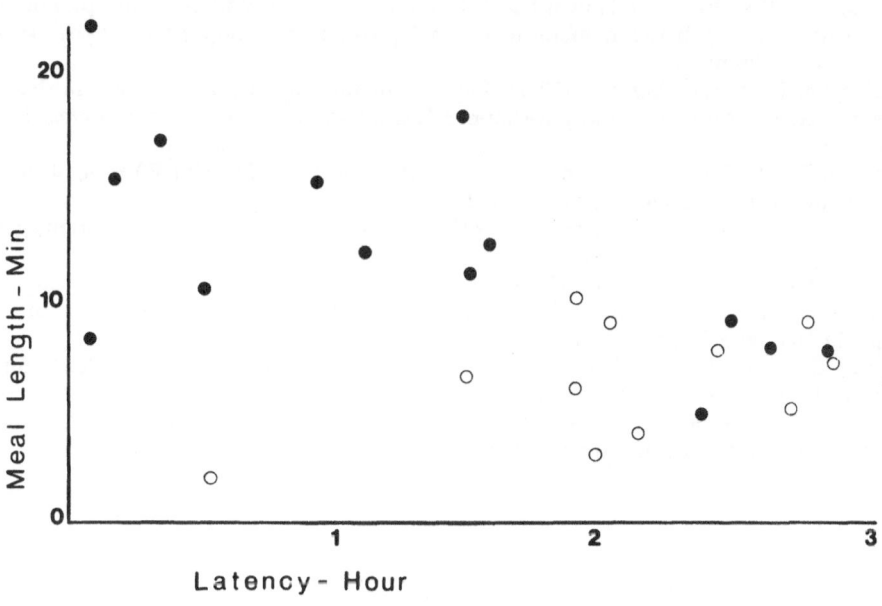

Figure 1. Meal size and latency to feeding by insects experiencing the same (open symbols) or novel (closed symbols) odors.

The data suggest that newly detected odors provide an arousal stimulus that leads to feeding activity and could be a mechanism for the switching between foods that characterized the foraging pattern in the field. In a typical field situation insects do encounter a series of items over relatively short periods (Raubenheimer & Bernays, in press), and it is conceivable that a chemical arousal effect is the mechanism involved in the "compulsive" switching. This may be an exaggerated example of the phenomenon of changing stimuli having a general arousal effect as reported elsewhere (e.g., Chapman, 1954; Simpson et al., 1988).

Learned aversions require a suitable timing between feeding on different items to allow an association to develop. In addition it would be efficient only if clear-cut differences occur in food items, and these should be limited in number since there are constraints on how many plant features could be learned. Similarly, memory constraints will limit the number of positive associations that can be learned, so that it may only be useful in situations where there are relatively few food items available. We conclude that the "compulsive" switching mechanism is an adaptation to the very complex environment in which *T. eques* lives, and work is in progress to address the role of different habitats on mechanisms of dietary mixing.

References

Bernays, E.A. & J.C. Lee (1988). Food aversion learning in the polyphagous grasshopper *Schistocerca americana* (Drury). *Physiol. Entomol.* **13**: 131-137.

Bernays, E.A. & D. Raubenheimer (1991). Dietary mixing in grasshoppers: changes in acceptability of different plant secondary compounds associated with low levels of dietary protein. *J. Insect Behav.* **4**: 545-556.

Bernays, E.A., K.L. Bright, J.J. Howard & D. Champagne (1992). Variety is the spice of life: frequent switching between foods in the polyphagous grasshopper *Taeniopoda eques*. *Anim. Behav.* (in press).

Champagne, D. & E.A. Bernays (1991). Phytosterol unsuitability as a factor mediating food aversion learning in the grasshopper *Schistocerca americana*. *Physiol. Entomol.* **16**: 329-336.

Chapman, R.F. (1954). Responses of *Locusta migratoria migratorioides* (R.&F.) to light in the laboratory. *British J. Anim. Behav.* **2**: 146-152.

Raubenheimer, D. & E.A. Bernays (1992). Feeding patterns in the polyphagous grasshopper *Taeniopoda eques*: a field study. *Anim. Behav.* (in press).

Simpson, S.J. & C. Simpson (1990). Mechanisms of nutritional compensation by phytophagous insects. In: E.A. Bernays (ed.), *Plant-Insect Interactions* Vol. II, pp. 111-160. Boca Raton: CRC Press.

Simpson, S.J. & P. White (1991). Associative learning and locust feeding: evidence for a "learned hunger" for protein. *Anim. Behav.* **40**: 506-513.

Simpson, S.J., M.S.J. Simmonds, A.R. Wheatly & E.A.Bernays (1988). The control of meal termination in the locust. *Anim. Behav.* **36**: 1216-1227.

Proc. 8th Int. Symp. Insect-Plant Relationships, Dordrecht: Kluwer Acad. Publ.
S.B.J. Menken, J.H. Visser & P. Harrewijn (eds), 1992

Semiochemicals isolated from the eggs of *Ostrinia nubilalis* as oviposition deterrent in three other moth species of different families

Denis Thiéry, Bruno Gabel and André Pouvreau
INRA-CNRS, Laboratoire de Neurobiologie Comparée des Invertébrés, Bures sur Yvette, France

Key words: *Aphomia sociella*, *Lobesia botrana*, Noctuidae, oviposition-deterring pheromones, *Sesamia nonagrioïdes*, Tortricidae

Epideictic pheromones that regulate the egg laying behaviour have recently been isolated and identified in the eggs of two phytophagous moths: *Ostrinia nubilalis* L. (Lepidoptera, Pyralidae) (Thiéry & Le Quéré, 1991) and *Lobesia botrana* Den. & Schiff. (Lepidoptera, Tortricidae) (Gabel & Thiéry, 1992 ; Thiéry et al., 1992). The five major constituents found in *O. nubilalis* were methyl esters of fatty acids: methyl palmitate, methyl linoleate, methyl palmitoleate, methyl oleate and methyl stearate. These compounds have been also identified in the extracts of *L. botrana* eggs (Thiéry et al., 1992).

We made the hypothesis that esters of fatty acids present in the eggs of different insect species, which regulate their oviposition, might not be specific. Our results document the activity of the compounds identified in the eggs of *O. nubilalis* on four moth species: (a) a phytophagous pyralid, *O. nubilalis*, (b) *Aphomia sociella* L., a pyralid predator of bumble bee larvae, (c) a noctuid corn pest, *Sesamia nonagrioïdes* Lef., and (d) a tortricid pest of vinegrapes, *L. botrana*.

Methods

The five methyl esters of fatty acids identified in the extracts of *O. nubilalis* eggs were blended in pure methanol according to the ratio measured by GC analysis (FID). The ratios of methyl palmitate, methyl palmitoleate, methyl linoleate, methyl oleate and methyl stearate were respectively 1.5, 1.1, 1.1, 1.3 and 0.1. The stock solutions of the synthetic blend contained four egg mass equivalents (*e.m.*) per 10^{-6} l of methanol. Different bioassays were used for each of the four species, and all the experiments were performed in climatic chambers on laboratory strains of insects. The numbers of eggs (or egg masses in *O. nubilalis*) laid on the treated and the non-treated areas were counted, and an index of deterrency (Di) was calculated: $Di = 100(A-B)/(A+B)$, where A and B are the numbers of eggs laid on the non-treated and treated areas (or plants), respectively. This index varies from 0% (no difference) to 100% (maximum avoidance).

O. nubilalis. The bioassay has been described in detail by Thiéry & Le Quéré (1991). Groups of females (2 days old) were allowed to oviposit for one night on an artificial substrate (waxed paper) divided into two equal areas: one impregnated with 90 dots of the blend in pure methanol (20 egg mass equivalents (*e.m.*)/dot, *i.e.*, 0.7 *e.m.* per cm²), the other left blank. Deposits of extracts were made with a syringe. The experiments were conducted with 62 females in 3 groups.

A. sociella. Isolated females (mated, n = 22) were allowed to oviposit for one night on an artificial oviposition substrate (paper) divided into two parts: one treated with droplets of the blend in pure methanol (3 *e.m.* per cm^2), the other left blank.

S. nonagrioïdes. Four potted plants of corn (high 15-20 cm) were offered to each group of 2 mated females (3 days old) for two consecutive nights. Two plants were sprayed with the blend of chemicals diluted in 10% of methanol in water, the other remaining blank. The equivalent amount of chemicals sprayed was about 10^4 *e.m.* per plant. Twenty replicates have been performed.

L. botrana. We used a bioassay described in detail elsewhere (Gabel & Thiéry, 1992). Fourteen groups of 5 mated females (3 days old) were allowed to oviposit for one night on an artificial substrate (cardboard). Twenty-three drops (20 *e.m.*/dot, *i.e.*, 5 *e.m.* per cm^2) of the blend in pure methanol were deposited on part of each cardboard by mean of a syringe, the remaining area was non-treated.

Results and discussion

Methanol used alone had no effect on the oviposition of the four species. In *O. nubilalis*, the synthetic blend of the five methyl esters of fatty acids reduced the number of egg masses laid on the treated areas to about 38%. This result is similar to that obtained with an identical dilution of a raw extract of *O. nubilalis* eggs. This confirms the biological activity of the five chemicals identified.

The deterrency index of *A. sociella*, was 35%. Of the 22 females, eight did not lay any eggs on the treated areas. Females of *S. nonagrioïdes* strongly preferred to oviposit on the non-treated plants. We obtained a deterrency index of more than 70%; eight females laid no eggs, and 14% of treated plants carried eggs. Twelve females (out of 40) laid their eggs outside the plants which was not observed in the controls, and is unusual behaviour in that species, possibly due to a dose effect. Data obtained with *A. sociella* and *S. nonagrioïdes* are still preliminary and need to be confirmed by further experiments.

The oviposition by *L. botrana* was reduced to 57%; in five of the replicates the reduction was more than 68%. The level of deterrency in response to the chemicals identified in *O. nubilalis* eggs was higher to that in response to raw extracts of *L. botrana* eggs (Gabel & Thiéry, 1992). This difference might be due to a dose effect.

Conclusions

The biological activity of rather simple compounds on the oviposition in four species raises interesting questions about the regulation of their oviposition behaviour. Our results suggest that non-specific compounds could enhance the avoidance for oviposition and that interspecific egg recognition may exist in some species of Lepidoptera. Experiments involving other insects species are currently in progress.

References

Gabel, B. & D. Thiéry (1992). Biological evidence of an oviposition deterring pheromone in *Lobesia botrana* Den. & Schiff. (Lepidoptera, Tortricidae). *J. Chem. Ecol.* **18**: 353-358.

Thiéry, D. & J.L. Le Quéré (1991). Identification of an oviposition-deterring pheromone in the eggs of the European corn borer. *Naturwissenschaften*, **78**: 132-133.

Thiéry, D., B. Gabel, P. Farkas & V. Pronier (1992). Identification of an oviposition regulating pheromone in the European grapevine moth, *Lobesia botrana* (Lepidoptera: Tortricidae). *Experientia* (in press).

Proc. 8th Int. Symp. Insect-Plant Relationships, Dordrecht: Kluwer Acad. Publ.
S.B.J. Menken, J.H. Visser & P. Harrewijn (eds), 1992

Roles of chemosensory organs in food discrimination by larvae of the tobacco hornworm, *Manduca sexta*

Gerrit de Boer
Depts of Entomology and Physiology and Cell Biology, University of Kansas, Lawrence, USA

Key words: Ablations, behavior, food selection, gustation, induction

The role of five different peripheral chemosensory organs in food discrimination by *Manduca sexta* (Johan.) larvae was examined by determining the competence of each organ and its necessity to mediate this behavior. Various host and non-host plant species were used in food choice tests to examine whether the same or different chemosensory organs mediate food discrimination.

Material and methods

Larvae were reared on leaves of tomato, *Lycopersicon esculentum* Mill. Different chemosensory organs were removed surgically, resulting in larvae having only one of the following organs remaining: the antennae, the maxillary palps, the epipharyngeal sensilla, the medial and lateral maxillary sensilla styloconica (De Boer, 1991). All operations were done using middle fourth instar larvae except for removal of the sensilla styloconica which was performed upon freshly molted fifth instar larvae. In addition, two groups of control larvae were used: unoperated larvae and larvae having all five chemosensory organs removed. Larvae were allowed to feed on leaves until the fifth larval molt. Food preferences were measured by offering a freshly molted fifth instar larva a choice between leaf discs and wetted glass fiber filter paper discs (De Boer & Hanson, 1987). Food consumption of each larva was measured visually after 50% of the total area of either choice substrate was eaten. Choice tests were repeated with at least 15 larvae. Choice scores of test larvae were compared with those of control larvae to determine the importance of individual chemosensory organs (Kruskal-Wallis test).

Results and discussion

The chemosensory organ(s) capable of mediating normal food discrimination (*e.g.*, by unoperated larvae) varies with the plant species sampled. The medial maxillary sensilla styloconica or the epipharyngeal sensilla are competent to mediate normal discrimination between the non-host *Canna generalis* Bailey and wet filter paper. In contrast, only the lateral maxillary sensilla styloconica are competent for the non-hosts *Vigna sinensis* Savi and *Raphanus sativus* L., and only the antennae for the non-host *Pelargonium hortorum* Bailey. Each of the five chemosensory organs is competent to mediate normal discrimination between the host *Solanum pseudocapsicum* L. and filter paper. However, only the maxillary palps are capable of mediating normal food discrimination for the host *Datura innoxia* L. and no individual chemosensory organ suffices for the host *L. esculentum*. For each of the above plant species, the remaining chemosensory organs mediate either a reduced or no food discrimination as compared with unoperated larvae. One exception was found: the maxillary palps mediate a food preference for the non-hosts *V. sinensis, R. sativus* and *P. hortorum,* which is opposite of the behavior shown by unoperated larvae. Apparently, the maxillary palps are not always capable to mediate information on the chemical quality of the plant. Thus, different chemosensory organs mediate normal food discrimination, depending on the plant species sampled. In addition, various chemo-sensory organs can provide conflicting input for feeding decisions: stimulation *vs* inhibition of feeding.

Although some chemosensory organs may be competent in mediating normal food discrimination, their necessity to elicit this behavior when other chemosensory organs are present (as in unoperated larvae) still needs to be determined. This was briefly studied by testing larvae having only the competent organs missing using *V. sinensis* and *P. hortorum* as test plants. Results show that only the lateral maxillary sensilla styloconica are required for normal feeding responses to *C. sinensis.* In contrast, although the antennae are the only organs competent to detect *P. hortorum,* combined input from three other organs are equally important: the epipharyngeal sensilla, the medial and lateral maxillary sensilla styloconica. Thus, each plant species seems to be detected by a unique combination of chemosensory organs. Some chemosensory organs are more important than others in mediating food discrimination.

Dietary experience of *M. sexta* larvae results in plasticity of food discrimination which appears to be, at least partly, due to changes in chemoperception. This was found with larvae reared on either *L. esculentum* or *V. sinensis* given a choice between the latter plant and filter paper (De Boer, 1991). Larvae reared on the former plant are deterred from eating *V. sinensis,* while larvae reared on the latter plant prefer *V. sinensis.* The function of the antennae to detect *V. sinensis* changes from incompetent by larvae reared on *L. esculentum* to competent by larvae reared on *V. sinensis.* In addition, the former group of larvae requires chemosensory input mediated by only the lateral maxillary sensilla styloconica while the latter group needs input from both the antennae and the maxillary palps. Thus, diet-induced changes in feeding responses to *V. sinensis* seem to involve a change in competence of the antennae and a shift in importance of various chemosensory organs to mediate food selection.

References

De Boer, G. (1991). Effect of diet experience on the ability of different larval chemosensory organs to mediate food discrimination by the tobacco hornworm, *Manduca sexta. J. Insect Physiol.* **37**: 763-769.
De Boer, G. & F.E. Hanson (1987). Differentiation of roles of chemosensory organs in food discrimination among host and non-host plants in larvae of the tobacco hornworm, *Manduca sexta. Physiol. Entomol.* **12**: 387-398.

Proc. 8th Int. Symp. Insect-Plant Relationships, Dordrecht: Kluwer Acad. Publ.
S.B.J. Menken, J.H. Visser & P. Harrewijn (eds), 1992

Sensory responses to the triterpenoid antifeedant toosendanin

Luo Lin-er and L.M. Schoonhoven
Dept of Entomology, Wageningen Agricultural University, Wageningen, The Netherlands

Key words: Deterrent receptor, gustation, *Pieris brassicae*, sensory inhibition

One of the basic problems in insect-plant relationships is how insects can perceive thousands of different antifeedant compounds, many of them occurring in plants in only low concentrations, according to Jermy (1983) "host plant specificity in phytophagous insects is determined mainly by the botanical distribution of plant substances inhibiting feeding". In several insect species so-called 'deterrent neurons' have been identified, which respond to a wide array of secondary plant compounds and upon stimulation, inhibit feeding behaviour. There is increasing evidence that some feeding deterrents may also affect the chemosensory system by inhibiting neurons that respond to the presence of feeding-stimulating compounds, such as sugars or amino acids (Schoonhoven *et al.*, 1992).

Several tree species belonging to the family Meliaceae have been found to contain a variety of terpenoids with strong antifeedant effects on many insect species. The bark of *Melia toosendan* S. et Z. contains toosendanin, a triterpenoid with a β-substituted furan ring (Shu & Liang, 1980), which strongly suppresses food intake in larvae of the large white cabbage butterfly, *Pieris brassicae* L. In this paper, we analyse the feeding inhibitory effect having studied the influence of toosendanin on four different taste neurons involved in food recognition in *P. brassicae* larvae.

Electrophysiological observations

Two sensilla styloconica located on the maxilla, each innervated by four chemoreceptory neurons, respond to various categories of plant compounds. The response spectrum of each cell can be determined using electrophysiological methods (for further details see Schoonhoven *et al.*, 1992). The medially located taste hair contains a neuron sensitive to several alkaloids and terpenoids, *e.g.*, strychnine and azadirachtin, which act as feeding deterrents. Toosendanin solutions at concentrations as low as 10^{-7} M or higher elicit phasic-tonic responses in this so-called 'deterrent cell'. Another deterrent cell located in the lateral sensillum styloconicum, is stimulated by flavonoids and is insensitive to toosendanin. The lateral sensillum contains three other neurons, which respond to sugars, amino acids and glucosinolates respectively. Toosendanin does not elicit action potentials in either of these neurons.

When the sugar sensitive neuron is stimulated with a mixture of sucrose and toosendanin, the latter compound appears to suppress the cell's normal response to sucrose. In the presence of 10^{-6} M toosendanin, neural activity elicited by 0.1 M sucrose shows a reduction of 45 per cent. The degree of inhibition at constant toosendanin levels is positively correlated with sucrose concentration, *i.e.*, the responses to high sucrose concentrations are reduced to a greater extent than those at lower sucrose levels.

Similarly, the glucosinolate sensitive neuron in the lateral taste hair is inhibited when its adeqate stimulus, *e.g.*, sinigrin, is mixed with toosendanin. However, in contrast to the sugar neuron, this degree of inhibition is negatively correlated with the concentration of sinigrin. When the amino acid sensitive neuron is tested with various concentrations of proline, either with or without toosendanin, it appears that the response of this cell to proline is not influenced by toosendanin.

Discussion

The presence of toosendanin in the food of *P. brassicae* is coded in a rather complex way: some neurons are stimulated, others are inhibited while a third category of neurons remains unaffected. The high efficacy of toosendanin as a feeding deterrent thus appears to be based upon its capacity to affect various chemoreceptors in different ways. Of particular interest is our observation that the inhibition of the sugar cell and the glucosinolate cell shows an opposite dependency of stimulus concentration. It may indicate that different types of molecular interactions are involved at the receptor membrane level. The fact that the amino acid neuron is not affected by toosendanin also contradicts the idea that toosendanin inhibits taste receptors via general and unspecific molecular mechanisms. Obviously different cell types vary in a rather subtle way according to their physico-chemical membrane characteristics and, consequently, react differently to secondary plant compounds like toosendanin.

The data presented show that to understand host-plant selection by phytophagous insects, the analysis of the perception mechanisms of natural deterrent compounds is just as important as the study of the sensory coding of phagostimulants.

References

Jermy, T. (1983). Multiplicity of insect antifeedants in plants. In: D.L. Whitehead & W.S. Bowers (eds), *Natural Products for Innovative Pest Management*, pp. 223-236. Oxford: Pergamon Press.

Schoonhoven, L.M., W.M. Blaney & M.S.J. Simmonds (1992). Sensory coding of feeding deterrents in phytophagous insects. In: E.A. Bernays (ed.), *Insect-Plant Interactions* Vol. 4. pp. 59-79. Boca Raton: CRC Press.

Shu, G.X. & X.T. Liang (1980). A correction of the structure of chuanliansu (toosendanin). *Acta Chim. Sin.* 38: 196-198.

Proc. 8th Int. Symp. Insect-Plant Relationships, Dordrecht: Kluwer Acad. Publ.
S.B.J. Menken, J.H. Visser & P. Harrewijn (eds), 1992

Insect antifeeding activity of some cardenolides, coumarins and 3-nitropropionates of glucose from *Coronilla varia*

J. Harmatha[1], J. Nawrot[2], K. Vokac[1], L. Opletal[3] and M. Sovova[3]
[1] Inst. of Organic Chemistry and Biochemistry, Cz. Acad. of Sci., Prague, Czechoslovakia
[2] Inst. of Plant Protection, Poznan, Poland
[3] Faculty of Pharmacy, Charles University, Hradec Kralove, Czechoslovakia

Key words: *Sitophilus granarius*, stored product pests, *Tribolium confusum*, *Trogoderma granarium*

The common crownvetch, *Coronilla varia* L. (Fabaceae), is a perennial herb growing on field boundaries and bushy slopes from lowlands to foothills. As this plant does not suffer from diseases, it is not attacked by pests and it is slightly poisonous for non-ruminants, it appears to be a promising source for obtaining various active compounds. To obtain new natural substances to regulate cardiac activity, tenths of compounds of *C. varia* were isolated in the course of recent phytochemical and pharmacological research.

We selected some of the compounds for testing feeding-deterrent activity towards three insect pest species on stored products (Table 1). The test method of Nawrot *et al.* (1986) was applied, using wheat wafer discs (1 cm in diameter) dipped into an ethanolic solution of the compound at 10 mg/ml concentration. Feeding of insects was recorded under three conditions: on pure food with non-treated discs (control), on food with one non-treated and one treated disc (choice test) and on food with treated discs only (no-choice test). After 5 days we calculated the coefficients of deterrency from the amount of food that had been consumed. These coefficients served as an index of activity expressed on a scale between 0 and 200. Strong antifeedants have an index of 150-200.

To test feeding-deterrent activity, we chose main constituents of coumarin type (1-3) and cardenolide type (4,5) together with 3-nitropropionic acid (10) and four of its glucose esters (6-9; Table 1). Comparisons were made with some commercially available coumarins and cardenolides and the relation between structure and activity was investigated. We discovered some highly efficient antifeedants among the structurally related compounds (Harmatha *et al.*, unpubl.).

The highest antifeeding activity was observed with 3-nitropropionic acid. Low activity of its glucose esters indicated that the free acid plays the main defensive role in the plant. As its concentration varies among different *C. varia* lines, resulting in differing susceptibility to pathogen/herbivore attack (Gustine, 1979), quantitative analyses need to be carried out in order to study the defensive mechanism. The low specificity of the effect on the three tested species implies a similar effect on other insect species that are ecologically more dependent on *C. varia*. The rather low activities of coumarin and cardenolide constituents probably only marginally contribute to the defence against storage pests.

3-Nitropropionic acid and its glucose esters were also found to be toxic to the cabbage looper, *Trichoplusia ni* Hübner, a non-pest of crownvetch (Byers *et al.*, 1977). However, experiments with chrysomelid beetles, *Chrysomela tremulae* Dejean (Pasteels *et al.*, 1982), containing 3-nitropropionic acid derivatives in the defensive secretion, demonstrated that *C. tremulae* has the capacity to biosynthesize the compounds *de novo* (Randoux *et al.*, 1991). More information is needed to assess the potential role of those compounds in defence, sequestration and other mechanisms related to plant-insect chemical interactions.

Table 1. Total coefficients of feeding deterrent activity of substances from *C. varia* on insect storage pests: *Sitophilus granarius* L. (A), *Tribolium confusum* Duv. (B) and *Trogoderma granarium* Everts (C)

Substances	Adults		Larvae	
	A	B	B	C
1. Daphnoretin	117.4	108.8	107.8	120.4
2. Scopoletin	151.8	140.8	64.3	88.4
3. Umbelliferone	74.4	142.4	108.2	85.1
4. Hyrcanoside	114.5	140.1	122.0	86.1
5. Deglucohyrcanoside	154.9	138.0	166.9	137.9
6. Karakin	132.2	119.5	89.4	108.0
7. Corollin	78.6	119.3	92.2	108.1
8. Coronillin	126.1	153.8	116.2	104.1
9. Coronarian	88.3	94.1	125.6	105.2
10. 3-nitropropionic acid	175.5	157.0	165.4	185.2

References

Byers, R.A., D.L. Gustine & B.G. Moyer (1977). Toxicity of ß-nitro-propionic acid to *Trichoplusia ni*. *Environ. Entomol.* **6**: 229-232.
Gustine, D.L. (1979). Aliphatic nitro compounds in crownvetch: A review. *Crop Science* **19**: 197-203.
Nawrot, J., E. Bloszyk, J. Harmatha, L. Novotny & B. Drozdz (1986). Action of antifeedants of plant origin on beetles infesting stored products. *Acta Entomol. Bohemoslov.* **83**: 327-335.
Pasteels, J.M., J.C. Braekman & D. Daloze (1982). Chemical defence in Chrysomelid larvae and adults. *Tetrahedron* **38**: 1891-1897.
Randoux, T., J.C. Braekman, D. Daloze & J.M. Pasteels (1991). *De novo* biosynthesis of 3-isoxazolin-5-one and 3-nitropropionic acid derivatives in *Chrysomela tremulae*. *Naturwissenschaften* **78**: 313-314.

Proc. 8th Int. Symp. Insect-Plant Relationships, Dordrecht: Kluwer Acad. Publ.
S.B.J. Menken, J.H. Visser & P. Harrewijn (eds), 1992

Seasonal variation in the importance of pollen volatiles on the reproductive biology of the sunflower moth

Jeremy N. McNeil[1] and Johanne Delisle[2]
[1] Dept of Biology, Laval University, Ste. Foy, P.Q., Canada
[2] Forestry Canada, Ste Foy, P.Q., Canada

Key words: Calling behaviour, daylength, dispersal, Homoeosoma electellum, temperature

Females of the sunflower moth, *Homoeosoma electellum* (Hulst), preferentially oviposit in plants with recently opened inflorescences (Teetes & Randolph, 1969; DePew, 1983), due to the presence of an oviposition stimulant present in the pollen (Delisle *et al.*, 1989). McNeil & Delisle (1989) demonstrated that at 25°C and 16L:8D, virgin females initiated calling behaviour (the emission of the sex pheromone) at a significantly younger age following emergence, spent more time calling, and had a higher rate of ovarian development than individuals held under the same conditions without pollen. While pollen is not used as a food source by sunflower moth adults, it is an essential resource for neonate larvae that are initially unable to feed on other flower parts due to the high levels of the major terpenoid (Rossiter *et al.*, 1986). However, any given host plant will only be available as a suitable oviposition site for a short time period following anthesis as the pollen load will decline due to the action of pollinators, wind and rain. Furthermore, in natural ecosystems host plants are patchily distributed so that the availability of suitable oviposition sites are also unpredictable in space. McNeil & Delisle (1989) proposed that the ability to initiate or delay reproduction depending on the presence or absence of pollen, was a life history strategy permitting sunflower moth females to cope with unpredictable habitat quality. In the presence of suitable hosts rapid maturation permits females to exploit temporarily available oviposition sites, while in the absence of pollen delayed maturation provided females the opportunity to disperse in search of suitable hosts, sometimes over considerable distances (Arthur & Bauer, 1981).

Effect of temperature and daylength. In the present study we wished to determine whether temperature and daylength, cues of predictable seasonal habitat deterioration, modified sunflower moth females' response to pollen. No significant difference was observed in the mean age of calling (2-3 days following emergence) when virgin females were held from emergence at 15°, 20° or 25°C in the presence of pollen. Similarly, in the absence of pollen the mean age of calling was about six days, regardless of whether females were placed at 15° or 25°C upon emergence. Furthermore, daylength had no significant effect on the age of first calling when females were held at 25°C in the presence of pollen. However, in the absence of pollen females held under 12L:12D called at a significantly younger age (3-4 days) than those at 16L:8D. Masaki (1978) reported that the rate of development in the univoltine cricket, *Teleogryllus emma* Ohmachi & Matsumura, varied with daylength and was greatest when the insect was reared under short day conditions. He proposed that this was an adaptation permitting the insect to reach sexual maturity at the appropriate time to produce diapausing eggs before the onset of winter. A similar situation may be occurring in the sunflower moth which overwinters as a diapausing prepupa. We propose that under fall conditions the energetic costs of dispersing in search of suitable hosts with a well developed egg complement are considerably less than those associated with a delay in the onset of reproduction until suitable hosts are found if, as the result of such delays, some progeny produced do not reach the prepupal stage before the onset of winter.

References

Arthur, A.P. & D.J. Bauer (1981). Evidence of the northerly dispersal of the sunflower moth by warm winds. *Environ. Entomol.* **10**: 523-528.

Delisle, J., J.N. McNeil, E. W. Underhill & D. Barton (1989). *Helianthus annuus* pollen, an oviposition stimulus for the sunflower moth, *Homoeosoma electellum*. *Entomol. exp. appl.* **50**: 53-60.

DePew, L.J. (1983). Sunflower moth (*Lepidoptera*: *Pyralidae*): Oviposition and chemical control of larvae on sunflowers. *J. Econ. Entomol.* **76**: 1164-1166.

Masaki, S. (1978). Seasonal and latitudinal adaptations in life cycles of crickets. In: H. Dingle (ed.), *Evolution of Insect Migration and Diapause*. New York: Springer Verlag.

McNeil, J.N. & J. Delisle (1989). Host plant pollen influences calling behavior and ovarian development of the sunflower moth, *Homoeosoma electellum*. *Oecologia* **80**: 201-205.

Rossiter, M., J. Gershenzon & M.T. Mabry (1986). Behavioral and growth responses of specialist herbivore, *Homoeosoma electellum*, to major terpenoid of its host, *Helianthus* spp. *J. Chem. Ecol.* **12**: 1505-1521.

Teetes, G.L. & N.M. Randolph (1969). Chemical and cultural control of the sunflower moth in Texas. *J. Econ. Entomol.* **62**: 1444-1447.

Proc. 8th Int. Symp. Insect-Plant Relationships, Dordrecht: Kluwer Acad. Publ.
S.B.J. Menken, J.H. Visser & P. Harrewijn (eds), 1992

Seasonal variation in plant chemistry and its effect on the feeding behaviour of phytophagous insects

W.M. Blaney[1] and M.S.J. Simmonds[2]
[1] *Dept of Biology, Birkbeck College, London, UK*
[2] *Jodrell Laboratory, Royal Botanic Gardens, Kew, Richmond, Surrey, UK*

Key words: Antifeedants, gustation, sensory physiology, *Spodoptera exempta*, *Spodoptera littoralis*, *Teucrium polium*

The behavioural and neurobiological effects of antifeedants in extracts from the leaves of *Teucrium polium* (ssp. *aureum* Schreber; Labiatae) were investigated with larvae of the oligophagous *Spodoptera exempta* Walker and the polyphagous *Spodoptera littoralis* Boisduval.

Antifeedant test. Plants were harvested each month from March to October, freeze dried and extracted in ethanol. These extracts were fractionated to provide a sugar

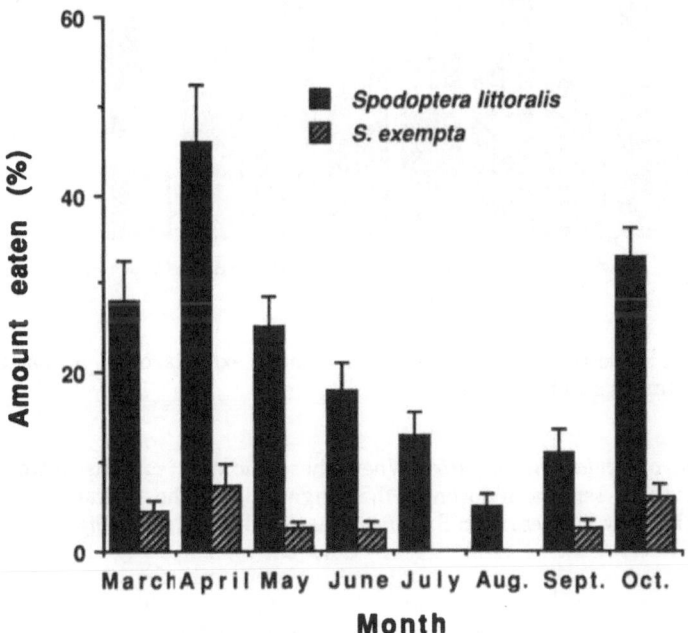

Figure 1. Amount eaten by stadium larvae of *S. littoralis* and *S. exempta* of glass-fibre discs treated with monthly ethanol extracts of *T. polium* (mean ± SEM, n = 20).

fraction and a terpenoid fraction. The terpenoid fractions were applied to glass-fibre discs and offered to final stadium larvae in no-choice bioassays so that the effects of any month-by-month changes in plant terpenoids could be monitored (Blaney *et al.*, 1990). When the degree of rejection at first contact, the duration of first meal, and the amount of extract-treated discs eaten were analysed it was seen that *S. exempta* was more selective than *S. littoralis*. The antifeedant effect of the plants varied significantly over the eight month period, being greatest in July and August (Fig. 1).

Variability in terpenoids and sugars. The amount of terpenoids and sugars present in each of the monthly extracts was measured (Fig. 2). Overall, the concentrations of sugars present in the plants were greater by an order of magnitude than those of the terpenoids. The concentration of both groups of compounds varied from month to month, the variability being much greater with the terpenoids than with the sugars. The highest concentration of terpenoids was found in plants harvested in August.

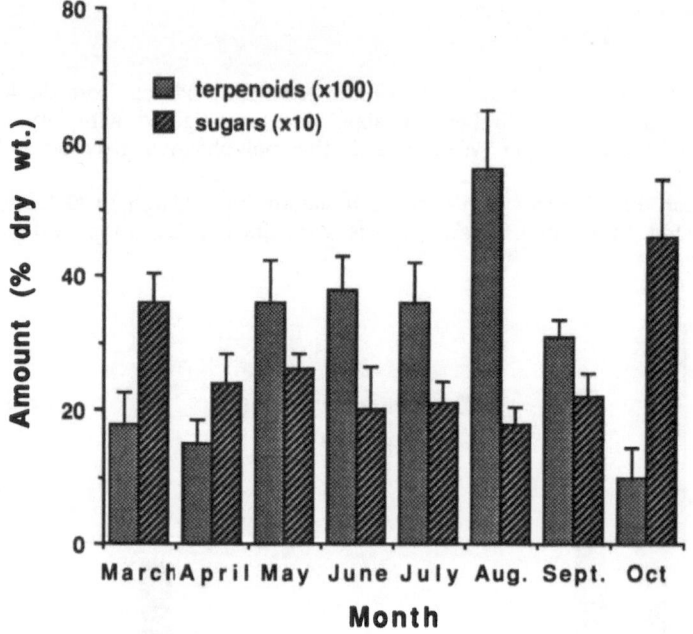

Figure 2. Amounts of terpenoids and sugars in ethanol extracts of *T. polium* taken at monthly intervals (mean ± SEM, n = 5).

Responsiveness of styloconic sensilla. When these monthly extracts were used to stimulate the maxillary sensilla styloconica the magnitude of the neural response was greater and its variability was less with *S. exempta* than with *S. littoralis* (Fig. 3).

Conclusions

Overall, with *S. exempta*, there were good correlations between behaviour, neural response and the levels of terpenoids in extracts of *T. polium*, but the picture was not so clear with *S. littoralis*.

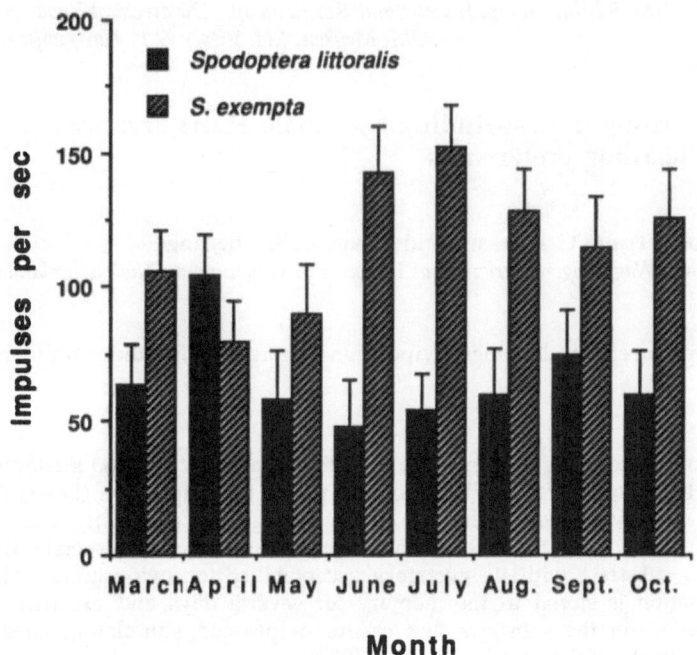

Figure 3. Total neural responses from lateral and medial maxillary sensilla styloconica of final stadium larvae of *S. littoralis* and *S. exempta* when stimulated with monthly extracts of *T. polium* (mean neural responses ± SEM, n = 10).

Reference

Blaney, W.M., M.S.J. Simmonds, S.V. Ley, J.C. Anderson & P.L. Toogood (1990). Antifeedant effects of azadirachtin and structurally related compounds on lepidopterous larvae. *Entomol. exp. appl.* 55: 149-160.

Proc. 8th Int. Symp. Insect-Plant Relationships, Dordrecht: Kluwer Acad. Publ.
S.B.J. Menken, J.H. Visser & P. Harrewijn (eds), 1992

Associative learning in host-finding by female *Pieris brassicae* butterflies: relearning preferences

Joop J.A. van Loon, Tjarda C. Everaarts and Renate C. Smallegange
Dept of Entomology, Wageningen Agricultural University, Wageningen, The Netherlands

Key words: Green substrates, landings, oviposition deterrent, oviposition stimulant, trainings

Butterflies are known to make extensive use of their visual, olfactory and gustatory senses when selecting both flowers for feeding and host plants for oviposition (Lewis & Lipani, 1990; Traynier, 1984). Moreover, in the case of the cabbage white butterfly, *Pieris rapae* L., Traynier (1979, 1984, 1986) has shown that females are able to associate the visual appearance of a substrate with its gustatory acceptability for oviposition. This visual-gustatory association is stored in the memory for several days and expresses itself in preferential landing on the substrate that carries oviposition stimulating chemicals. *P. rapae* failed to learn about deterrents (Traynier, 1987).

Female butterflies of *P. rapae* and its relative *Pieris brassicae* L. prefer Crucifers as hosts for oviposition. Oviposition strategies may differ, however. While *P. rapae* females lay their eggs singly and perform a new flight for each egg, *P. brassicae* lays its eggs in clusters of 30-50 eggs. We thought that it might be interesting to compare for their capacity to learn associatively these species. Two questions were addressed: (1) is *P. brassicae* capable of visual-chemosensory associative learning and, if so (2) is *P. brassicae* capable of relearning? In other words, can a preference resulting from a positive association be turned into a preference for a previously less preferred alternative, by a subsequent negative association using deterrents?

Methods

Adult eclosion groups of butterflies were kept 'naive' (*i.e.*, no contact with compounds known to stimulate or deter oviposition, and no green substrates). Females (10-13) and males (5-6) were confined in circa 0.5 m³ cages in a greenhouse at 23°C with supplementary light from mercury vapour lamps. They were allowed to drink *ad lib.* from a 10% sucrose solution. Behaviour of individual females (recognizable by numbers on their hindwings) were scored during 60 min; results on landings will be presented here. Substrates offered for landing were darkgreen (DG) or lightgreen (LG) cardboard circles (radius 5 cm), were present only during the observations.

Two different 4-day training-testing sequences were used. On day 1 of the sequential training schedule, the LG substrate sprayed with oviposition stimulant sinigrin (250 nMol/cm²) was offered to group A (LG+). Group B was presented with the combination of sinigrin and DG (DG+). On day 2, females were offered a choice between LG and DG substrates (untreated) and first and total landings were scored. On day 3, a second association was offered to groups A and B; the colour offered on day 1 was offered again but now treated with a compound deterring oviposition, namely, helveticoside (5 nMol/cm^2) (treatments LG- and DG-). On day 4, the preference test of day 2 was repeated. A third group was kept naive up to day 2 to assess any possible naive colour preferences. In the simultaneous schedule, on day 1, females were either offered LG+ and DG- together (group C) or LG- and DG+ together (group D). On day 2, a dual choice test (LG vs DG, both lacking any chemical) was carried out for both groups. For each individual female, the substrate on which they performed their first landing and the total number of landings during 60 min were scored. On day 3, group C was offered LG- and DG+ together and group D was presented with LG+ and DG-, the reciprocal situation of day 1. On day 4, the dual choice test of day 2 was repeated.

Results and discussion

Of the eight associations offered, two resulted in preference changes for first landings and six for total landings. Six of these eight changes could be categorized as the result of associative learning. The naive (innate) preference for LG could be changed into a weak but significant preference for DG for total landings in just one case. Notably, this change was still possible after reinforcement on LG. The indiscriminate total landings of naive females could be changed into preferential landing on either DG or LG by both training schedules. A preference for DG could be induced and readily turned into a preference for LG or an indifference to both shades of green offered.

We conclude that *P. brassicae*, like *P. rapae*, can learn to associate visual substrate characteristics with chemosensory evaluation. Furthermore, we found that a preference for a substrate can readily be turned into a preference for an alternative substrate, which indicates the re-learning ability as described above. *P. brassicae* demonstrated that it can learn as a result of experience with deterrents, unlike *P. rapae*. We used a definitely more powerful deterrent (naturally occurring in Cruciferae) than was used by Traynier (1987). The sequential and simultaneous trainings produced different results; the latter schedule was more effective in modifying landing preferences. The simultaneous schedule is considered to represent the normal situation better.

The functional significance of such flexible learning capacities may be the optimal use of time and/or energy by increasing the frequency of landings on substrates suitable for oviposition. The situation studied, however, is highly artificial. Although the capacity for learning and re-learning is interesting, its significance in nature is difficult to assess (Szentesi & Jermy, 1990).

References

Lewis, A.C. & G.A. Lipani (1990). Learning and flower use in butterflies: hypotheses from honey bees. In: E.A. Bernays (ed.), *Insect-Plant Interactions* 2, pp. 95-110. Boca Raton: CRC Press.

Szentesi, Á. & T. Jermy (1990). The role of experience in host plant choice by phytophagous insects. In: E.A. Bernays (ed.), *Insect-Plant Interactions* 2, pp. 39-74. Boca Raton: CRC Press.

Traynier, R.M.M. (1979). Long term changes in the oviposition behaviour of the cabbage butterfly, *Pieris rapae*, induced by contact with plants. *Physiol. Entomol.* 4: 87-96.

Traynier, R.M.M. (1984). Associative learning in the ovipositional behaviour of the cabbage butterfly, *Pieris rapae*. *Physiol. Entomology* 9: 465-472.

Traynier, R.M.M. (1986). Visual learning in assays of sinigrin solution as an oviposition releaser for the cabbage butterfly, *Pieris rapae*. *Entomol. exp. appl.* 40: 25-33.

Traynier, R.M.M. (1987). Learning without neurosis in host finding and oviposition by the cabbage butterfly, *Pieris rapae*. In: V. Labeyrie, G. Fabres & D. Lachaise (eds), *Insects-Plants. Proc. 6th Int. Symp. Insect-Plant Relationships*, pp. 243-247. Dordrecht: Junk.

Proc. 8th Int. Symp. Insect-Plant Relationships, Dordrecht: Kluwer Acad. Publ.
S.B.J. Menken, J.H. Visser & P. Harrewijn (eds), 1992

Circadian stability of olfaction in *Lobesia botrana*

F. Marion-Poll and B. Gabel
INRA-CNRS, Laboratoire de Neurobiologie Comparée des Invertébrés, Bures sur Yvette, France

Key words: Electroantennograms, European grapevine moth, olfaction, plant volatiles, Tortricidae

A circadian modulation of the behavioural activity of a number of insects such as Lepidoptera has already been described. This modulation is possibly induced by a combination of factors affecting both sensory perception and central nervous system excitability. Short-term and long-term variations of sensitivity in insect taste receptors in relation to feeding has been documented (Blaney *et al.*, 1986). Moreover, long-term variations of olfactory sensitivity has been demonstrated in relation to age or reproductive state in some insect genera (Blaney *et al.*, 1986). Although a diel variation in olfactory sensitivity to sex pheromones has not been found in Lepidoptera (Worster & Seabrook, 1989), electrophysiological experiments on nocturnal insects are often performed with insects reared in an inverted photoperiodic cycle.

We have investigated whether a circadian rhythm affects the olfactory sensitivity to plant odours in adult *Lobesia botrana* Den. et Schiff. females as measured by the electroantennogram method (EAG). Three experiments were conducted in order to test variations in: (1) the absolute and relative EAG responses to different compounds, (2) the dose-response curve of a single compound, (3) the adaptation rate.

Material and methods

Biological material. L. *botrana* females were obtained from a strain raised in the laboratory on a semi-artificial diet. Two-day-old mated females were isolated in individual glass tubes under a normal day-light regime (light: 1.00-17.00 h) at 25°C and 70% RH until use for the experiment. The experiments were performed in the laboratory under artificial light at 23°C.

Experimental set up and recording. EAG responses were obtained from head-antenna preparations. The electrical potential recorded between a reference electrode inserted into the head and an electrode slipped over the intact tip of one antenna, was digitized at 25 Hz together with the stimulus command. The preparation was continuously flushed with moistened air (50 ml/s). Odorants were delivered by an air flow (1 ml/s, duration 1.6 or 2 s) through a Pasteur pipette containing a filter paper loaded with 20 µl of chemicals diluted in paraffin oil. The hexanol, heptanol, octanol, terpineol and benzaldehyde came from Aldrich (USA). Thujone was extracted and purified from tansy flowers. Responses were evaluated by the maximal amplitude deflection reached during a stimulation and by the half-recovery time (Kaissling, 1972) using a custom computer program.

Results

Absolute and relative amplitudes. In the first experiment, six compounds were delivered during 1.6 s at approximately regular intervals (35-40 s). The order of presentation of the stimuli was kept constant: paraffin oil (20 µl), hexanol 10^{-4}, 10^{-3}, 10^{-2}, 10^{-1} (dilution v:v), and thujone, heptanol, octanol, benzaldehyde, terpineol and hexanol (10^{-2} dilutions). During each two hour period, 5-7 individuals were tested.

The absolute EAG amplitudes showed an irregular distribution with, however, no statistical significant difference related to the time period (as tested by analysis of variance). The relative amplitudes (computed either relative to the response to a single compound or to the mean response to all compounds) were distributed along parallel lines except for the period of time, 1:00-5:00 a.m., when the responses to hexanol 10^{-1} were noticeably more similar to hexanol 10^{-2} than during any of the other periods of time.

Dose-response curves. In the second experiment, the inter-stimulation time was kept constant (40 s) and the number of stimulations reduced. The stimulation order was: paraffin oil, hexanol 10^{-4}, 10^{-3}, 10^{-2}, 10^{-1}. Ten insects were tested for each period of time.

The dose response curves showed no statistically significant differences related to the time period, both on the absolute EAG amplitudes and on the half-recovery time of the responses.

Adaptation to high stimulus load. The third experiment was designed to investigate a possible variation in the adaptation rate. Hexanol 10^{-1} was delivered as five consecutive 1.6 s stimulations spaced by 4.8 s. The five stimulations were drawn from a single Pasteur pipette which was then discarded. A minimum of ten individuals were tested for each time period.

The average responses to consecutive stimulations decreased from roughly 4 mV to 2 mV. A statistically significant decrease of the EAG amplitudes was observed during the periods before and after the dark phase.

Discussion

The electroantennogram responses of adult female moths sampled from a stock culture were found to be stable over time. The inter-individual variability in the absolute EAG amplitudes and in the half-recovery times was so high that a consistent variability associated with time of day could not be detected. This result corresponds with previous experiments performed on male insects stimulated with sex pheromonal compounds (Worster & Seabrook, 1989).

References

Blaney, W.M., L.M. Schoonhoven & M.S.J. Simmonds (1986). Sensitivity variations in insect chemoreceptors: a review. *Experientia* **42**: 13-19.

Kaissling, K.-E. (1972). Kinetic studies of transduction in olfactory receptors of *Bombyx mori*. In: D. Schneider (ed.), *Int. Symp. Olfaction and Taste* **IV**, pp. 207-213. Stuttgart: Wiss. Verlagsges.

Worster, A.S. & W.D. Seabrook (1989). Electrophysiological investigation of diel variations in the antennal sensitivity of the male spruce budworm moth *Choristoneura fumiferana* (Lepidoptera: Tortricidae). *J. Insect Physiol.* **35**: 1-5.

Proc. 8th Int. Symp. Insect-Plant Relationships, Dordrecht: Kluwer Acad. Publ.
S.B.J. Menken, J.H. Visser & P. Harrewijn (eds), 1992

Whitefly preference-performance relationships

Dave Crawshaw Thomas
Dept of Entomology, Wageningen Agricultural University, Wageningen, The Netherlands

Key words: Host-plant acceptance, host-plant adaptation, life history parameters, *Trialeurodes vaporariorum*

For the past two and a half years I have conducted a series of experiments with the greenhouse whitefly (*Trialeurodes vaporariorum* (Westwood)), to ascertain to what extent and at what rate it can adapt to various host plants. This adaptation was measured by monitoring the changes in egg laying, survivorship and development time of a given population on several different host plants over several consecutive generations. Furthermore, under the conditions of a choice test, I tested whether such differences in performance on the various host plants tested are in anyway related to the whitefly's preference.

Method

The experiments to assess the adaptation of a whitefly population to different host plants are described in Thomas (1992). In these experiments the adaptation of a population originating from gerbera to four cucumber varieties was assessed over nine consecutive generations, and the adaptation of a population from tomato var. Moneymaker to two gerbera varieties, aubergine and *Lycopersicon hirsutum* C.H. Mull *cv*. Glabratum was assessed over 12 consecutive generations.

The method for the preference testing was as follows. Three whitefly populations were taken from cucumber, gerbera and sweet pepper, respectively, such that they were conditioned to these host plants. Replicated parallel choice experiments were conducted under standardized conditions in a greenhouse. For each population, 50 female whiteflies (in the case of sweet pepper somewhat less due to the scarcity of whiteflies) were released from a small glass vial and allowed to fly freely to the three host plants offered: cucumber, gerbera and sweet pepper. The plants were of a similar size and equidistant from the release point. The number of whiteflies per host plant was counted at 1 and 24 h after releasing the whiteflies.

Results

The egg laying of the whiteflies on all plants tested remained highly variable (CV circa 50%) over consecutive whitefly generations, although there was a clear rank order in the number of eggs laid: eggplant > cucumber > gerbera > tomato > sweet pepper. Except for that upon sweet pepper, the immature mortality of all populations decreased over successive generations. For cucumber and eggplant this decrease was minimal (a few %) and for gerbera it was more pronounced (10 - 20%). The mortality showed a clear rank order of eggplant < cucumber < gerbera = tomato < sweet pepper. The development time decreased over successive generations of the whiteflies as did the variability in the development time. The rank order of development time was eggplant = cucumber < gerbera = tomato.

The preference tests demonstrated that the initial choice of host plant exhibited by the whiteflies was random. However, 24 h after being released, they exhibited a clear rank order of preference of cucumber > gerbera > sweet pepper, this rank order of preference being invariate, independent of the host plant origin of the whitefly population tested. Although the overall rank order of preference was invariate, raising a whitefly population upon a particular host plant increased the preference of the population for that host plant, relative to the other host plants offered.

Discussion

The results clearly show a correlation between preference and performance in the greenhouse whitefly. Data from the adaptation experiments showed a rank order of performance of eggplant > cucumber > gerbera > tomato > sweet pepper. This same rank order was seen in part in the preference experiments where the rank order of preference was cucumber > gerbera > sweet pepper. This correlation between preference and performance is in good agreement with other data for this same species, as reviewed by Van Lenteren and Noldus (1990). The results indicate that a whitefly population can readily adapt to a host plant. A similar result was found by Gould (1979) for spider mite. A whitefly population will tend to accept all hosts offered to it in accordance with the rank order of preference, promoting its polyphagous habit by ensuring that to a greater or lesser extent, all available hosts acceptable to the whitefly population concerned are utilised. The genetic basis of this preference - performance correlation is as yet unknown, but an investigation into this would be an appropriate step in this line of research.

Acknowledgements. I thank Piet Huisman, Joop van Lenteren, Henk Smid, Nanni Volmer and Wim Veldhuis for their assistance during this project.

References

Gould, F. (1979). Rapid host range evolution in a population of the phytophagous mite *Tetranychus urticae* Koch. *Evolution* 33: 791-802.
Thomas, D.C. (1992). Host plant adaptation in the glasshouse whitefly IV: Finding an appropriate methodology for assessing host plant adaptation. *Proc. Exper. & Appl. Entomol.*, N.E.V. Amsterdam 3: 201-207.
Van Lenteren, J.C. & L.P.P.J. Noldus (1990). Behavioural and ecological aspects of whitefly-plant relationships. In: D. Gerling (ed.), *Whiteflies: Bionomics, Pest Status and Management*, pp. 47-89. Wimborne England: Intercept Ltd.

Proc. 8th Int. Symp. Insect-Plant Relationships, Dordrecht: Kluwer Acad. Publ.
S.B.J. Menken, J.H. Visser & P. Harrewijn (eds), 1992

The role of salts in the feeding behaviour of locusts

Sheila Trumper
Dept of Zoology, University of Oxford, UK

Key words: Learning, micronutrient, neophilia

The extent to which phytophagous insects regulate their intake of more than 30 essential macro- and micronutrients is not known. Several studies have followed the effect of dilution of all nutrients and the mechanisms of regulation of protein and carbohydrate have been studied in depth in a number of insects and in *Locusta migratoria* (L.) in particular (Simpson & Simpson, 1990). However, the study of the role of micronutrients in the feeding behaviour of insects has been largely neglected.

Pretreatment assays. I was interested in whether fifth instar *L. migratoria* nymphs could regulate their salt intake. The key point about regulation is that it occurs in response to current nutritional state and is not simply an inflexible, innate response (Simpson & Simpson, 1990). Inducing a salt deficiency, then, ought to lead to an altered response to a diet containing salt if regulation is taking place. Variations on the artificial diet developed by Dadd (1961) were used. In the assays described here, insects were pretreated for 24 h on the control diet (termed +salt) or a salt-free diet (-salt) on days 1-2 of the fifth instar. Fig. 1 shows the relative amounts of each diet eaten during the pretreatment period. Significantly, ($P < 0.01$), more +salt diet was eaten than -salt.

Choice of complex diets. Following a 24-h pretreatment period on +salt or -salt diet insects were observed with a choice of these diets. Insects that had been salt-deprived ate for less time on either diet in the choice test despite being 'hungrier' (Fig. 2). This is suggestive of a generalised learned aversion to the artificial diets in -salt pretreated insects.

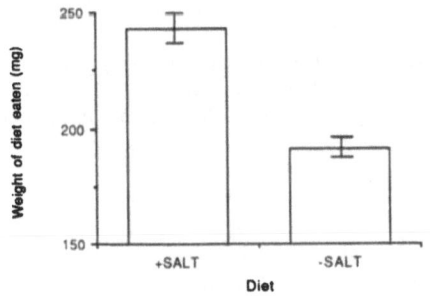

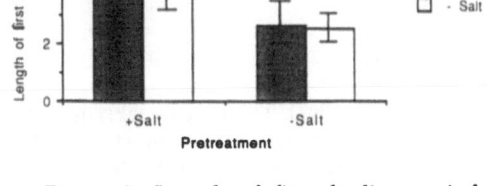

Figure 1. Weight of diet eaten during pretreatment period (mean ± SE).

Figure 2. Length of first feeding period after pretreatment (mean ± SE).

Choice of glass microfibre discs. After the 24 h pretreatment period insects were presented with a choice of four glass fibre discs (Whatman GF/A). All discs were impregnated with 5% dry weight of sucrose to ensure feeding. In addition to the sugar two of the 4 discs were impregnated with NaCl (0.5, 1 or 2%, respectively).

The insects from both pretreatment groups showed a significant (P < 0.05) preference for the salt-free discs compared to salt-containing ones. Salt-deprived insects ate more (P < 0.05) of the discs overall than controls (Fig. 3). Covariate analysis shows that the response was over and above that due to hunger. This result is suggestive of neophilia - an enhanced reaction to novel food. Note the difference to insects tested on artificial diets in the choice period (Fig. 2).

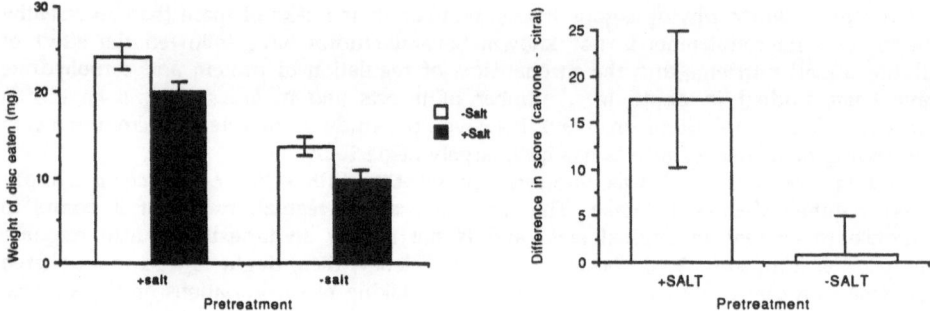

Figure 3. Weight (mg) of each disc eaten after pretreatment (mean ± SE).

Figure 4. Difference in score (carnivor-citral for each pretreatment group (mean ± SE).

Odour pairing during pretreatment. On the day after moulting the insects were pretreated on either +salt or -salt diet. Wicks impregnated with 50 µg carvone in 10 µl hexane were mounted on pins just above the diet. On the afternoon of day 2 all diets were replaced by cellulose (indigestible to locusts) and the insects were provided with a choice of carvone or citral. Insects were scored according to the distance they moved towards the odour source.

Insects conditioned with the +salt-carvone pairing scored significantly higher (P < 0.05) for carvone compared to citral in the behavioural test. -Salt pretreated insects showed no such distinction between the two odours (Fig. 4).

Thus we have tentative evidence of a plastic response to salts possibly involving neophilia, learned aversion and associative learning. Work is continuing on the relative roles of these behaviours and how they affect the regulation of other nutrients such as protein.

References

Dadd, R.H. (1961). The nutritional requirements of locusts. *J. Insect Physiol.* 6: 1-12.
Simpson, S.J. & C.L. Simpson (1990). The mechanism of nutritional compensation by phytophagous insects. In: E. Bernays (ed.), *Insect-Plant Interactions.* Vol.II, pp. 111-160. Boca Raton: CRC Press.

Proc. 8th Int. Symp. Insect-Plant Relationships, Dordrecht: Kluwer Acad. Publ.
S.B.J. Menken, J.H. Visser & P. Harrewijn (eds), 1992

Computer-aided analysis of chemosensory data

P. Roessingh[1], A. Fritschy[2], E. Städler[1], L. Pittenger[3] and F.E. Hanson[3]
[1] Eidg. Forschungsanstalt, Wädenswil, Switzerland
[2] Umwelt-Informatik, Zürich, Switzerland
[3] University of Maryland Baltimore County, Baltimore, USA

Key words: Chemoreception, electrophysiology

Electrophysiology is a powerful technique to determine which environmental stimuli are detected by the insect sense organs. However, it often results in large amounts of primary data which can be difficult to digest. This aspect was particularly onerous in the past when the only method of data acquisition and analysis was to photograph the oscilloscope screen and count the spikes by hand. The laboratory computer has begun to replace this technology by permitting the direct acquisition of data (thus precluding the need for a tape recorder), display, printing, counting spikes and analysing data (Frazier & Hanson, 1986), as well as data logging and management.

Computer-aided analysis

Several software packages that run on personal computers have been developed especially for recording from insect chemosensilla. For the IBM PC/XT/AT family of computers, perhaps the most widely used is 'Sapid Tools' (Smith et al., 1990), written mostly in BASIC. A second one is written in FORTRAN (Marion-Poll & Tobin, 1989). Another package was developed by Piesch and colleagues at the University of Regensburg that uses a PC with a transputer card to provide very high speed analysis. For the Macintosh, two software packages are under development; "SpikeAnalysis" (Roessingh, Fritschy & Städler) written in PASCAL, and "Spike Data Processor" (Pittenger & Hanson) written in object oriented C++, both utilize the user-friendly aspects of the Macintosh platform.

These packages are similar in overall function, and a flowchart outlining the procedure is depicted in Fig. 1. First, the data are accepted in analog form as voltages from the neurophysiological amplifier and converted to digital information by an analog-to-digital conversion card added to the computer. Normally this occurs at 10,000 samples per second, so that individual spikes are preserved with high fidelity.

Once the data are in the computer or on disk in digitized form, the next step is to "filter" the spikes from the baseline noise by an algorithm that looks for peaks and troughs having certain characteristics. The spikes are then classified on the basis of amplitude or total spike shape, the latter using a computation-intensive template matching procedure. The underlying rationale is that each sensory neuron has an action potential that is recorded as a unique spike shape. The identification of these spikes can be difficult, however, the amplitude classification method confuses spikes of the same amplitude but of different shape, and the shape-based method of classification becomes unreliable if spikes change in shape with time or with different stimuli. Thus new ways of classifying spikes are needed and are currently being investigated (e.g., using neural

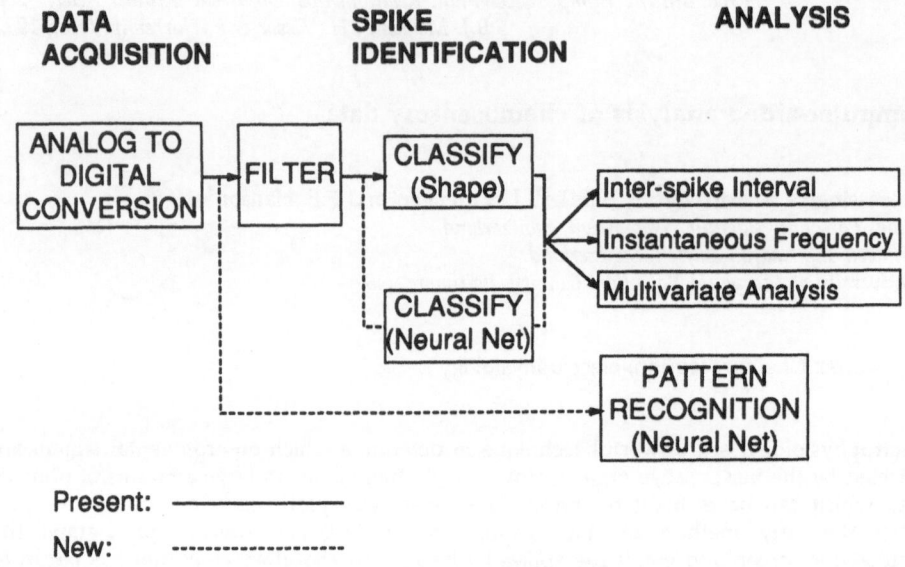

| DATA ACQUISITION | SPIKE IDENTIFICATION | ANALYSIS |

Figure 1. Flowchart illustrating the procedures used in computer-aided data analysis.

networks). During all of these procedures, feedback to the operator via video displays is very important to avoid artifacts of misclassification.

Usually the desired end point is the ability to analyze spike trains to determine their information content. Some of these packages have internal routines for this, or it can be done by exporting the preprocessed data to commercial statistical and graphical packages that provide powerful and flexible analytical tools. New approaches to the analysis of spike train data are being sought, such as the use of neural networks as suggested in a companion paper in this volume (Hanson *et al.*, this vol., pp. 173-175).

References

Frazier, J.L. & F.E. Hanson (1986). Electrophysiological recording and analysis of insect chemosensory responses. In: J.R. Miller & T.A. Miller (eds), *Insect-Plant Interactions*, pp. 285-330. New York: Springer.

Marion-Poll, F. & T.R. Tobin (1989). A microcomputer program to drive and analyze electrophysiological experiments on insect olfactory receptor neurons. *Chem. Senses* 14: 218.

Smith, J.J.B., B.K. Mitchell, B.M. Rolseth, A.T. Whitehead & P.J. Albert (1990). Sapid tools: microcomputer programs for analysis of multi-unit nerve recordings. *Chem. Senses* 15: 253-270.

Proc. 8th Int. Symp. Insect-Plant Relationships, Dordrecht: Kluwer Acad. Publ.
S.B.J. Menken, J.H. Visser & P. Harrewijn (eds), 1992

Analysis of sensory information using neural networks

F.E. Hanson[1], J.P. Stitt[2] and J.L. Frazier[2]
[1] *Dept Biology, University of Maryland, Baltimore County, Baltimore, USA*
[2] *Dept Entomology, Pennsylvania State University, University Park, USA*

Key words: Chemoreception, Lepidoptera, *Manduca sexta*, neural networks

Feeding and oviposition decisions are made by the insect central nervous system based on neural input from chemoreceptors responding to phytochemical stimuli. There are four or more chemosensory neurons in most gustatory sensilla, and each substance or plant elicits a response from more than one. The mixture of neuronal responses represents a "pattern" that is repeatable when stimulating with the same substance or plant on the same sensillum. Presumably patterns elicited by acceptable substances or plants are different from those of unacceptable ones. In attempting to understand what these input patterns mean to the animal, we have used a cybernetic approach to analyze the information content of the sensory inputs to help us understand the types of information needed for a behavioral decision. The eventual goal is to have the computer find patterns in the data that the insect uses or might use to differentiate between host plants and non-host plants. The specific question we address here is whether neural network algorithms can be "trained" with raw electrophysiological recordings as inputs, and, if so, whether they can then "recognize" similarities in other recordings submitted as "test" data.

Methods

Following is a report of an initial attempt using a Kohonen network algorithm (Eberhart & Dobbins, 1990; Pao, 1989; Wasserman, 1989; programmed on a Macintosh IIfx by J.S.) having a large input layer capable of accepting raw digitized data, namely action potentials from a gustatory sensillum on the maxilla of the tobacco hornworm, *Manduca sexta* (Johan.). The procedure is to "train" the neural network using responses of the same sensillum to repeated stimulations by a substance or extract. We used 4096 samples of digitized data (acquired at 10 KHz) as the input for each record. The trained software is then challenged by a test record, *e.g.*, a response to a different substance or from a different sensillum. The program compares each test response against a training set using multidimensional analysis, and returns a value between 1 (very similar) and -1 (very different). This "fuzzy logic" type of output permits an operator to adjust a threshold level to force a decision if a bipolar (yes/no) decision is needed.

Results and discussion

The results show that the computer can be trained to discriminate among recordings from different sensilla and/or stimuli. When trained with ten responses to inositol on the lateral styloconicum (Fig. 1A), the algorithm rated the five test responses as follows: (1) responses to inositol by the lateral styloconica are very similar; (2) responses to inositol by

A. Trained with Lateral Inositol

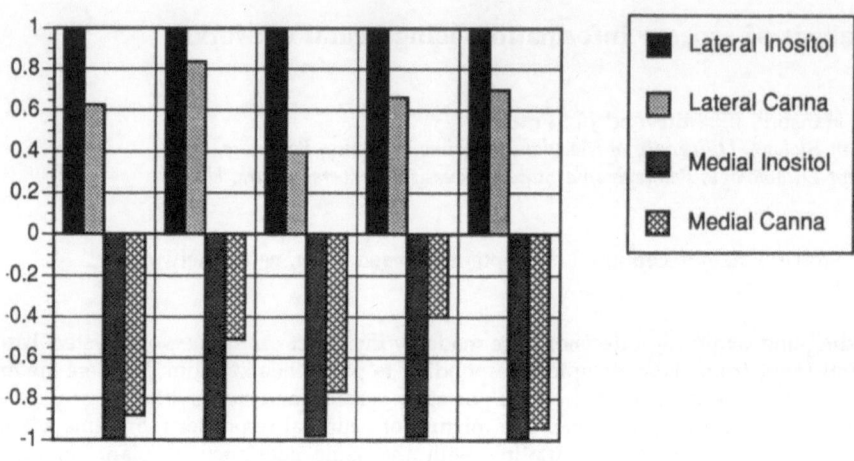

B. Trained with Lateral Canna Extract

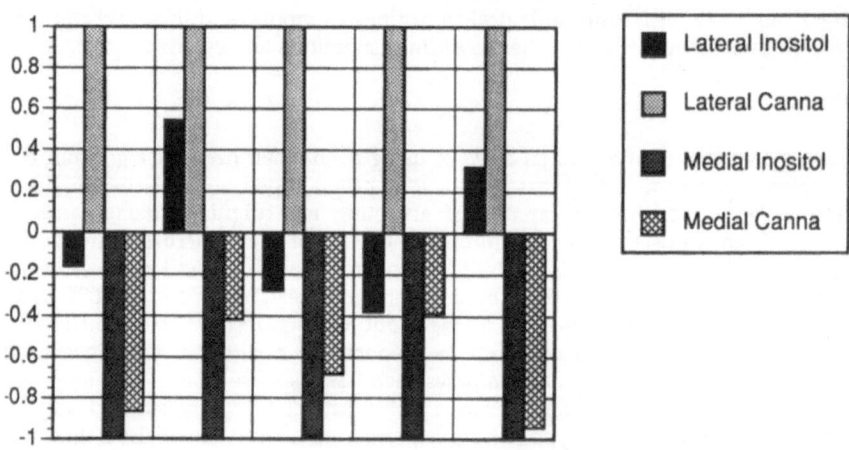

Figure 1. Results of testing the trained network with five test sets, each comprised of a separate recording from responses to each of the following categories: (1) inositol or (2) canna extract on the lateral styloconicum; (3) inositol or (4) canna extract on the medial styloconicum. Values graphed indicate the degree of similarity between the test recording and the training set: the higher the positivity, the higher the similarity. The neural network algorithm was trained using responses from the lateral styloconicum to (A) inositol and (B) canna extract.

the medial styloconica are very different; (3) responses to canna extract by the lateral styloconica are somewhat similar; and (4) responses to canna extract by the medial styloconica are very different. When trained with responses to canna extract on the lateral styloconicum (Fig. 1B), it recognized canna responses on the lateral as being very similar, but the other test responses as being different or very different. Human observers generally agreed with such evaluations after visual inspection of the training set and test sets.

This research will be continued by testing responses from several sensilla and several host and non-host plant extracts, as well as by testing the effects of these extracts on feeding behavior.

References

Eberhart, R.C. & R.W. Dobbins (1990). *Neural Network PC Tools*. San Diego: Academic Press.

Pao, Y. (1989). *Adaptive Pattern Recognition and Neural Networks*. Reading, MA: Addison-Wesley.

Wasserman, P.D. (1989). *Neural Computing: Theory and Practice*. New York: Van Nostrand Reinhold.

Proc. 8th Int. Symp. Insect-Plant Relationships, Dordrecht: Kluwer Acad. Publ.
S.B.J. Menken, J.H. Visser & P. Harrewijn (eds), 1992

Azadirachtin treatment and host-plant selection

A. Jennifer Mordue and Munira Nasiruddin
Dept of Zoology, University of Aberdeen, Aberdeen, UK

Key words: Analogues, antifeedant, Lepidoptera, locusts, sub-lethal effects

Azadirachtin is a most effective insect antifeedant and growth regulator (IGR) and has substantial potential in insect control. Insects may refuse to eat azadirachtin or may feed and ingest sufficient quantities to disrupt their physiology and behaviour. The well documented effects of delayed growth, moult aberrations, death at the moult and sterility of the adult are due to a malfunctioning endocrine system. In addition, epithelial, midgut, fat body and muscle cells, are also known to be affected by azadirachtin treatment. Such sub-lethal events cause deficiencies in intermediary metabolism and muscle tone and may be important in the overall toxicity of azadirachtin. This paper examines the effects of azadirachtin and some analogues at sub-lethal doses in polyphagous and oligophagous locusts, and in lepidopterous pests of brassicas in the field.

Methods

Locusts were treated with azadirachtin by injection: 22,23-dihydroazadirachtin, 22-α-bromo-22,23-dihydro-23-α,ß-ethoxyazadirachtin, dihydrodeacetylazadirachtin or 11-deoxy-azadirachtin at doses of 0.1, 1.0 or 10 µg/g. Growth parameters were measured in *Schistocerca gregaria* Forskål from 2nd instar to mature adult stage and in *Locusta migratoria* L. from day 2 of the 5th instar for 16 days. Antifeedancy experiments were carried out using choice tests (Blaney *et al.*, 1990). Field experiments were carried out in Scotland, grid reference: NT 366 729 using guidelines in EPPO/OEPP 1984, No. 83 for caterpillars on brassicas. Four replicates of control, or cypermethrin treatments, and four concentrations of azadirachtin (50, 250, 500 & 750 ppm) were carried out. Plots were assessed for crop damage and caterpillar number and species prior to and after treatment. Crop damage was assessed as the number of holes in each leaf.

Results and discussion

The graminivorous oligophagous species *L. migratoria* is deterred from feeding by the presence of plant secondary compounds and avoids such compounds which are found mainly in the Dicotyledons. The polyphagous species *S. gregaria* gives a variable response to such compounds from an extreme deterrence to azadirachtin through to phago-stimulation and shows a more detailed chemoreceptor response and also a greater physiological tolerance of such compounds (Cottee *et al.*, 1988). *L. migratoria* is deterred by azadirachtin at 25 ppm and by the structurally related compounds dihydroazadirachtin (at 50 ppm), 11-deoxyazadirachtin (at 50 ppm) and bromoethoxyazadirachtin (at 100 ppm), *i.e.*, over a fourfold dose range. *S. gregaria*, however, is deterred by azadirachtin at 0.05 ppm (a 500-fold increase in sensitivity *vs L. migratoria*) and by the analogues at 0.1,

0.5 and 5.0 ppm respectively (*i.e.*, over a range of 100 fold). Therefore, *S. gregaria*, is both more sensitive to azadirachtin and more discriminatory between closely related analogues, perhaps reflecting significant differences in receptor specificity between the two species.

Injection of fifth instar *L. migratoria* with 10 µg/g azadirachtin causes a high incidence of mortality both before and during the adult moult, markedly reduces growth rate, and food intake (as measured by faeces production), and increases instar length. Dihydroazadirachtin has similar effects to azadirachtin with dihydrodeacetylazadirachtin and bromo-ethoxyazadirachtin having decreasing effectiveness. The latter causes a significantly lower growth rate than in controls but only slightly increased mortalities, all occurring in the adult stage after a normal moult. Bromoethoxy-treated insects have smaller dry weights than those of control insects 16 days after treatment (Table 1). These results confirm those of Blaney *et al.* (1990) with *Spodoptera* and *Heliothis* spp. Hydrogenation of the $C_{22,23}$ positions does not significantly alter the biological effectiveness of azadirachtin. Increasing the steric bulk of the dihydrofuran ring at the $C_{22,23}$ positions by the addition of bromo- or ethoxy- groups markedly reduces both its antifeedancy and IGR properties, whereas changes at C_{11} or C_3 on the decalin ring are also important and reduce the molecule's effectiveness down to intermediate levels.

Table 1. The physiological effects of compounds on fifth instar ♂♂ *Locusta migratoria* treated with 10 µg/g body wt on day 2 of the instar. (Chemical injected into the haemolymph, n = 6-11)

Compound	% wet wt. gain (day 0-2)	Cumulative % mortality (day 0-16)	Ave. faeces production (mg/insect/ day)	Instar length (days)	Time of death	Dry wt. (g) at day 16
Control	41.8[a]	0	71.7	8.6 ± 0.2[a]	0	0.567 ± 0.03[a]
Azadirachtin	5.9[b]	100[a] (by day 13)	30.2	12.7 ± 0.3[b]	Before or during ecdysis	-
Dihydro-azadirachtin	2.2[b]	90[b]	52.8	14.1 ± 1.1[b]	"	-
Dihydro-desacetylazad.	8.7[b]	66.7[b]	29.4	12.5	"	0.245
Bromoethoxy-azadirachtin	25.5	50[b]	61.3	7.7 ± 0.5	As adult	0.432 ± 0.06[b]

a-b: P ≤ 0.01 (Mann Whitney *U*-test for wt. gain & mortalities; Wilcoxon signed rank test for instar length)

The ED_{50} for azadirachtin for all biological effects is in the region of 2 µg/g. Topical application of sub-lethal levels of azadirachtin and analogues to 2nd instar *S. gregaria* enabled long term effects to be monitored. Doses of 0.1 and 1.0 µg/g revealed the same trends, although at a decreased level, as for the 10 µg/g dose in *L. migratoria* with reduced growth and feeding and increased mortalities occurring within the first few days of treatment. Insects which survived regained normal growth and feeding and at adult maturity were not different from the controls in terms of maturation rate, dry body weight, fat body weight or ovary weight. Thus sub-lethal effects of azadirachtin and related analogues may be of a temporary nature.

Preliminary field studies showed that crop damage is reduced by azadirachtin spraying of the plants. However, importantly the caterpillars were not killed and remained on the plant. Presumably the insects had ingested enough azadirachtin to depress their feeding and disrupt growth in a similar manner to that seen in the locust nymphs above.

It is clear that complex inter-relationships exist between antifeedancy and sub-lethal behavioural effects (lack of feeding, lower mobility) after ingestion. Such relationships may also vary between oligophagous and polyphagous species. A fuller understanding of the dynamics of deterrency/IGR effects in a variety of species would assist in the design of more successful semiochemicals for pest management.

Acknowledgements. We thank Prof. S.V. Ley and Prof. E.D. Morgan for supplies of azadirachtin and related molecules, the British Council for a studentship to Ms Nasiruddin and Ms G. Davidson for assistance with fieldwork. The field study was carried out together with Dr R. McKinlay, SAC, University of Edinburgh.

References

Blaney, W.M., M.S.J. Simmonds, S.V. Ley, J.C. Anderson & P.L. Toogood (1990). Antifeedant effects of azadirachtin and structurally related compounds on lepidopterous larvae. *Entomol. exp. appl.* 55: 149-160.

Cottee, P.K., E.A. Bernays & A.J. Mordue (Luntz) (1988). Comparisons of deterrency and toxicity of selected secondary plant compounds to an oligophagous and a polyphagous acridid. *Entomol. exp. appl.* 46: 241-247.

Proc. 8th Int. Symp. Insect-Plant Relationships, Dordrecht: Kluwer Acad. Publ.
S.B.J. Menken, J.H. Visser & P. Harrewijn (eds), 1992

The effects of azadirachtin on feeding by *Myzus persicae*

A.J. Nisbet[1,2], J.A.T. Woodford[2] and R.H.C. Strang[1]
[1] Dept of Biochemistry, University of Glasgow, Glasgow, UK
[2] Dept of Zoology, Scottish Crop Research Institute, Invergowrie, Dundee, UK

Key words: Allochemicals, antifeedant, aphids, systemic

Azadirachtin, a tetranortriterpenoid isolated from the seeds of the neem tree (*Azadirachta indica* A. Juss), has strong antifeedant effects on a wide range of insects, but there is confusing evidence for its efficacy against aphids. In initial tests, azadirachtin applied systemically, (20 ppm + 0.02% Tween 80) did not deter settling and feeding by *Aphis fabae* (Scop.), but was strongly antifeedant against locusts (*Schistocerca gregaria* Forskål) (A.J. Nisbet, unpubl. results).

Materials and methods

Azadirachtin was purified from de-fatted, ground neem seeds by successive flash column chromatography. Purity was assessed by 200 MHz ^{1}H NMR, HPLC, and analytical TLC.

Electrical recording of aphid feeding. Roots of 3 week old *Nicotiana clevelandii* Gray seedlings were immersed for 36 h in test solutions containing azadirachtin + 0.02% Tween 20 + 2% ethanol in distilled water to allow systemic uptake. An apterous adult *Myzus persicae* (Sulz.) was connected by a fine gold wire to a DC amplifier and an "Electrical Penetration Graph" (EPG) of its feeding behaviour (Tjallingii, 1988) was recorded for 9 h.

Artificial diet studies. Azadirachtin was incorporated into an artificial diet (Griffiths *et al.*, 1975) and five apterous adult aphids were allowed feeding access to sachets of this diet for 48 h. After this period, the amount of aphid excrement (honeydew) was quantified by reacting the collected material with ninhydrin solution (1% in acetone/acetic acid) and reading the absorbancy of a methanolic solution of the reacted material at 500 nm.

Results

M. persicae probed more frequently on seedlings treated systemically with azadirachtin. The aphids also initiated more phloem feeds, as indicated by the number of sustained E(pd) patterns, on treated plants. The mean duration of each of these individual, sustained E(pd) patterns was reduced from 416 min in the control group, to 163 min in the 300 ppm treated group, and to 32 min in the 500 ppm treated group. The mean duration of total E(pd) pattern over a 9 h recording period was decreased from 456.2 min on the control seedlings, to 321.9 min on 300 ppm treated seedlings, and to 152.9 min on the 500 ppm treated seedlings.

Honeydew production by aphids feeding on diet treated with 100 ppm azadirachtin was reduced by more than 80% compared with that of aphids feeding on control diets. At 500 and 1000 ppm azadirachtin concentration, the mean amount of honeydew produced

179

by the aphids was less than 10% of that produced by the control group, and was not significantly different from that produced by aphids which were starved during the honeydew collection period.

Conclusions

Azadirachtin applied to plant roots increases the frequency of aphid probing on the foliage, but decreases the duration of discrete phloem feeds and the total time for which aphids feed in the phloem. When presented in a diet solution, azadirachtin, at a concentration of 100 ppm, reduces aphid feeding by more than 80%.

Acknowledgements. Alasdair Nisbet would like to acknowledge the receipt of a post-graduate studentship from the Scottish Office Agriculture and Fisheries Department.

References

Griffiths, D.C., A.R. Greenway & S.L. Lloyd (1975). An artificial diet with improved acceptability to three strains of *Myzus persicae* (Sulz.) (Hemiptera, Aphididae). *Bull. Entomol. Res.* **65**: 349-358.
Tjallingii, W.F. (1988). Electrical recording of stylet penetration activities. In: A.K. Minks & P. Harrewijn (eds), *Aphids, Their Biology, Natural Enemies and Control*, Vol. 2B, pp. 95-108. Amsterdam: Elsevier.

Proc. 8th Int. Symp. Insect-Plant Relationships, Dordrecht: Kluwer Acad. Publ.
S.B.J. Menken, J.H. Visser & P. Harrewijn (eds), 1992

Effects of the plant-derived antifeedant polygodial on aphid host selection behaviour

Glen Powell[1], Jim Hardie[1] and John A. Pickett[2]
[1] *AFRC Linked Research Group in Aphid Biology, Imperial College, Silwood Park, UK*
[2] *AFRC Institute of Arable Crops Research, Rothamsted Experimental Station, Harpenden, UK*

Key words: Electrical recording, *Myzus persicae*, video

The drimane antifeedant polygodial has been shown to reduce settling of aphids on treated leaf halves and inhibit acquisition of persistent, semipersistent and nonpersistent viruses (Griffiths *et al.*, 1989). Tethering aphids to a fine gold wire enables their stylet activities to be electrically recorded, but may also influence behaviour (Tjallingii, 1988). The behaviour of untethered aphids on plants has been investigated by direct observation and video recording, but the value of these approaches depends on the use of reliable criteria to judge the initiation and termination of stylet penetrations. This study was designed to assess the reliability of video-recorded antennal movements for estimating plant penetration by *Myzus persicae* (Sulz.), and to investigate the behavioural effects of polygodial.

Methods

The 'electrical penetration graph' signal from tethered adult apterous aphids, connected to the DC electrical recording system (Tjallingii, 1988), was recorded simultaneously with videoed antennal and body movements (Hardie *et al.*, 1992) on Chinese cabbage (*Brassica pekinensis* L.) seedlings. The behaviour of untethered aphids was then continuously video recorded during 15 min access to cabbage seedlings, following various treatments with 0.1% (±)-polygodial or 50% ethanol solvent.

Results and discussion

The durations of stylet penetrations, estimated as periods of body and antennal immobility, were significantly correlated with durations of electrical contacts (r = 0.97; P < 0.001; slope = 1.01; n = 74). The behavioural effects of various treatments on free aphids were then investigated by quantifying four behavioural parameters (Table 1); timed data were normalised by $\log_{10}$ transformations. Polygodial treatment of a seedling did not significantly affect these parameters, confirming previous reports (Hardie *et al.*, 1992). It is therefore not clear how application of polygodial to leaves immediately inhibits acquisition of potato virus Y (Griffiths *et al.*, 1989), a process which occurs when aphid stylets puncture epidermal cell membranes (Powell, 1991).

After 24 h exposure to polygodial, applied to either green paper or a leaf surface, aphids made fewer penetrations of longer duration when compared to solvent pre-treated controls. Such prolonged exposure to polygodial also caused a delay before a first penetration was initiated, but the difference was only significant following pre-treatment on green paper. Similar observations have been made with *Aphis fabae* (Scop.) (Hardie *et al.*, 1992).

Table 1. The immediate effects of polygodial *vs* solvent treatment of cabbage seedlings on the behaviour of starved *M. persicae* (1) and the effects of polygodial 24 h pre-treatments presented on cabbage leaves (2) or green paper (3). Video-assessed stylet penetration behaviour was recorded on untreated cabbage seedlings in protocols 2 and 3. Values in the table are means ± SE

Treatments		Penetration parameters			
24 h pre-treatment	Treatment of cabbage seedling	Number	Log time (s) to first	Log mean duration (s)	Log mean total duration (s)
1 Starved	polygodial	7.3±1.4 NS	0.84±0.06 NS	1.35±0.03 NS	2.85±0.02 NS
Starved	ethanol	4.9±0.4	0.75±0.04	1.35±0.05	2.83±0.06
2 Cabbage leaf: polygodial	none	2.0±0.2 **	1.63±0.10 NS	1.68±0.14 *	2.87±0.02 NS
Cabbage leaf: ethanol	none	3.8±0.6	1.39±0.07	1.39±0.07	2.86±0.03
3 Green paper: polygodial	none	2.5±0.3 ***	1.38±0.07 ***	1.79±0.10 ***	2.88±0.02 NS
Green paper: ethanol	none	4.6±0.4	0.74±0.04	1.34±0.04	2.88±0.01

n = 15; NS = non-significant difference; *P < 0.05; **P < 0.01; ***P < 0.001, Student t-tests.

References

Griffiths, D.C., J.A. Pickett, L.E. Smart & C.M. Woodcock (1989). Use of insect antifeedants against aphid vectors of plant virus disease. *Pestic. Sci.* **27**: 269-276.
Hardie, J., M. Holyoak, N.J. Taylor & D.C. Griffiths (1992). The combination of electronic monitoring and video-assisted observations of plant penetration by aphids and behavioural effects of polygodial. *Entomol. exp. appl.* **62**: 233-239.
Powell, G. (1991). Cell membrane punctures during epidermal penetrations by aphids: consequences for the transmission of two potyviruses. *Ann. Appl. Biol.* **119**: 313-321.
Tjallingii, W.F. (1988). Electrical recording of stylet penetration activities. In: A.K. Minks & P. Harrewijn (eds), *Aphids. Their Biology, Natural Enemies and Control.* Vol. B, pp. 95-108. Amsterdam: Elsevier.

Proc. 8th Int. Symp. Insect-Plant Relationships, Dordrecht: Kluwer Acad. Publ.
S.B.J. Menken, J.H. Visser & P. Harrewijn (eds), 1992

The geometry of feeding: a new way of looking at insect nutrition

S.J. Simpson, D. Raubenheimer and P.G. Chambers
Dept of Zoology, University of Oxford, UK

Key words: Compensatory feeding, dietary selection, locust

Meeting the nutritional needs of an insect can be seen as a problem of multidimensional geometry. At any given time there is a requirement by the tissues for a particular quantity and mix of nutrients. This constitutes a "nutritional target" in an n-dimensional s'.&' e, where n is the number of nutrients. Providing the tissues with nutrients involves two stages: feeding and post-ingestive processing. Just as there is a nutritional target, there is also an "intake target", which is the optimal amount and balance of nutrients that need to be ingested so that post-ingestive processes can operate at maximal efficiency. An insect eating a nutritionally homogeneous food is confined to a "rail" in multi-dimensional space. The intake target is only achievable if it lies on the rail. A choice of foods containing different proportions of nutrients expands the nutritional space available to the insect from a line (rail) to a volume. Even then, it may be that the intake target is still unreachable.

Methods

We have explored such concepts by giving locusts artificial diets containing different concentrations and proportions of carbohydrate and protein. Two-dimensional plots of protein and carbohydrate intake and growth were used to infer the positions of the intake and nutritional targets. These estimates were then used to test hypotheses about the functional rules employed by locusts to regulate their intake of the two classes of nutrient.

Results and discussion

Both protein and carbohydrate intake were regulated equally efficiently. In no-choice tests locusts reached the closest point on their nutritional rail to the intake target ("least squares optimization") (Fig. 1a-b). They "jumped rails" by differentially utilizing ingested food and showed a remarkable ability to reach their "growth target" (Fig. 1c). Differential utilization could, in theory, be achieved without any active nutritional response. Only on extremely unbalanced diets did insects have an extended instar duration. If given a choice of diets enclosing a plane containing the intake target, then insects were able to reach the target by varying the proportion of the two diets eaten (Fig. 1d).

Such concepts and experiments provide a new framework within which to investigate the proximate and ultimate causes of insect feeding. Complex behavioural decisions like those in Fig. 1 can be made on the basis of nutrient-specific feedbacks, acting in part via direct modulation of gustatory responsiveness, and also through learned associations (Abisgold & Simpson, 1988; Simpson & White, 1990; Simpson *et al.*, 1991). Work is beginning on aphids and caterpillars under similar dietary regimes and it is hoped that general principles of nutritional regulation will emerge.

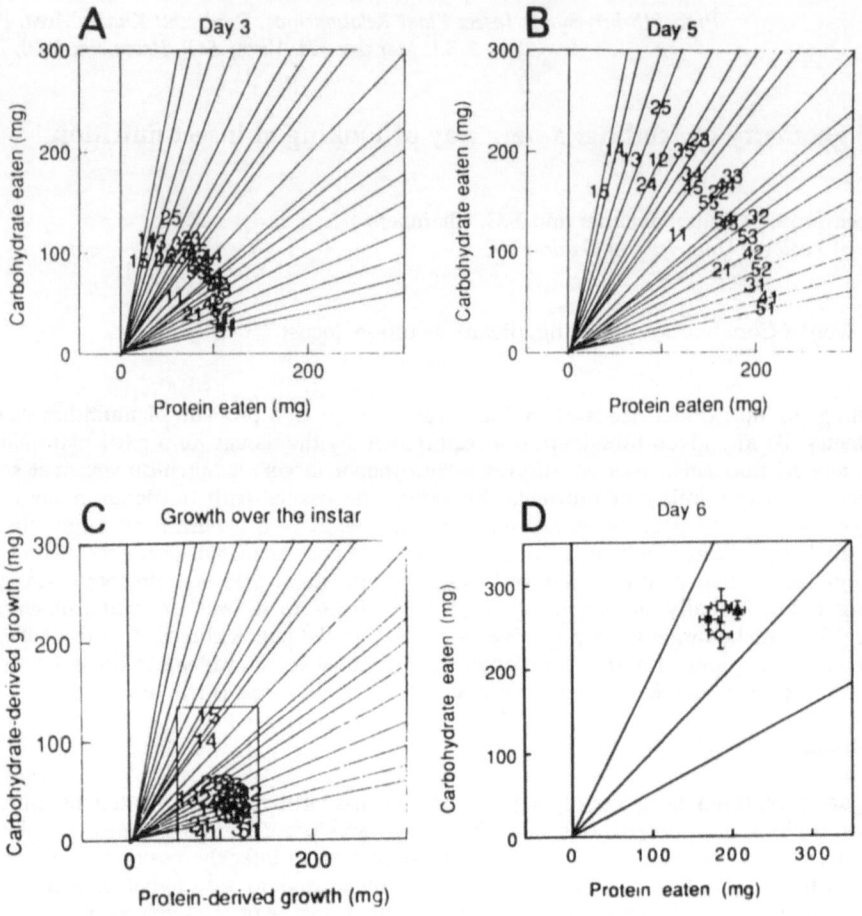

Figure 1a-c. Results from an experiment in which locusts were fed one of 25 artificial diets, containing one of five levels each of protein and digestible carbohydrate. 1a-b show the cumulative intake of digestible carbohydrate and protein over the first 3 and 5 days of the 5th stadium. Each two-figure number is the mean of 8 locusts. Multiplying the first digit by seven gives the percentage dry weight of protein in the diet, and seven times the second digit gives the percentage of digestible carbohydrate. The radiating lines represent intake rails. The strikingly arc-shaped arrays are very close to what would be expected if the insects were exhibiting least-squares optimization with respect to the intake target, with the target lying close to the point reached by insects fed on diets containing equal concentrations of protein and carbohydrate. This fits well with estimates based on growth, respiration and post-ingestive efficiencies for protein and carbohydrate. Fig. 1c shows protein-and carbohydrate-derived growth across the stadium.

Figure 1d. Mean intake (± SE, n = 10) of protein and carbohydrate by locusts given a choice of two foods over days 1-6 of the 5th stadium. The treatments were choices of diets 12 and 21 (circle); 24 and 21 (star); 12 and 24 (triangle); 24 and 42 (square). Note how the insects end up at the same point. This corresponds well with earlier estimates of the rail bearing the intake target.

References

Abisgold, J.D. & S.J. Simpson (1988). The effect of dietary protein levels and haemolymph composition on the sensitivity of the maxillary palp chemoreceptors of locusts. *J. exp. Biol.* 135: 215-229.

Simpson, S.J. & P.R. White (1990). Associative learning and locust feeding: evidence for a "learned hunger" for protein. *Animal Behav.* 40: 506-513.

Simpson, S.J., M.S.J. Simmonds, W.M. Blaney & S. James (1991). Modulation of chemo-sensitivity and the control of dietary selection in the locust. *Appetite* 17: 141-154.

Proc. 8th Int. Symp. Insect-Plant Relationships, Dordrecht: Kluwer Acad. Publ.
S.B.J. Menken, J.H. Visser & P. Harrewijn (eds), 1992

A common chemical mechanism for insect-plant communication

Dale M. Norris and Shaohua Liu
University of Wisconsin, Madison, USA

Key words: Electrochemistry, plasma membrane, protein, redox, sulfhydryl-disulfide

During interactions among insects, plants and their environments, chemical communication (*i.e.*, the exchange of chemical energy units as information) occurs. In our laboratory, we have shown that such information exchange involving insects is dependent on sulfhydryl-disulfide (*i.e.*, -SH/-S-S-) oxidative-reductive electrochemistry by receptor-energy transducer proteins in the involved dendritic neuronal membrane (Rozental & Norris, 1973; Norris, 1985, 1986). Our findings have shown further that the electrochemical action of a given amount (moles) of a chemical messenger on an insect biotype's chemosensory system is highly positively correlated with its effect on the live insect's behavioral response whether the former is (1) measured polarographically as the maximum induced millivolt shift (mV Max) in the U½ value for the sulfhydryl-disulfide receptor-transducer protein; or (2) determined neurophysiologically as the maximum percentage (%) of mV inhibition in a standardized electroantennogram (EAG) elicited by a given amount of the excitant, amyl acetate (Norris, 1986). These experimentally determined correlations are each so high (r > 0.95) that either is adequate for deducing the involved energy code for the information transfer as expressed here in the equation, LogY = 3.40 - 0.112 LogX. In this particular derived equation, "X" equals the messenger (*e.g.*, moles of chemical) input (afferent) variable and "Y" equals the primary output (efferent) polarographic variable, mV Max. In these findings the transfer of energy as information from chemical messenger to whole insect behavior thus was interpreted readily in quantitative mathematical terms.

Plant cell communication. Early findings on chemical responses by plant cells to stresses (*i.e.*, messages) in their environments suggested that sulfhydryl groups might also be involved in this reception. Our group subsequently showed that classical sulfhydryl reagents, iodoacetic acid (IAA) and N-ethylmaleimide (NEM), effectively elicited soybean cells and whole plants to alter their chemical defenses against *Trichoplusia ni* (Hübner) larval feeding. These findings led us to test further the hypothesis that plant cell chemical communication with stresses in their environments (including herbivorous insects) was also dependent on sulfhydryl-disulfide oxidative-reductive electrochemistry by receptor-energy transducer proteins in the plasma membrane of the cell. Subsequent experiments to test this hypothesis have all yielded data supporting it. Not only the classical sulfhydryl reagents, IAA, NEM and p-mercuribenzenosulfonic acid (PMBS), functioned as messengers to alter plant defensive chemistries, but so did the antioxidants, α-tocopherol (vitamin E) and L-ascorbic acid (vitamin C), and the analogous *T. ni* herbivory (Neupane & Norris, 1992). Thus, an effective messenger for plant cell communication with its environment does not have to be a classical sulfhydryl reagent; however, such a messenger apparently must be a molecule, ion or free radical-yielding entity (*e.g.*, herbivore) capable of effectively altering the redox electrochemistry of the involved sulfhydryl-disulfide receptor-energy transducer protein in the plant cell plasma membrane.

The proposed common chemical mechanism for insect and plant communication with each other and with their environments. Thus, both insect and plant chemical communications with each other and with their other environments are dependent on sulfhydryl-disulfide oxidative-reductive electrochemistry at the protein receptor-energy transducer level in the involved plasma membrane. Similar sulfhydryl-disulfide protein receptor-energy transducer macromolecules have been isolated from the plasma membrane of the involved insect chemosensory cell or plant cell. In addition, common chemicals, *e.g.*, the antioxidants L-ascorbic acid and α-tocopherol, serve as effective messengers to both the insect chemosensory system and the plant cell defensive chemical system. Because common chemicals serve as messengers to both insects and plants then it may be concluded that insects and plants can communicate with each other by the sulfhydryl-disulfide redox electrochemical information-encoding mechanism which can be decoded experimentally in millivolt (mV) units using either of the proven electrochemical techniques, dropping-mercury-electrode (d.m.e.) polarography or EAGs. Based on the dual findings of d.m.e. polarography and EAGs we now know that the essential encoding of the message from the messenger chemical, ion or free radical, necessary to elicit a predictable whole insect or plant response, occurs at the plasma membrane level. The current understandings of energy transduction across plasma membranes fully support our here-summarized findings regarding the sulfhydryl-disulfide dependent common chemical communication mechanism for insect and plant interactions with each other and their environments. We have termed this dynamic communication scheme the environmental energy exchange (EEE) code (Norris, 1986; Neupane & Norris, 1992).

Our elucidated information-encoding properties of the EAG explain why scientists have been so successful in using this experimental technique to identify active messenger molecules in insect chemical (*e.g.*, pheromone) communication.

The validity of our long-term findings on insect-plant chemical communications was further supported very recently by the report of Raina *et al.* (1992) that a proven hormonal messenger to plants regarding environmental stress, ethylene, serves also as a messenger to induce pheromone production in *Helicoverpa* (*Heliothis*) *zea* (Boddie) females. Ethylene is a messenger involved with sulfhydryl-disulfide protein electrochemistry in plasma membrane.

References

Neupane, F.P. & D.M. Norris (1992). Antioxidant alteration of *Glycine max* defensive chemistry: Analogy to herbivory elicitation. *Chemoecology* 3 (in press).

Norris, D.M. (1985). Electrochemical parameters of energy transduction between repellent naphthoquinones and lipoprotein receptors in insect neurons. *Bioelectrochem. Bioenerget.* 14: 449-456.

Norris, D.M. (1986). Anti-feeding compounds. In: C. Haug & H. Hoffmann (eds), *Chemistry of Plant Protection.* Berlin: Springer Verlag.

Raina, A.K., T.G. Kingan & A.K. Mattoo (1992). Chemical signals from host plant and sexual behavior in a moth. *Science* 255: 592-594.

Rozental, J.M. & D.M. Norris (1973). Chemosensory mechanism in American cockroach olfaction and gustation. *Nature* 244: 370-371.

Genetics and Evolution

Proc. 8th Int. Symp. Insect-Plant Relationships, Dordrecht: Kluwer Acad. Publ.
S.B.J. Menken, J.H. Visser & P. Harrewijn (eds), 1992

Genetics and the phylogeny of insect-plant interactions

Douglas J. Futuyma
Dept of Ecology and Evolution, State University of New York, Stony Brook, New York, USA

Key words: Asteraceae, constraints, genetic correlation, *Ophraella*, specialization

Introduction

Two salient features of phytophagous insects are the subjects of this paper: the host specificity (specialization) that all species display, often to an extraordinary degree; and the phylogenetic conservatism of host associations, manifested in the frequent association of related insects with related plants. Both features are frequently attributed to the genetically determined behavioral and physiological responses of insects to plants' secondary compounds. Phylogenetic conservatism of diet is presumed to arise from the greater ease with which insects will adapt to plants that are chemically similar to their ancestral hosts than those that are dissimilar (Ehrlich & Raven, 1964). This hypothesis, a specific instance of Darwin's presumption that *natura non facit saltum*, assumes the existence of lineage-specific, historically determined developmental or genetic constraints. Although we can readily appreciate such constraints in the evolution of morphology -- we do not expect beetles to evolve a lepidopteran proboscis in a single step -- we know so little of the neurobiology and biochemistry of insects that we cannot readily specify which plants might lie within the adaptive scope of an insect lineage, and which do not. Likewise, host plant specialization, exemplified by cases of related species that have diverged in host affiliation and therefore have abandoned an ancestral host in favor of another, is often attributed to constraints on the capacity to process different plant allelochemicals, a constraint that is frequently termed a "trade-off" (Futuyma & Moreno, 1988; Jaenike, 1990).

In both cases, it is implicitly assumed that an insect species lacks adequate selectable variation to adapt, without concomitant "penalties," to a greater range of host plants. Such a conclusion seems to conflict with the apparent abundance of genetic variation in most characters of most species that have been examined (Lewontin, 1974; Mousseau & Roff, 1987), including responses of insects to plants (Futuyma & Peterson, 1985). But although it is widely believed that "the simplest possible evolutionary constraint, *viz.* lack of genetic variation, would appear not to be important" (Barker & Thomas, 1987, p. 6), phylogenetic conservatism and the ecological and geographic limits of every species suggest that limitations on genetic variation do exist (Bradshaw, 1991). Unlike most ecological genetic investigations, which describe genetic variation in features a species possesses (*e.g.,* variation in preference of an insect for various of its natural hosts), exploration of the limits to evolution may require study of features that have *not* evolved -- for example, the genetic capacity to adapt to potential food plants that an insect does *not* use.

Genetic correlations

Although paucity of genetic variation is seldom considered a severe constraint, negative genetic correlations, owing to pleiotropy, have been considered a likely constraint on breadth of host utilization. A trade-off in performance on different plants, which may provide selection for specialized preference, would be reflected as a negative correlation in the performance of genetically different families across different plants. The majority of such studies have provided little evidence for negative correlations that might explain the selective advantage of specialization (Futuyma & Moreno, 1988; Jaenike, 1990; Via, 1990; Futuyma & Keese, in press). However, most such studies face two limitations (Rausher, 1988). If performance only on natural hosts is examined, trade-offs may not be evident precisely because these hosts are similar enough to be included in the diet without trade-offs. Furthermore, if only two hosts are used, variation in overall "vigor" may confer a positive correlation that obscures an underlying trade-off. The positive correlation may be "factored out" if each brood is tested on more than two plants.

Futuyma & Philippi (1987) measured growth of clonal genotypes of the polyphagous geometrid *Alsophila pometaria* Harris on three species of Fagaceae and one of Aceraceae. With reference to performance on one host as a standard of general "vigor," only two of nine correlations among three plants, with measurements taken twice during development in the laboratory and once in the field, were significant: one was negative and one was positive. Jaenike (1990) applied a more suitable statistical analysis (principle components analysis) to these data, and found a significant negative correlation between growth on *Acer* L. versus the three Fagaceae. (However, this method forces some variables to carry negative signs, so the analysis may not be definitive.) Jaenike concluded that it is premature to dismiss the role of trade-offs in the evolution of diet breadth. Some studies of aphid genotypes have also found evidence of trade-offs in performance on different hosts (*e.g.*, Via, 1991), although others have not (*e.g.*, Weber, 1985).

Tests for genetic correlations in a context in which they should have the greatest potential for explaining specialization have been performed by Mark C. Keese in our laboratory, using sister species of the chrysomelid beetle genus *Ophraella* Wilcox. Phylogenetic analysis of this genus (see below) indicates that the association of *O. notulata* (Fabr.) with *Iva frutescens* L. (Asteraceae, Ambrosiinae) has evolved from association with *Ambrosia artemisiifolia* L., the host of *O. slobodkini* Futuyma. Negative genetic correlations in performance on these plants might explain why speciation entailed a host shift rather than expansion of host range.

Keese has reared larvae from wild-caught females of both species on both plants. Two replicate groups of 5 larvae per brood were reared on leaf fragments of each species, and scored for survival and time to pupation. Significant variation among families was detected in both species for development time, and in *O. slobodkini* for survival. The family X host interaction was also significant for developmental time in *O. slobodkini* (Fig. 1). Genetic correlations (of family means on the two hosts) were generally positive (Table 1), although none was statistically significant. Thus these data do not provide evidence for trade-offs. Keese has found that the mean performance (measured as feeding and developmental rates, survival, and eclosion weight) of *O. notulata* on the ancestral host (*Ambrosia*) is far better than that of *O. slobodkini* on the derived host (*Iva*). In view of the relatively high fitness of *O. notulata* on *Ambrosia* and the lack of evidence for a trade-off, Keese suggests that ecological factors rather than physiological trade-offs have been responsible for the complete host shift in this species.

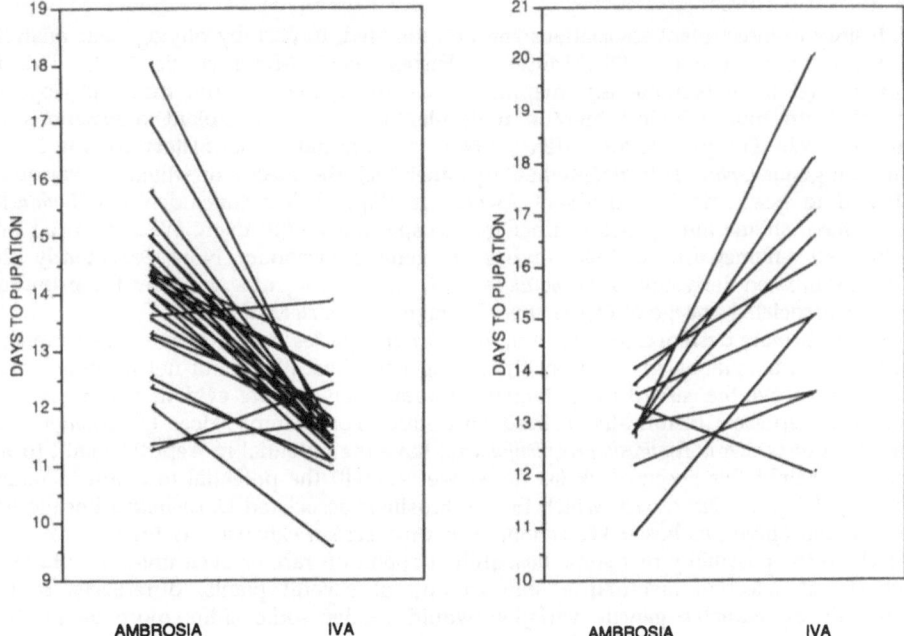

Figure 1. Mean development time (days from hatch to pupation) of *Ophraella notulata* (left) and *O. slobodkini* (right) on foliage of *Ambrosia artemisiifolia* and *Iva frutescens* in the laboratory. Each line connects the means of the progeny of one wild-caught female from populations in northern Florida, divided between the plants. In each species, mean time to pupation is lower on the natural host. In both species, there is significant variation among broods and significant brood x host interaction, but the correlation in development time on the two hosts is not statistically significant.

Table 1. Family mean correlations in performance of *Ophraella notulata* and *O. slobodkini* on *Iva frutescens* and *Ambrosia artemisiifolia*. N is the number of broods. Correlations were calculated from the mean survival in two replicate groups per brood, and from the mean of the two groups' mean time to pupation. None of the correlation coefficients is significantly different from zero.

	Ophraella notulata		*Ophraella slobodkini*	
	N	r	N	r
Survival	34	0.140	20	0.246
Development time	32	-0.273	10	0.115

Genetic variation and its constraints

The history of insect-plant associations can be estimated, in part, by phylogenetic analysis of extant species (Miller, 1987; Mitter & Farrell, 1991; Mitter *et al.*, 1991). This is accomplished by parsimoniously mapping onto an estimate of the insect phylogeny, derived from morphological and/or molecular data, the host-plant associations as characters. We (Futuyma & McCafferty, 1990) have estimated this history for the North American genus *Ophraella* (Coleoptera: Chrysomelidae), the species of which are variously restricted to genera in four tribes of Asteraceae (Fig. 2). We concluded that *Ophraella* species have shifted among hosts rather than cospeciating with them. We have used this phylogenetic framework for our studies of genetic variation. Note particularly the well-substantiated derivation of *O. notulata*, with its host *Iva frutescens*, from the primarily *Ambrosia*-associated lineage of *O. communa* Le Sage and *O. slobodkini*.

Our studies are designed to determine if *Ophraella* species harbor genetic variation that would enable adaptation to certain of their congeners' host plants but not to others, and, should this prove the case, if the potential for adaptation is more evident with respect to plants that represent historically realized host shifts. For example, does *O. communa*, the major host of which is *Ambrosia artemisiifolia* L., have the potential to "repeat" a shift to *Iva frutescens*? Might this species lack (as far as we can tell) the potential to adapt to plants such as *Solidago* or *Eupatorium*, which the Ambrosiinae-associated *O. communa* lineage has not included among its hosts? We cannot, of course, screen exhaustively for rare variants, or exclude the possibility that some host shifts depend on rare or even unique mutations. However, if selection favoured a shift to any of several plants, differences in the magnitude of available genetic variation would render some shifts more likely than others. Our research program is based on the assumption that, as for most phenotypic traits, those involved in host shifts have a polygenic basis (Futuyma & Peterson, 1985) with a long-persistent pattern of variation (Lande, 1975).

Among response variables, we have focused on two critical steps in establishment on a new host: feeding behavior (by adults and especially hatchling larvae) and larval survival. We have no data on long-distance orientation and almost none on oviposition. Feeding is scored as the area of leaf disc consumed in 24 h, without choice (the natural host invariably elicits greater feeding if a choice is offered). Hatchlings, from eggs laid on the natural host, were removed for feeding trials within 24 h of hatching. Adults, which in some trials did and in others did not have post-eclosion experience of their natural host, were usually deprived of food for 24 h before testing. Larval survival, scored at variable periods after placement of hatchlings on test material, was scored either in dishes to which fresh leaf material was introduced every two days, or on cuttings inserted into water. The most extensive data (Futuyma *et al.*, submitted) are for *O. communa*, a multivoltine species, with an egg-to-egg generation time of *ca* one month, that in eastern North America feeds only on *Ambrosia artemisiifolia*. In most of the experiments, we have assayed for quantitative genetic variation by using a half-sib design (Falconer, 1981), mating each male to two or more virgin females and individually testing several progeny of each female. Significant variance among sires is interpreted as evidence of additive genetic variance (generally considered the most available to selection). Variance among dams can include additive and non-additive genetic variance, as well as variance arising from maternal effects and common environment (*e.g.*, of the eggs or hatchlings in a single clutch). For a few tests of *O. communa*, and for tests of several other species, we have data only for progenies of wild-caught females; these include an unknown mixture of full- and perhaps half-sibs.

194

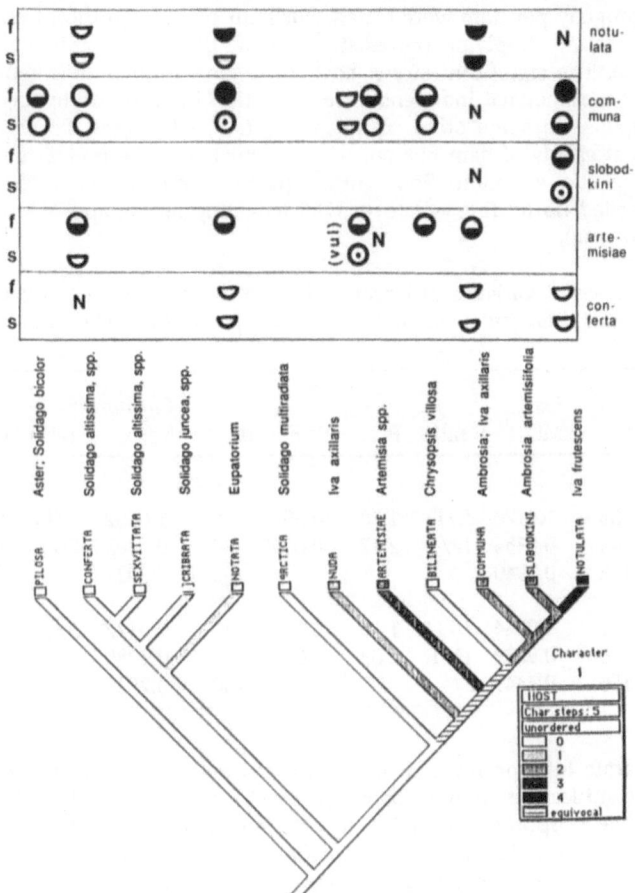

Figure 2. Estimated phylogeny of *Ophraella* species (named at the branch ends), with principal hosts arrayed above each species. Shading on branches of the cladogram represents the most parsimonious interpretation of shifts among tribes of Asteraceae, given as characters 0 (Astereae), 1 (Eupatorieae), 2 (Ambrosiinae), 3 (Anthemideae). *Iva frutescens* (Ambrosiinae) is marked as character 4 to emphasize derivation of association with this plant (in *O. notulata*) from association with *Ambrosia*. The diagram above the host names summarizes results of tests for genetic variation in five species of *Ophraella* (named at the right) for feeding response (f) and larval survival (s) on various test plants below the symbols. A filled circle indicates evidence for genetic variation (significant variation among both sires and dams), a half-filled circle evidence for significant variation among dams only, an unfilled symbol failure to find evidence of genetic variation. N indicates the natural host of a species. The pairs of symbols above *Artemisia* for *O. communa* represent tests on *A. carruthii* and *A. vulgaris* respectively. *O. artemisiae* Futuyma was tested on *A. vulgaris*.

195

Full descriptions of the experiments will be published elsewhere. An example of our results is one test of hatchling feeding responses to two hosts of other *Ophraella species* (Table 2). For the test on *Iva*, each of 55 sires was mated to an average of 1.8 dams, and a mean of 2.95 progeny per dam were tested. For both plants, significant effects of both sire and dam were found, implying the existence of additive and probably of non-additive genetic variance. The significant sire x host interaction implies that genetic variation in response to *Iva* is somewhat independent of variation in response to *Eupatorium*. In tests of feeding response to some other plant species (*e.g.*, *Chrysopsis*, *Solidago bicolor* L.), we found significant effects of dam but not of sire, which we interpret as evidence for some non-additive genetic variation. Some other plants (*Artemisia carruthii* Wood, *Solidago altissima* L.) elicited no feeding whatever (288 hatchling larvae and 653 adults have been tested on *S. altissima*).

Table 2. Analyses of variance of larval and adult *O. communa* feeding scores on each of two test plants, *Iva frutescens* and *Eupatorium perfoliatum*. The variate is log (units consumed +1)

Source	df	*Iva* MS	F ratio	F	P>F	df	*Eupatorium* MS	F ratio	F	P>F
F_1 larvae										
Sire	54	1.4176	S/D	1.70	<0.05	54	1.0352	S/D	1.99	<0.025
Dam	44	0.8354	D/E	2.12	0.0003	44	0.5203	D/E	1.72	0.0066
Error	193	0.3940				198	0.3021			
F_1 adults										
Sire	49	1.0744	S/D	1.19	>0.25	49	0.4177	S/D	1.05	>0.25
Dam	25	0.9022	D/E	2.03	0.005	25	0.3985	D/E	1.41	0.108
Error	150	0.4449				150	0.2827			

In four separate tests on *Iva* and one on *Eupatorium*, survival from first instar to 8-11 days varied significantly among dams. Table 3 presents results of one such test. Compared to the proportion surviving to eclosion on the natural host *Ambrosia* (0.56), survival on *Iva* was modest (0.14) and on *Eupatorium* was very low (0.002: 1/576). On *Eupatorium* as on some other plants, mortality was complete or almost so despite some feeding and growth, suggesting an effect of toxicity or nutritional inadequacy. Mortality before pupation was complete not only on those plants that elicited no feeding (*A. carruthii*, *S. altissima*), but also on several that elicited slight to substantial feeding (*S. bicolor*, *Chrysopsis villosa* Nutt., *Artemisia vulgaris* L.). Thus, on none of these plants is there evidence of genetic variation in capacity for larval survival.

These results are related to the inferred phylogenetic history of host associations in Fig. 2, in which the upper and lower rows of symbols in the horizontal panel marked "communa" summarize the evidence on genetic variation in feeding response and larval survival, respectively, to the hosts of the *Ophraella* species arrayed below. In *O. communa*, evidence of genetic variation in feeding response is perhaps more pronounced in respect to the hosts of closely allied species. It is also germane to note that a phenetic analysis of overall similarity in secondary compounds (Futuyma & McCafferty, 1990, Fig. 8F) showed that *Iva* is most similar to *Ambrosia*, *Artemisia* is rather less similar, and *Solidago* and *Eupatorium* both differ markedly. (No information on the secondary chemistry of *Chrysopsis* was found, but as a member of the Astereae, it may be expected to resemble *Solidago*.)

Table 3. Larval survival, to 10 days, on cuttings of *Iva frutescens* and *Ambrosia artemisiifolia*). Data are $\sqrt{x + 0.5}$. Analyses of variance for both hosts together and each separately

Source	df	MS	F ratio	F	P>F
Overall analysis					
Sire	24	0.5517	S/D	0.98	>0.50
Dam	35	0.5630	D/E	2.74	<0.001
Host	1	44.6956	H/S*H	131.69	<0.001
Sire * Host	24	0.3394	S*H/(D*H)+E	1.61	<0.05
Dam * Host	35	0.2275	D*H/E	1.11	0.334
Error	119	0.2054			

	Iva					Ambrosia				
	df	MS	F ratio	F	P>F	df	MS	F ratio	F	P>F
Sire	24	0.4630	S/D	1.04	>0.25	24	0.4476	S/D	1.29	>0.10
Dam	35	0.4434	D/E	2.06	0.007	35	0.3474	D/E	1.77	0.025
Error	59	0.2151				60	0.1957			

Although the ancestral association of the genus *Ophraella* appears to have been with *Solidago* or related genera in the tribe Astereae, the species of *Solidago* used in these experiments are the hosts of *Ophraella* species only distantly related to *O. communa*; we found evidence in *O. communa* of genetic variation in larval feeding response to *S. bicolor* (although mean response was very low) but no capacity for survival. A more extreme deficiency of genetic variation was displayed on *S. altissima*, which evoked no feeding.

Artemisia (tribe Anthemideae) and *Chrysopsis* (tribe Astereae) are hosts of rather close relatives of *O. communa*. It is not certain whether these associations evolved from a lineage that fed on Ambrosiinae (which would make the association of *O. communa* plesiomorphic) or on Astereae. *O. communa* displayed genetic variation in feeding response to *Artemisia vulgaris*, a European species on which *O. artemisiae* can be reared successfully through a complete life cycle, but a small sample (n = 90) showed no response to *A. carruthii*, a natural host of *O. artemisiae*. The most abundant evidence of genetic variation, including additive genetic variation, was displayed in response to *Iva frutescens*, which moreover supported rather high survival through eclosion and also elicited oviposition by some females (Futuyma et al., submitted). Chemically more similar and more closely related to *Ambrosia* than any of the other test plants, *I. frutescens* is an immediately derived host association, relative to *Ambrosia*. Although isolated populations of *O. communa* in the desert of eastern California feed on a related plant, *I. axillaris* Pursh (the host also of *O. nuda* Le Sage), this plant differs substantially from *I. frutescens* in secondary chemistry, habitus, and habitat, and undoubtedly is, relative to *Ambrosia*, in these populations.

The most conspicuous conflict with the expectation that a lineage might not harbor genetic variation in response to plants with which it has had no evolutionary experience is the response to *Eupatorium*, the autapomorphic (uniquely derived) host association of a distantly related species (*O. notata* (Fabr.)). This plant also supports a low level of larval survival, unlike all other test plants except *I. frutescens* as noted above.

Fig. 2 also summarizes preliminary analyses of similar screens for genetic variation in four other species of *Ophraella*. No evident relationship of genetic variation to the phylogenetic position of several test plants is displayed by either *O. conferta* (Le Conte), which fed on none of them, or *O. artemisiae*, which displays a significant dam effect (indicative of non-additive genetic variance) on all. *Ophraella notulata*, like its close relative *O. communa*, has a genetically variable feeding response to *Eupatorium*, and also to its presumptive ancestral host, *Ambrosia*, but, like *O. communa*, does not respond to *Solidago altissima*. Thus a similar pattern of genetic variation has been retained since the speciation event that separated *O. communa* from the *O. notulata* lineage. This divergence may be estimated, from Nei's genetic distance (data on 19 enzyme loci in Futuyma & McCafferty, 1990), to have occurred 1.7 to 6.2 million years ago, using respectively the calibrations suggested by Nei (1987) and Thorpe (1982). This provides some justification for the presumption underlying this research, that genetic variation in ancestral populations can be estimated from their contemporary descendants.

As noted above, Mark Keese has found evidence of genetic variation in feeding responses to and survival on each other's host in the sister (and sibling) species *O. notulata* (host: *I. frutescens*) and *O. slobodkini* (host: *Ambrosia artemisiifolia*). It is interesting and perhaps not surprising that the species with the derived host shows higher performance, by all measures, on its ancestral plant than does the species with the ancestral host on the derived plant.

Discussion

This research on *Ophraella* has been aimed at providing a broad picture of the relationship between the availability of genetic variation and the evolutionary history of host affiliation. At a more mechanistic level, it is very crude. An insect confronts in a potential host plant a great panoply of compounds that may affect its physiology and behavioral responses. In general we know little, and for *Ophrella* nothing at all, about how many of these compounds influence the likelyhood of a host shift, but the greater the number of compounds that require independent genetic change, the less likely adaptation to the plant may be. For some compounds, an insect may simply not possess the requisite, modifiable biochemical machinery, and so could not exhibit genetic variation. Although we cannot say that a proposed host shift could never occur, paucity of genetic variation, revealed by genetic screening, may at least indicate that some shifts are unlikely, relative to others.

Overall chemical similarity among plants need not predict host shifts, which could be based on the response to one or a few compounds, as appears likely in *Yponomeuta* Latreille (Menken *et al.*, 1992). In *Yponomeuta* and some other groups, moreover, shifts to distantly related, chemically rather dissimilar plants have occurred (Jermy, 1984). Nevertheless, overall chemistry appears to be at least loosely correlated with host shifts in *Ophraella* (Futuyma & McCafferty, 1990), as well as with feeding responses to congeners' hosts and with the pattern of genetic variation in *O. communa*.

Although many investigators have documented some capacity of insects to feed and survive on certain plants that they do not naturally utilize, this study is among the first (see Thompson, 1988; James *et al.*, 1988; Karowe, 1990) to investigate genetic variation in these capacities, and so to probe the possible potential for evolution of host shifts. It also represents the first attempt to relate genetic variation explicitly to the phylogenetic history of host shifts.

Summary

We report some results of screening for genetic variation in behavioural responses and survival of species of *Ophraella* (Coleoptera: Chrysomelidae) to host plants of their congeners, within a phylogenetic context. We wish to determine if the presence or absence of detectable genetic variation in these features bears a relationship to the apparent history of host shifts. Among our observations are:

(1) Confronted with certain host plants used by congeneric species, even rather large samples of some species of *Ophraella* display no evidence of genetic variation in capacity to feed and/or to survive. Paucity of genetic variation may therefore make some imaginable host shifts relatively unlikely.

(2) At least in *O. communa*, there may exist a loose correlation between the genetic potential for adaptation to a plant, and the plant's overall chemical similarity to the species' natural host. The major exception is *Eupatorium*.

(3) A genetically variable capacity to perform on *Solidago* may have been lost in the lineage that today is associated with Ambrosiinae, inasmuch as the remote ancestor of this lineage is presumed to have fed on *Solidago* or related plants. The pattern of presence vs. apparent absence of genetic variation in response to various plants is similar in *O. communa* and *O. notulata*, both of which feed on Ambrosiinae. Thus a similar pattern of constraint *vs* genetic potential for adaptation has persisted since the divergence of these species, implying long-term "phylogenetic conservatism" of adaptive potential.

(4) Genetic variation may be more evident in *O. communa* for plants that represent realized host shifts than for those that do not. *Eupatorium* is the conspicuous exception.

(5) Our limited data do not provide evidence for trade-offs, in the form of negative genetic correlations, as possible constraints on host utilization by *Ophraella*.

(6) In one cluster of three closely related species, the mean performance of the species with the derived host association on its ancestral host is greater than the performance of species with the ancestral host association when they are placed on the derived host.

Our evidence that in some instances genetic variation is very low or absent suggests the existence of genetic constraints that can account for the phylogenetic conservatism of host affiliation that is so conspicuous in many groups of insects. Whether or not standing levels of genetic variation can serve to predict the realized history of host shifts -- whether or not we can show that genetic constraints have at least loosely steered the course of evolution -- is not yet certain, but some of our data suggest that this may be the case.

Acknowledgements

I am grateful to M. Keese, S. Scheffer, D. Funk, T. Morton, and J. Walsh for assistance, and to M. Keese for permission to cite his unpublished work. I acknowledge support by the U. S. National Science Foundation (BSR 8516316, BSR 8817912). This is Contribution No. 820 in Ecology and Evolution from the State University of New York at Stony Brook.

References

Barker, J.S.F. & R.H. Thomas (1987). A quantitative genetic perspective on adaptive evolution. In: V. Loeschcke (ed.), *Genetic Constraints on Adaptive Evolution*, pp. 3-23. Berlin: Springer-Verlag.

Bradshaw, A.D. (1991). The Croonian Lecture, 1991. Genostasis and the limits to evolution. *Phil. Trans. R. Soc. Lond.* B 333: 289-305.

Ehrlich, P.R. & P.H. Raven (1964). Butterflies and plants: a study in coevolution. *Evolution* 18: 586-608.

Falconer, D.S. (1981). *Introduction to Quantitative Genetics*. Second edition. New York: Longman Press.

Futuyma, D.J. & M.C. Keese (1992). Evolution and coevolution of plants and phyto-phagous arthropods. In: G.A. Rosenthal and M.R. Berenbaum (eds), Herbivores: Their Interactions with Secondary Plant Metabolites (Second Edition), Volume II: *Evolutionary and Ecological Processes*. San Diego: Academic Press.

Futuyma, D.J. & S.S. McCafferty (1990). Phylogeny and the evolution of host plant associations in the leaf beetle genus *Ophraella* (Coleoptera: Chrysomelidae). *Evolution* **44**: 1885-1913.

Futuyma, D.J. & G. Moreno (1988). The evolution of ecological pecialization. *Annu. Rev. Ecol. Syst.* **19**: 207-233.

Futuyma, D.J. & S.C. Peterson (1985). Genetic variation in the use of resources by insects. *Annu. Rev. Entomol.* **30**: 217-238.

Futuyma, D.J. & T.E. Philippi (1987). Genetic variation and covariation in responses to host plants by Alsophila pometaria (Lepidoptera: Geometridae). *Evolution* **41**: 269-279.

Jaenike, J. (1990). Host specialization in phytophagous insects. *Annu. Rev. Ecol. Syst.* **21**: 243-273.

James, A.C., J. Jakubczak, M.P. Riley & J. Jaenike (1988). On the causes of monophagy in Drosophila quinaria. *Evolution* **42**: 626-630.

Jermy, T. (1984).Evolution of insect/host plant relationships *Amer. Nat.* **124**: 609-630.

Karowe, D.N. (1990). Predicting host range evolution: colonization of Coronilla varia by Colias philodice (Lepidoptera: Pieridae). *Evolution* **44**: 1637-1647.

Lande, R. (1975). The maintenance of genetic variability by mutation in a polygenic character with linked loci. *Genet. Res.* **26**: 221-234.

Lewontin, R.C. (1974). *The Genetic Basis of Evolutionary Change*. New York: Columbia University Press.

Menken, S.B.J., W.M. Herrebout & J.T. Wiebes (1992). Small ermine moths (*Yponomeuta*): their host relations and evolution. *Annu. Rev. Entomol.* **37**: 41-66.

Miller, J.S. (1987). Host-plant relationships in the Papilionidae (Lepidoptera): Parallel cladogenesis or colonization? *Cladistics* **3**: 105-120.

Mitter, C. & B. Farrell (1991). Macroevolutionary aspects of insect/plant interactions, In: E. Bernays (ed.), *Insect-Plant Interactions*, Vol. 3, pp. 35-78. Boca Raton: CRC Press.

Mitter, C., B. Farrell & D.J. Futuyma (1991). Phylogenetic studies of insect-plant interactions: insights into the genesis of diversity. *Trends Ecol. Evol.* **6**: 290-293.

Mousseau, T.A. & D.A. Roff (1987). Natural selection and the heritability of fitness components. *Heredity* **59**: 181-197.

Nei, M. (1987). *Molecular Evolutionary Genetics*. New York: Columbia University Press.

Rausher, M.D. (1988). Is coevolution dead? *Ecology* **69**: 898-901.

Thompson, J.N. (1988). Variation in preference and specificity in monophagous and oligophagous swallowtail butterflies. *Evolution* **42**: 118-128.

Thorpe, J.P. (1982). The molecular clock hypothesis: Biochemical evolution, genetic differentiation and systematics. *Annu. Rev. Ecol. Syst.* **13**: 139-168.

Via, S. (1990). Ecological genetics and host adaptation in herbivorous insects: the experimental study of evolution in natural and agricultural systems. *Annu. Rev. Entomol.* **35**: 421-446.

Via, S. (1991). The genetic structure of host plant adaptation in a spatial patchwork: demographic variability among reciprocally transplanted pea aphid clones. *Evolution* **45**: 827-852.

Weber, G. (1985). Genetic variability in host plant adaptation of the green peach aphid, *Myzus persicae*. *Entomol. exp. appl.* **38**: 49-56.

Proc. 8th Int. Symp. Insect-Plant Relationships, Dordrecht: Kluwer Acad. Publ.
S.B.J. Menken, J.H. Visser & P. Harrewijn (eds), 1992

Host-race formation in a leaf-mining moth

Michael Auerbach[1] and Robert Fleischer[2]
[1] Dept of Biology, University of North Dakota, Grand Forks, USA
[2] National Zoological Park, Smithsonian Institution, Washington, USA

Key words: Allozyme variation, host preference, Phyllonorycter, Populus

Larvae of the leaf-mining moth, *Phyllonorycter salicifoliella* Chambers (Lepidoptera: Gracillariidae) feed within leaves of three *Populus* species in Itasca State Park, north-central Minnesota, USA. Since 1985, leaf-miner densities have been consistently high on *Populus tremuloides* Michx, intermediate on *P. balsamifera* L., and low on *P. grandidentata* Michx. Despite this pattern, a previous study found no difference in egg to adult survivorship among miners on the three hosts, although larval growth rate and pupal mass did vary among hosts (Auerbach & Alberts, 1992).

Adult moths mate on host trees following termination of diapause in early spring. Females prefer to oviposit on young, still-expanding leaves. Although population-wide oviposition spans about 3 weeks, the vast majority of eggs are deposited over a 4-5 day period. In most years, oviposition is well synchronized with availability of preferred foliage on *P. tremuloides* and *P. balsamifera*, but not with leaf availability on *P. grandidentata*, which flushes leaves 3 weeks after the other host species.

Materials and methods

We used horizontal starch gel electrophoresis to test if populations of miners were genetically differentiated among host species. We were particularly interested in whether miners associated with the phenologically-delayed host, *P. grandidentata*, differed genetically from miners on other hosts. In 1989 we examined allozyme variation among females ovipositing on, larvae feeding on, and adults reared from each host species. Analyses were repeated in 1990 with adults reared from host plants in at least two locations, to test for variability between sites and years.

Results and conclusions

Seven polymorphic loci that resolved well for adults, and 7 for larvae (5 same as adults), were selected from among 18 loci we initially screened. There were no significant differences in allele frequencies between miners associated with *P. tremuloides* and *P. grandidentata*. Thus, phenological differences among hosts do not appear to partition leaf-miner populations. However, adults reared on *P. tremuloides* and *P. grandidentata* differed significantly from ones on *P. balsamifera* in allele frequencies at 6 loci (*Me*, *Pgm-1*, *Pgm-2*, *Mdh*, *Est-2*, *Lap*). The same results were obtained with females ovipositing on these hosts. In most cases, allele frequencies were fixed or nearly fixed for alternative alleles. Similar differences were found among larvae at 5 loci (*Me*, *Pgm-1*, *Pgm-2*, *Mdh*, *Est-4*). These patterns were consistent from year to year and across sites, including ones with mixed stands of host species.

These results provide strong evidence for the existence of host-associated, genetically distinct races of *P. salicifoliella*, as has been found for other phytophagous insects (Waring *et al.*, 1990). Clearly, variation in host preference is fundamental for segregation of *P. salicifoliella* races, since the same genetic differences were observed in ovipositing females and other developmental stages. Variation in habitat (host) preference is most likely to produce genetic divergence when mating occurs on or near the preferred resource (Diehl & Bush, 1989), as is true in this case.

Throughout its extensive range in North America, *P. salicifoliella* occurs only on members of two genera in the Salicaceae, *Populus* and *Salix*. Phenolic glycosides unique to these taxa presumably form at least part of the basis for host discrimination. Identity and concentration of phenolic glycosides vary greatly among the three *Populus* species (Auerbach & Alberts, 1992). We suspect that genetically-based differences in perception of and response to varying phenolic glycoside profiles cause the segregation of host-associated genotypes and, as a result, assortative mating.

We conducted an additional experiment in 1990, to test if host races had evolved host-associated specialization. Ovipositing females were collected from *P. tremuloides* and *P. balsamifera* and transferred into sleeve cages covering branches of the same (control) or alternative host species. Females transferred singly would not oviposit on an 'incorrect' host, but we were able to induce oviposition by transferring multiple females into the same cage. Egg to adult survivorship did not vary between miners reared on their 'correct' host (43.5% and 43.9% for *P. tremuloides* and *P. balsamifera* respectively). Survivorship for miners on the 'wrong' host was 48.2% and 13.6% for switches from *P. tremuloides* to *P. balsamifera* and the converse. Thus, miners associated with *P. tremuloides* survived equally as well on *P. balsamifera*, whereas those associated with *P. balsamifera* had greatly reduced survivorship on the other host. Although sample sizes were small, these results provide evidence that individuals associated with *P. tremuloides* are preadapted for colonization of *P. balsamifera*, while those on *P. balsamifera* appear to have evolved some degree of specialization for this host.

References

Auerbach, M. & J.D. Alberts (1992). Occurrence and performance of the aspen blotch miner, *Phyllonorycter salicifoliella*, on three host-tree species. *Oecologia* **89**: 1-9.
Diehl, S.R. & G.L. Bush (1989). The role of habitat preference in adaptation and speciation. In: D. Otte & J.A. Endler (eds), *Speciation and Its Consequences*, pp. 345-365. Sunderland: Sinauer.
Waring, G.L., W.G. Abrahamson & D.J. Howard (1990). Genetic differentiation among host-associated populations of the gallmaker *Eurosta solidaginis* (Diptera: Tephritidae). *Evolution* **44**: 1648-1655.

Proc. 8th Int. Symp. Insect-Plant Relationships, Dordrecht: Kluwer Acad. Publ.
S.B.J. Menken, J.H. Visser & P. Harrewijn (eds), 1992

Host-race formation in the two-spotted spider mite (*Tetranychus urticae*)

Jan Bruin[1], Tetsuo Gotoh[1,2], Maurice W. Sabelis[1], Steph B.J. Menken[3] and
Wil E. van Ginkel[3]
[1] *University of Amsterdam, Dept of Pure and Applied Ecology, Amsterdam, The Netherlands*
[2] *Ibaraki University, Faculty of Agriculture, Ami, Ibaraki, Japan*
[3] *University of Amsterdam, Institute of Taxonomic Zoology, Amsterdam, The Netherlands*

Key words: Allozymes, assortative mating, host-plant preference, speciation

The two-spotted spider mite, *Tetranychus urticae* Koch (Acari: Tetranychidae), has been reported to be found on numerous host-plant species throughout the world. It is an important pest of many agricultural crops, including tomato and cucumber, which are often grown simultaneously, in neighbouring compartments, or alternately, in the same compartment, in greenhouses in the Netherlands. In both situations, host shifts are likely to occur. The two host plants are generally thought to differ in suitability for the spider mites; the surface of tomato leaves and stems are covered with glandular hairs that can cause high mortality, due to entrapment and intoxication (*e.g.*, Carter & Snyder, 1985). Cucumber does not have such hairs. Hence, spider mites are subject to severe selection to colonize the crop occasionally available.

Several studies have demonstrated *T. urticae*'s ability to rapidly adapt to new hosts (*e.g.*, Gould, 1979; Fry, 1989). Whether this colonization success leads to the formation of host races is unknown. In this paper we address this question to the tomato-strain and the cucumber-strain of *T. urticae* (for a more detailed report on part of this study see Gotoh *et al.*, in prep.). We assessed the following three characteristics: mate choice by males; host-plant choice by females; the genetic composition of both strains.

A selection experiment was carried out to attempt to mimic host shifts: mites of the cucumber-strain were transferred to tomato leaves and kept on tomato for successive generations.

Materials and methods

Mate choice. On a leaf disc, a single male was offered a choice between two females in their final moulting stage, one female from each strain. The mites were observed for 30 min at 5-min intervals. When a male had guarded one female for more than 10 min it was considered to have made a choice. This test was done on both tomato and cucumber.

Host-plant choice. Two leaf discs, one from each host plant, were interconnected by a piece of waxed cocktail pricker. Adult female mites were placed individually on the bridges and observed after 24 h. The host on which the mites had settled was considered to be the preferred.

Genetic composition. Preliminary experiments revealed that the enzyme phosphoglucose isomerase (*Pgi*) showed high activity and, unlike most other enzymes, could easily be detected electrophoretically in individual mites. For details, see Gotoh *et al.* (in prep.).

Selection experiment. Females of the cucumber strain were reared on cut tomato leaves. Every three weeks ca. 120 adult females were transferred to fresh tomato leaves (ca. 20 females/leaf), and allowed to oviposit for five days. Thereafter the females were removed. At 10-generation intervals we assessed: 1) the genetic composition at the *Pgi* locus, and 2) the ovipositional rate [as a measure of host-plant quality] of females of the different strains, on tomato (one female/leaf disc). The latter was done to monitor (the rate of) adaptation to the new host.

Results

Mate choice. Males of the tomato strain preferred females of their own strain to females of the cucumber strain. Males of the cucumber strain showed no preference. This pattern was evident on both tomato and cucumber leaves.

Host-plant choice. Females of both strains preferred their original host to the new host (tomato strain: n = 52, P < 0.001; cucumber strain: n = 80, 0.001 < P < 0.01; binomial test).

Genetic composition. Mites of the cucumber strain were highly polymorphic at the *Pgi* locus; evidence of five alleles was found. The tomato strain appeared to be fixed for the most common allele of the cucumber strain.

Selection experiment; genetic composition. After 19 generations, the variation in the cucumber strain-on-tomato did not show any evidence of directional selection at the *Pgi* locus towards fixation of the tomato strain allele.

Selection experiment; ovipositional rate. Females of the "parental" cucumber strain performed relatively poorly on tomato leaves. However, females of the cucumber strain that had been kept on tomato for 10 or 20 generations performed as well as females of the tomato strain.

Discussion and conclusions

We conclude that the tomato and the cucumber strain of the two-spotted spider mite are biologically distinct, since 1) the tomato-mites mate assortatively, and 2) both strains prefer the host they originated from. Both traits favour the formation of host races. The most likely evolutionary scenario is that a host shift had occurred in the past, from cucumber to tomato, the more hostile host. Recurrent bottlenecks in subsequent generations may have increased the effects of drift on the genetic composition of the population. Indeed, we found the tomato strain to be monomorphic at the *Pgi* locus for the most common allele in the cucumber strain.

Our selection experiment showed no evidence of selection by tomato on the *Pgi* locus of the mites that originated from cucumber. It did however show that adaptation to tomato can occur within 10 generations.

References

Carter, C. & J.C. Snyder (1985). Mite responses in relation to trichomes in *Lycopersicon esculentum* x *L. hirsutum* F2 hybrids. *Euphytica* **34**: 177-185.

Fry, J.D. (1989). Evolutionary adaptation to host plants in a laboratory population of the phytophagous mite *Tetranychus urticae* Koch. *Oecologia* **81**: 559-565.

Gould, F. (1979). Rapid host range evolution in a population of the phytophagous mite *Tetranychus urticae* Koch. *Evolution* **33**: 791-802.

Proc. 8th Int. Symp. Insect-Plant Relationships, Dordrecht: Kluwer Acad. Publ.
S.B.J. Menken, J.H. Visser & P. Harrewijn (eds), 1992

Variation in the suitability of *Barbarea vulgaris* (Cruciferae) for the flea beetle *Phyllotreta nemorum*

Jens Kvist Nielsen
Chemistry Dept, Royal Veterinary and Agricultural University, Frederiksberg, Denmark

Key words: Inheritance, plant defences, sex chromosomes

The flea beetle, *Phyllotreta nemorum* L., (Coleoptera: Chrysomelidae) is an oligophagous species feeding on a restricted number of plants belonging to the plant family Cruciferae (Nielsen, 1989). Eggs are laid in the soil close to the host plants, and the first instar larvae have to climb the plant to find a site to initiate a leaf mine. Full-grown third instar larvae leave the leaf mines and pupate in the soil.

Plants which are acceptable under field conditions are usually highly acceptable to both larvae and adults in laboratory tests (Nielsen, 1989). Laboratory tests with one plant species, *Barbarea vulgaris* R.Br., were very variable (Nielsen, unpubl.). Although the larvae readily initiated leaf mines in this species, sometimes the larvae developed well, and sometimes they did not. Some larvae left the mines after a few hours; others died in the mine. Therefore, *B. vulgaris* seemed to contain defences similar to those found in *Bunias erucago* L., *Matthiola parviflora* (Schousb.) R.Br., and *Reseda alba* L. (Nielsen, 1989). As a result of this unpredictability, *B. vulgaris* was judged to be an unlikely host plant for *P. nemorum* under natural conditions. It was therefore surprising to find large numbers of *P. nemorum* larvae on *B. vulgaris* at a single site, Ejby. This observation started the investigations on the variability of the plant, *B. vulgaris* as well as the insect, *P. nemorum*.

Materials and methods

Seeds of *B. vulgaris* were collected at different sites in Denmark, and grown in small field plots at the Agricultural Experimental Station, Højbakkegaard, Taastrup. Detached leaves from the field plots were presented to first instar larvae in the laboratory, and survival after three days was measured to estimate the suitability of the leaves for *P. nemorum* (Nielsen, 1989). Leaves with larvae were then transferred to plastic vials containing a layer of moist peat about 3 cm deep. Full-grown third instar larvae left the leaf mines and pupated in the peat. Sex ratios were determined of adult beetles emerging from the peat.

Two populations of *P. nemorum* were used. The T-population was collected from radish, *Raphanus sativus* L. at the Agricultural Experimental Station in Taastrup, and the E-population was collected from *B. vulgaris* in Ejby. The inheritance of the ability to survive in *B. vulgaris* was studied by crossing virgin females from the T-population with males from the E-population and *vice versa* (F_1). Mating pairs were kept separately. Beetles of the F_1 generation were used in further crossing experiments with the T-population.

Results and discussion

The results demonstrated a seasonal variability in the suitability of *B. vulgaris* for the T-population. Rosette leaves, although unsuitable during the summer period, became suitable in the autumn and remained so until spring. In spring, the plants grew and flowered in April and May, when the older (lower) leaves were more suitable than the younger (upper) leaves. There were only minor differences between plants originating from different localities. Some variation between years may have been caused by environmental effects (*e.g.*, temperature, water relations), but detailed analyses have not yet been made. All plant stages were suitable for the E-population (Nielsen, unpubl.).

F_1 larvae developed well in *B. vulgaris*, and the ability to survive in this plant seemed to depend on the presence of one or more dominant genes in the E-population. The T-population was supposed to have the corresponding recessive genes. When the F_1 generation was backcrossed with the T-population, about 50% of the progeny survived on *B. vulgaris*. This was the expected frequency if the E-population had a single dominant gene. The sex ratios of the survivors from backcrosses involving F_1 males suggested that the most abundant gene(s) was (were) sex-linked. Males whose father was from the E-population produced an excess of sons with the gene which allowed the larvae to survive in *B. vulgaris*. F_1 males whose mother was from the E-population produced mainly daughters with this ability. F_1 females mated with T-males produced equal numbers of males and females with the ability to survive in *B. vulgaris*. This mode of inheritance would occur if the major genes for survival in *B. vulgaris* are located on the X- and Y-chromosomes, but there may be other possibilities (Nielsen, in prep.). The male is the heterogametic sex in *P. nemorum* and other beetles (Segarra & Petitpierre, 1990). If the two major genes were located on the X- and Y-chromosomes, they probably originated from different mutations, since crossing-over is supposed not to occur between sex chromosomes in beetles (Smith & Virkki, 1978).

The ability to survive in *B. vulgaris* is thought to be an extension of the host range in the E-population, and not a restriction in the T-population. This is because most Danish populations are susceptible to the defences of *B. vulgaris* in the same way as the T-population (Nielsen, unpubl.). The adaptation of *P. nemorum* to *B. vulgaris* may have been facilitated by the variability of the plant, which is sometimes undefended even against susceptible populations. The adaptation has progressed to where the insect has developed a tolerance to the plant defences, but not yet any preference for the new host plant (Nielsen, unpubl.).

Acknowledgements. This study was supported by the Danish Agricultural and Veterinary Research Council, the Danish Natural Science Research Council, and the Carlsberg Foundation.

References

Nielsen, J.K. (1989). Host plant relations of *Phyllotreta nemorum* L. (Coleoptera: Chrysomelidae). II. Various defensive mechanisms in plants and their role in defining the host plant range. *J. appl. Entomol.* **107**: 193-202.

Segarra, C. & E. Petitpierre (1990). Chromosomal survey in three genera of Alticinae (Coleoptera: Chrysomelidae). *Cytobios* **64**: 169-174.

Smith, S.G. & N. Virkki (1978). Coleoptera. In: B. John (ed.), *Animal Cytogenetics*, Vol. 3: *Insecta 5*. Berlin & Stuttgart: Gebrüder Borntraeger.

Proc. 8th Int. Symp. Insect-Plant Relationships, Dordrecht: Kluwer Acad. Publ.
S.B.J. Menken, J.H. Visser & P. Harrewijn (eds), 1992

Plant secondary chemistry and the evolution of feeding specialization in insect herbivores: a different perspective

David N. Karowe
Dept of Biology, Virginia Commonwealth University, Richmond, Virginia, USA

Key words: Genetic correlations, *Orgyia leucostigma*, trade-offs

Many hypotheses to explain the evolution of feeding specialization among insect herbivores assume a trade-off in fitness across host plant species, *i.e.*, that adaptation to one host (or host taxon) necessarily results in loss of adaptation to at least some other hosts (or host taxa). As potentially powerful and intuitively appealing as this "Jack of all trades is master of none" hypothesis may be, we presently know little about how common fitness trade-offs across hosts are, or about which mechanisms might give rise to such fitness trade-offs (Futuyma & Peterson, 1985; Jaenike, 1990). In this study, I tested the hypothesis that differences among plant species in seconday chemistry give rise to fitness trade-offs across host plant species, *i.e.*, that adaptation to one type of secondary compound necessarily results in loss of adaptation to other types of secondary compounds.

Materials and Methods

The white-marked tussock moth, *Orgyia leucostigma* (J.E. Smith), is a generalist lymantriid which, as a species, is recorded from over 60 tree species in North America. I divided the larvae from each of 40 full-sib *O. leucostigma* families across 10 artificial diets: control diet (BioServ Douglas Fir Tussock Moth Diet), and nine other diets consisting of the control diet plus a single naturally-occurring secondary compound. Five diets contained an alkaloid (0.01% berberine, 0.5% caffeine, 0.5% nicotine, 1.0% quinine, 0.5% scopolamine), one diet contained coumarin (0.05%), two diets contained a phenoloic compound (0.1% quercetin, 0.1% rutin), and one diet contained a terpenoid (0.5% α-pinene). Concentrations were chosen such that growth of *O. leucostigma* was decreased (*i.e.*, the chemical acted in a defensive capacity) but mortality was not greatly increased.

I measured egg-to-pupa growth rate for all individuals on each diet, and used ANOVA to determine whether this measure of fitness differed significantly among families. Males and females were analyzed separately, because females have an additional larval instar. To correct for differences among families in general vigor (Futuyma & Philippi, 1987), I subtracted family mean growth rate on the contol diet from family mean growth rate on each of the chemical-containing diets. To determine whether chemically-mediated fitness trade-offs exist, I calculated correlation coefficients for corrected family mean growth rate across each pair of chemical-containing diets.

Results

ANOVA revealed that *O. leucostigma* harbors significant genetic variation for growth rate on all 10 diets. Full-sib heritability estimates were significantly different from zero in all cases except females on the coumarin diet, and ranged from 0.13 to 0.69.

Of the 90 possible genetic correlations across chemical-containing diets (45 for each sex), 26 were significant and positive; none was significant and negative. Significant positive genetic correlations were as common among diets containing chemicals belonging to different classes (*e.g.*, an alkaloid and a phenolic compound) as among diets containing chemicals belonging to the same class (*e.g.*, two alkaloids).

Discussion

These results reveal that, for *O. leucostigma*, adaptation to one type of secondary compound does not result in loss of adaptation to other secondary compounds, or even to other classes of secondary compound. In fact, the opposite appears to be true. To the extent that *O. leucostigma* is typical of insect herbivores, it appears that interspecific diversity in secondary chemistry does not give rise to fitness trade-offs across plant species (even among plant species which contain different classes of secondary chemicals), and therefore does not drive the evolution of feeding specialization among insect herbivores.

The absence of evidence for chemically-mediated fitness trade-offs across hosts does not argue that fitness trade-offs do not occur. Indeed, there exist many alternative mechanisms which plausibly may give rise to fitness trade-offs across plant species. For example, it is conceivable that adaptation to the nitrogen quantity and/or quality of a particular plant species may necessarily result in loss of adaptation to plant species with different nutritional profiles. Alternatively, selection for phenological synchrony with one plant species may result in loss of synchrony, and therefore lower fitness, on phenologically dissimilar plant species (but see M. Auerbach, this volume). In addition, the limited behavioral repertoire of insect herbivores may give rise to fitness trade-offs across plant species if, for instance, the efficiency with which any one host is located is compromised by simultaneous searching for additional hosts.

Finally, it is important to note that while these results suggest that chemically-mediated fitness trade-offs do not drive the evolution of feeding specialization among insect herbivores, they do not suggest that plant chemistry is unimportant in host specialization. It is possible that plant secondary chemistry functions primarily in reinforcing specialization once it has evolved. Sixty-four of the 90 possible pairwise comparisons in this study revealed no significant genetic correlation in fitness across diets. This suggests that, in the context of selection for increased adaptation to *e.g.*, berberine, genetic variation for fitness on *e.g.*, nicotine may be selectively neutral. Genetic drift then may result in the loss of genetic variation necessary for adaptation to nicotine, particularly if host shifts commonly occur in small, relatively isolated populations. Once such variation is lost, the presence of nicotine in non-host plants may provide strong selection for maintenance of host specialization.

References

Futuyma, D.J. & S.C. Peterson (1985). Genetic variation in the use of resources by insects. *Annu. Rev. Entomol.* **30**: 217-238.

Futuyma, D.J. & T.E. Philippi (1987). Genetic variation and covariation in responses to host plants by *Alsophila pometaria* (Lepidoptera: Geometridae). *Evolution* **41**: 269-279.

Jaenike, J. (1990). Host specialization in phytophagous insects. *Annu. Rev. Ecol. Syst.* **21**: 243-273.

Proc. 8th Int. Symp. Insect-Plant Relationships, Dordrecht: Kluwer Acad. Publ.
S.B.J. Menken, J.H. Visser & P. Harrewijn (eds), 1992

Population genetical evidence for host-race formation in *Yponomeuta padellus*

Léon E.L. Raijmann and Steph B.J. Menken
Institute of Taxonomic Zoology, University of Amsterdam, Amsterdam, The Netherlands

Key words: Allozyme variation, gene flow, speciation, Lepidoptera, Yponomeutidae

The idea of speciation in sympatry (*i.e.*, individuals of two diverging populations being physically capable of encountering one another with moderately high frequency), attracts increasing interest. Geographical barriers are usually considered to be a prerequisite for genetic differentiation and speciation. Many examples provided evidence for this allopatric speciation model strongly advocated by Mayr (1963). Nevertheless, two key observations supported the idea of sympatric speciation. First, the species richness of phytophagous insects in small isolated areas could not be explained purely by the allopatric model. Second, the shift of phytophagous insects onto introduced plants and the rapid adaptation to those hosts, resulted in the idea of sympatric speciation through host-race formation. Research into host races (*i.e.*, populations of a species partially reproductively isolated from other conspecific populations as a direct consequence of adaptation to a specific host), therefore, plays an important role in discussing the probability of speciation in sympatry.

The small ermine moth *Yponomeuta padellus* (L.) was chosen to test the hypothesis of sympatric speciation through host-race formation for two reasons: 1) The results of an integrated study examining the phylogeny, insect-host plant relationships, population structure and speciation in the genus *Yponomeuta* suggest that host shifts and subsequent sympatric speciation have occurred in the evolution of this genus (Menken *et al.*, 1992). 2) *Y. padellus* feeds on plants from various genera of the Rosaceae. Populations on various host species possibly represent host races (Raijmann, 1992).

In this paper, we focus on a sympatric situation in Zuidoostbeemster, in the western part of the Netherlands, consisting of two populations infesting *Crataegus* spp. and *Prunus cerasifera* Ehrh., respectively. The gene flow level between these two populations was estimated indirectly, and compared with the intraspecific gene flow level of five *Crataegus*-infesting populations (the one in Zuidoostbeemster and four others which were scattered over the Netherlands). The main question was to investigate whether these two sympatric populations in Zuidoostbeemster are genetically more differentiated than the geographic *Crataegus* populations.

Methods and materials

In 1990, 5th instar larvae were collected in the field and they were reared on their own food plant in the laboratory. Adults were deep frozen and stored until they were used for allozyme analysis, when allele frequencies were also calculated. They were tested for homogeneity with a G-test. If significant differences between the populations in allele frequencies for one or a few enzyme loci were found, and host-plant specific selection was

209

not involved (Menken *et al.*, 1992), then this indicated a reduced gene flow level between the populations. We determined the inbreeding coefficient *Fst*, *i.e.*, the among population variance in allele frequencies (Wright, 1965; modified by Weir & Cockerham, 1984). From these parameters, we calculated the gene flow levels represented by *Nm* values (where *N* is the effective population size and *m* the migration ratio, *i.e.*, the average number of individuals exchanged between populations per generation). *Nm* values were also calculated according to Slatkin (1985), who demonstrated an almost linear relationship between *Nm* and ln *p(1)*, the average frequency of private alleles (alleles found in only one population).

Results and conclusions

Based on 16 polymorphic loci, the *P. cerasifera* population revealed nine host-specific alleles compared with the Zuidoostbeemster *Crataegus* population, whereas the latter had two host-specific alleles, although frequencies were low (0.01 - 0.09). These alleles, however, may have been absent due to sampling error. Therefore, given the sample size, the probability of not detecting a rare allele in one host was calculated, assuming that the rare allele frequency was the frequency observed on the other host in the same generation. It was found improbable that three out of nine *P. cerasifera* specific alleles ($P \leq 0.01$) and both *Crataegus* specific alleles ($P < 0.01$) would not be detected in the other population. If those nine *P. cerasifera* specific alleles were compared with any of the five other *Crataegus* populations, there were still four restricted to this population. The absence of three alleles could be explained by sampling error. Nevertheless, we can assume that the presence of the hydroxybutyrate dehydrogenase (Z)-allele ($P < 0.01$) indicates that this *Prunus cerasifera* population and these five *Crataegus* populations do not represent a panmictic population.

Forty-five individuals (both females and males) were tested for allele frequency homogeneity. At least 11 and no more than 16 polymorphic loci were compared. It appeared that 30% of all polymorphic loci detected in the five *Crataegus* populations were heterogeneous. The sympatric populations in Zuidoostbeemster had almost twice as many heterogeneous loci.

The average intra-*Crataegus Fst* value, *i.e.*, the gene flow level calculated among the five *Crataegus* populations was 0.021 with a corresponding *Nm* value of 14. The gene flow level between the sympatric Zuidoostbeemster populations, however, was ten times lower: the *Fst* value being 0.155 corresponding with a *Nm* value of about 1.4.

Finally, Slatkin's method of investigating levels of gene flow based on the presence and frequency of private alleles, was used to test the above-mentioned results. The difference in gene flow level between the intra-*Crataegus* and the *Crataegus* - *P. cerasifera* comparison was not as obvious as it was according to the *F*-statistics, though the average level of gene flow among *Crataegus* populations exhibited a somewhat higher value than the *Crataegus* - *P. cerasifera* gene flow level (*Nm* values were 12 and 16, respectively).

In summary, it can be concluded that sympatric populations of *Y. padellus* on different hosts (*i.e.*, *P. cerasifera* and *Crataegus*) show substantial differences in allelic distribution, thus supporting the idea of host-race formation. However, the consistency of this pattern has still to be tested in space and time.

References

Mayr, E. (1963). *Animal Species and Evolution*. Cambridge, Massachusetts: Harvard University Press.

Menken, S.B.J., W.M. Herrebout & J.T. Wiebes (1992). Small ermine moths (*Yponomeuta*): their host relations and evolution. *Annu. Rev. Entomol.* **37**: 41-66.

Raijmann, L.E.L. (1992). Genetical population structure of the small ermine moth *Yponomeuta padellus*. *Proc. Exper. & Appl. Entomol.* **3**: 94-98.

Slatkin, M. (1985). Rare alleles as indicators of gene flow. *Evolution* **39**: 53-65.

Weir, B.S. & C.C. Cockerham (1984). Estimating F-statistics for the analysis of population structure. *Evolution* **38**: 1358-1370.

Wright, S. (1965). The interpretation of population structure by F-statistics with special regard to systems of mating. *Evolution* **19**: 395-420.

Proc. 8th Int. Symp. Insect-Plant Relationships, Dordrecht: Kluwer Acad. Publ.
S.B.J. Menken, J.H. Visser & P. Harrewijn (eds), 1992

Latitudinal trends in oviposition preferences: ecological and genetic influences

J.M. Scriber
Dept of Entomology, Michigan State Univ., East Lansing, USA

Key words: Defoliation, *Papilio glaucus*, *P. canadensis*, sex-linkage

The acceptability of host plants by ovipositing insects can be determined by a balance of various internal and external excitatory factors (Miller & Strickler, 1984). However, insects are generally assumed to be more "hard-wired", with regard to sensory physiology, genetics, detoxification systems and historically constrained phylogenies of host selection due to chemical constraints on their evolution (Feeny, 1991). The divergence of insect preferences for certain plants results from physiological, biochemical, and behavioral adaptations to plant availability, acceptability, and suitability, which in turn are determined by the interactions of the plant nutrients and allelochemicals, the associated community of natural enemies and/or competitors and various abiotic factors such as microclimate or seasonal thermal unit accumulations affecting latitudinal voltinism and feeding specialization patterns (Scriber & Lederhouse, 1992).

We have been examining the genetic basis of differential oviposition preferences in *Papilio* species with the use of hand-paired hybrids and multi-choice oviposition arenas and find that the patterns of host plant preference are generally: 1) repeatable for an individual female on a day-to-day basis, 2) not induced by previous host plant exposure, 3) latitudinally variable with a "step-cline" in the Great Lakes hybrid zone and 4) probably sex-linked on the paternal ("X") chromosome for the *P. glaucus* L. tuliptree preference pattern and for the *P. canadensis* R&J quaking aspen preference pattern (Table 1).

In addition, a local and significant host-plant shift from aspen (*Populus tremuloides* Michx) to ash (*Fraxinus americana* L.) was recently observed in Michigan populations of *P. canadensis*. These "ash shifts" correlated precisely with the areas of heavy defoliation of the Salicaceae, Betulaceae, Rosaceae and Tiliaceae by gypsy moths (*Lymantria dispar* L.), forest tent caterpillars (*Malacosoma disstria* Hübner) or eastern tent caterpillars (*M. americanum* Fabr.) which have occurred simultaneously in Michigan for the last 3-5 years. Ash leaves are basically immune to defoliation by these three species and represent one of the only remaining resources for *P. canadensis* in heavily defoliated areas (Fig. 1). In Michigan, during 1991, more than 620,000 acres exhibited serious defoliation, and another 215,000 acres were sprayed with the microbial pesticide *Bacillus thuringiensis* Berliner. We are attempting to differentiate between 1) direct competitive displacement due to defoliation and 2) indirect competition due to selection of ash for "enemy-free space" or 3) indirect competition due to induced chemical resistance in refoliated aspen, birch, cherry and basswoods as causes for the observed local shifts in *Papilio* oviposition patterns.

Table 1. Three choice oviposition preference tests with *Papilio* spp. Data are percentage of total eggs (mean ± SE)

	Females (n)	*Liridendron tulipifera*	*Prunus serotina*	*Populus tremuloides*
P. rutulus (OR)	5	45.8 ± 11.0	29.0 ± 12.6	25.0 ± 10.4
P. canadensis (AK)	14	52.1 ± 6.6	21.5 ± 4.5	26.4 ± 5.6
P. canadensis (nMI & CAN)	8	36.5 ± 8.7	30.5 ± 6.9	33.0 ± 7.5
P. canadensis (cMI, Isabella Co.)	12	36.9 ± 6.9	30.9 ± 9.5	35.2 ± 7.8
P. glaucus (OH, KY, IL, WV, GA)	39	74.0 ± 2.8	18.4 ± 2.3	7.4 ± 1.4
P. glaucus (sMI, PA, GA)	9	82.0 ± 4.1	12.4 ± 3.3	5.6 ± 1.5
P. glaucus (FL)	3	86.7 ± 7.9	8.4 ± 4.2	4.8 ± 4.8
Hybrid (c x g)	5	84.1 ± 6.9	7.5 ± 6.7	8.4 ± 4.2
Hybrid (g x c)	1	37.1	31.9	31.0

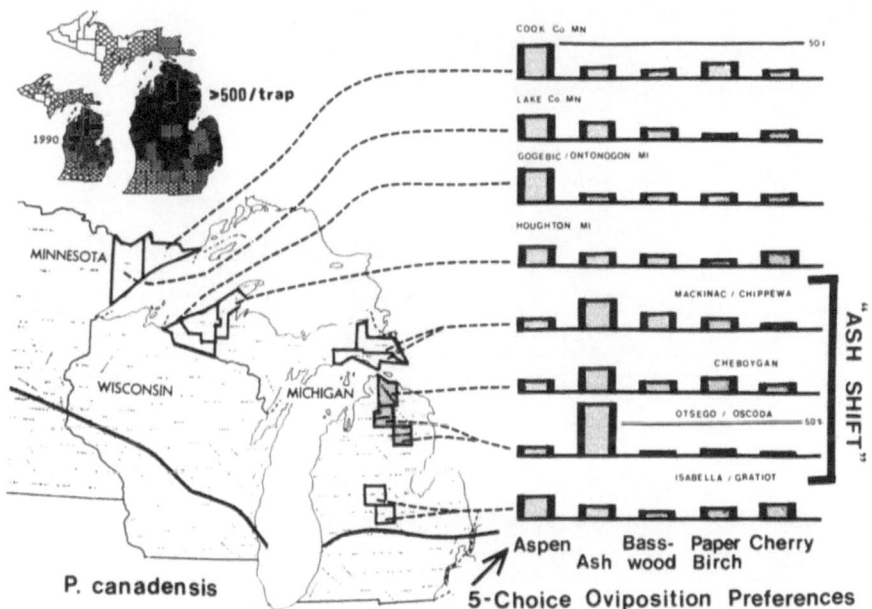

Figure 1. 5-choice oviposition preferences of various *P. canadensis* populations. Note the shift from aspen to white ash in north central Michigan where defoliations by gypsy moths (and forest tent caterpillars) have been most intense.

References

Feeny, P. (1991). Chemical constraints on the evolution of swallowtail butterflies. In: P.W. Price, T.M. Lewinsohn, G.W. Fernandes & W.W. Benson (eds), *Plant-Animal Interactions: Evolutionary Ecology in Tropical and Temperate Regions*, pp. 315-340. New York: John Wiley & Sons.

Miller, J.R. & K.L. Strickler (1984). Finding and accepting host plants. In: W. Bell & R.T. Carde (eds), *Chemical Ecology of Insects*, pp. 127-158. New York: Chapman and Hall.

Scriber, J.M. & R.C. Lederhouse (1992). The thermal environment as a resource dictating geographic patterns of feeding specialization of insect herbivores. In: M.R. Hunter, T. Ohgushi & P.W. Price (eds), *Effects of Resource Distribution on Animal-Plant Interactions*, pp. 429-466. San Diego: Academic Press.

Proc. 8th Int. Symp. Insect-Plant Relationships, Dordrecht: Kluwer Acad. Publ.
S.B.J. Menken, J.H. Visser & P. Harrewijn (eds), 1992

Role of any single herbivore species in the evolution of willows

J. Tahvanainen, H. Roininen and R. Julkunen-Tiitto
Dept of Biology, University of Joensuu, Joensuu, Finland

Key words: Chemical defence, coevolution, phenolic glycosides, specialization

Willows (*Salix* spp.) are an ancient group of deciduous shrubs and trees which has undergone an intensive phase of speciation during and after pleistocene glaciation. The majority of willow species, in total more than 300, occur in the northern temperate and boreal zone. Willow species vary greatly in their morphology and ecology: some are tiny dwarf shrubs and others grow to mid-sized trees; leaves can be glabrous or hairy; habitats occupied by different willow species range from flooded river banks to rocky hills, and their geographical distribution and local abundance are highly variable.

Phenolic secondary metabolites, especially lower-molecular phenolic glycosides, are characteristic of willows. More than 10 different phenolic glycosides are found in highly variable combinations and concentrations in different willow species substantially affecting the rate and type of herbivory on these plants (Julkunen-Tiitto, 1989).

A conspicuously high number of herbivorous species - mammals, birds and insects - regularly feeds on willows. The estimated number of insect species attacking willows in Northern Europe exceeds 500. Although all these mammals and the majority of insects are polyphagous, some insect taxa such as sawflies contain mostly oligophagous or strictly monophagous species. Feeding by such a high number of herbivorous species can be a selective force resulting in adaptive responses in willow plants. Our aim here is to decipher how individual herbivorous species and herbivory as a whole might affect the evolution of willows.

Mode of host utilization by different herbivores

The scale and resolution of host plant utilization by different herbivorous animals varies greatly. Mammals being selective generalists, exploit individual or entire stands of willow and are active year around. The pattern of host exploitation by polyphagous insects resembles that of mammals but is spatially and temporally more restricted. Specialized insects, on the other hand, usually utilize only a minute fraction of potentially suitable plant biomass: for example, bud- and shoot-galling sawflies attack only certain types of shoots of their host species within a very limited period of time.

Mammalian herbivores preferably browse on willow species with a low content of deterrent phenolic glycosides, selecting weakly defended mature plants or plant parts. Insect species associated with willows containing high amounts of phenolic glycosides tend to be more specialized than those feeding on willows with a low content of these defensive metabolites. Some generalist insects appear to prefer stressed, slow-growing plants or plant parts while highly specialized insects, such as galling sawflies, mostly attack vigorously growing, juvenile shoots of their host willow (Roininen, 1991). Population densities of most polyphagous insects seem to fluctuate strongly and randomly whereas the numbers of many specialized insects seem to be more stable. Moreover, many herbivorous insects are so rare that they may not even have an impact on their host. It thus seems obvious that highly variable patterns and intensities of host exploitation occur among willow-feeding herbivores. This apparently results in contrasting selective pressures on willows and complex plant-mediated interactions between different herbivorous animals. For example, the browsing by the mountain hare (*Lepus timidus*), which prefers mature shoots of weakly defended willow species, stimulates the growth of juvenile shoots in the following growing season. Such shoots are unpalatable to hares but provide a source of high-quality buds for the bud-galling sawfly, *Euura mucronata* whose repeated attack on the same willow clone improves the quality of that clone for the subsequent generations of the same insect species (resource regulations). Similar complicated interactions may exist among other willow-feeding animals, and may have an intricate ecological and evolutionary influence on the willow-herbivore association.

Evolution of willow-herbivore interaction

The coevolutionary model predicts that plant defences and herbivore adaptations evolve as a result of reciprocal interactions of a pair of plant and herbivore species (Ehrlich & Raven, 1964). According to Jermy's (1976) model of sequential evolution, herbivorous animals adapt individually to the properties of their host plants, but the feeding by herbivores does not have any marked effect on the evolution of the plant. Many herbivorous species, such as specialized leaf beetles and galling sawflies, are intricately adapted to chemical and other characteristics of certain species of willow. This definitely indicates that willow hosts exert an evolutionary impact on herbivores. But it is more difficult to decipher what the role of herbivory in the evolution of willows is.

Great variability in the feeding patterns among hundreds of willow-feeding animals does not allow any real coevolutionary relationships between individual species of willow and herbivore to become established. On the other hand, the overall impact from herbivory must be strong enough as to cause adaptive responses in willows. For instance, the great variability in the quantity and quality of secondary chemicals, shown to affect the behaviour and performance of many herbivores, is difficult to see just as a random consequence of plant metabolism. We suspect that numerous herbivorous animals make up a composite, diffuse selective force (herbivore load) to which individual willow species may respond in a highly unpredictable manner. Since most of the willow-feeding herbivores are polyphagous, favouring plants and plant parts with a low chemical defence, the diffuse herbivore pressure ought to be stronger on weakly defended willow species. Consequently, there should be a general trend in willow evolution towards defences, such as phenolic glycosides, which are effective against generalist herbivores. Interestingly, many willow species, contain low amounts of phenolic glycosides and are consequently palatable to most mammalian and generalist insect herbivores, *e.g.*, *Salix phylicifolia* L., *S. caprea* L., *S. cinerea* L. and *S. aurita* L., which belong to the youngest group of the genus. Presumably these willows have speciated after the last glaciation; they seem to have effectively colonized pioneer plant communities devoid of established herbivore fauna. Gradually, herbivore communities have been building up resulting in increasing herbivore pressure on these weakly defended pioneer willows.

References

Ehrlich, P.R. & P.H. Raven (1964). Butterflies and plants: A study in coevolution. *Evolution* **18**: 586-608.

Jermy, T. (1976). Insect-host-plant relationship - coevolution or sequential evolution? *Symp. Biol. Hung.* **16**: 109-113.

Julkunen-Tiitto, R. (1989). Distribution of certain phenolics in Salix species (Salicaceae). *Univ. Joensuu, Publications in Sciences* **15**: 1-29.

Roininen, H. (1991). The ecology and evolution of the host plant relationships among willow-feeding sawflies. *Univ. Joensuu, Publications in Sciences* **23**: 1-21.

Host-Plant Resistance and
Application of
Transgenic Plants

Proc. 8th Int. Symp. Insect-Plant Relationships, Dordrecht: Kluwer Acad. Publ.
S.B.J. Menken, J.H. Visser & P. Harrewijn (eds), 1992

Potential of plant-derived genes in the genetic manipulation of crops for insect resistance

Angharad M.R. Gatehouse, Vaughan A. Hilder, Kevin Powell, Donald Boulter and John A. Gatehouse
Dept of Biological Sciences, Plant Molecular Biology Group, University of Durham, Durham, UK

Key words: Antimetabolites, lectins, protease inhibitors, transgenic plants

Introduction

The plant kingdom provides a rich and diverse source of secondary compounds. A protective role against various pests, pathogens and competitors for many of these has been established and in recent years the utilisation of such compounds in crop protection, either by conventional plant breeding or by genetic engineering has been, and is being, investigated (Gatehouse *et al.*, 1990).

Since the major criteria on which crops have been selected in the past have been based primarily on high and dependable yields, nutritional value (including low mammalian toxicity) and, where necessary, adaptation to certain environmental conditions, few cultivated species have retained the degree of resistance exhibited by their wild relatives (Feeny, 1976). Mankind has sought to redress this balance, to some extent, by a heavy reliance upon the use of chemical insecticides, but despite an annual expenditure of approximately 6 billion U.S. dollars it is estimated that 37% of all crop production is lost world-wide to pests and diseases, with 13% lost to insects. Chemical control methods often result in the rapid build-up of resistance by insect pests to such compounds, and it is not uncommon for examples of resistance in a major pest to be noted within the first year of field use (Metcalf, 1986). Furthermore, the indiscriminate use of pesticides has, in some cases, exacerbated the problem of insect herbivory - where elimination of a wide range of predatory species along with the primary pests has resulted in secondary pests becoming primary pests with even more devastating effects. An example of this is illustrated by the brown plant hopper (*Nilaparvata lugens* Stål), a major field pest of rice (Heinrichs & Mochida, 1983).

Although complete elimination of chemical control agents is neither a realistic nor feasible possibility in the near future, a significant reduction in their use is both necessary and desirable. The enlightened agronomist is, therefore, looking towards an integrated pest control programme comprising a combination of practices which might include the judicious use of pesticides, crop rotation, field sanitation, use of pest-free seeds, but above all exploiting inherently resistant plant varieties (Meiners & Elden 1978).

Breeding crops for resistance offers many advantages over an almost exclusive reliance upon chemicals, the most obvious ones being:
(i) it provides season long protection,
(ii) insects are always treated at the most sensitive stage,
(iii) protection is independent of the weather,
(iv) it involves no application costs
(v) it protects plant tissues which are difficult to treat using insecticides. For example

chemical insecticides are inadequate at controlling larvae of *Ceutorhynchus assimilis* Marsh. (pest of oil seed rape) since they attack the developing ovules within the pods,

(vi) only crop-eating insects are exposed,

(vii) the material is confined to the plant tissues expressing it and therefore does not leach into the environment,

(viii) the active factor is biodegradable and choice of suitable genes/gene products can ensure it is not toxic to man and animals,

(ix) consumer acceptability. In recent years there has been much concern over the presence of pesticide residues in food crops; inherently resistant crops should offer the consumer the alternative of produce containing a well defined and characterised gene product as opposed to unspecified pesticide residues. However, there is also concern regarding the use of genetically modified plants in many Western countries, and much attention should be paid to public debate stressing the benefits.

(x) considerable financial savings (Fig. 1). These pest-resistant crops might thus offer economic advantage to the farmer which should ultimately benefit the consumer, depending on the prices to be set for resistant crop seed.

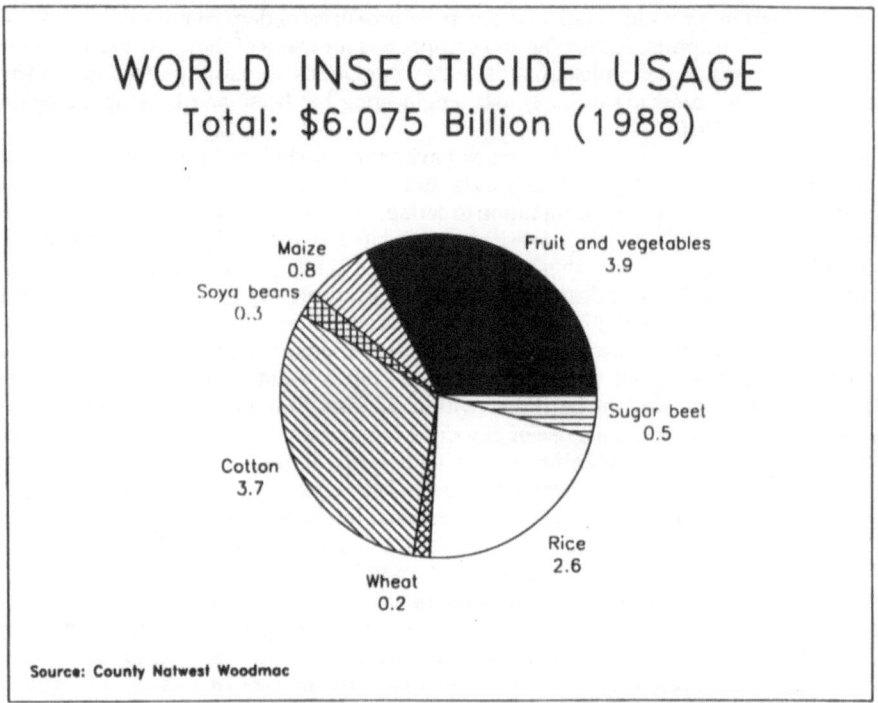

Figure 1. Total world insecticide usage, divided according to crops (1988 figures).

Plant genetic manipulation as part of a breeding programme can make a significant contribution in the production of insect-resistant crops, and offers two major advantages over using conventional plant breeding alone. It enables the desired gene(s) to be transferred to the recipient plants without the co-transfer of undesirable characteristics, thereby greatly

222

speeding up the development time of new varieties, and also allows the transfer of genes across incompatibility barriers - genes can be introduced from sources which are wholly unavailable to conventional plant breeding. Constraints to this new technology include the inability to transform and regenerate some specific crop cultivars, the identification and production of 'useful' genes for transfer, regulatory barriers, and consumer acceptability of such crops. These, however, are being eroded as it is now possible to successfully transform and regenerate a large number of different crop species of importance in the developed and developing world. Also, the number of different 'insecticidal' genes which have been transferred to foreign plant species to confer varying levels of resistance is gradually increasing.

This paper attempts to address the progress to date in the isolation and transformation of insect resistance genes of plant origin. The reader is also referred to a recent review by Gatehouse *et al.* (1991a).

Potential candidates in engineering insect resistance

Protease inhibitors. A protective role for protease inhibitors was suggested by Ryan and co-workers (Green & Ryan, 1972) when they demonstrated that damage to the leaves of certain plant species (notably tomato and potato), either by insect feeding or mechanical wounding, induced the synthesis of protease inhibitors. This induction occurs not only in the attacked leaf but throughout the plant. Production of these inhibitors was thought to be controlled by an oligosaccharide wound hormone, proteinase inhibitor-inducing factor (PIIF), which is released from the damaged leaves and transported to other leaves where it initiates synthesis and accumulation of inhibitors (Shumway *et al.*, 1976; Walker-Simmons & Ryan 1977; Brown *et al.*, 1985). More recent results have, however, suggested a more complex signalling mechanism with either oligosaccharides (local responses) or an 18-amino acid peptide, systemin (systemic responses) stimulating the production of jasmonic acid, which

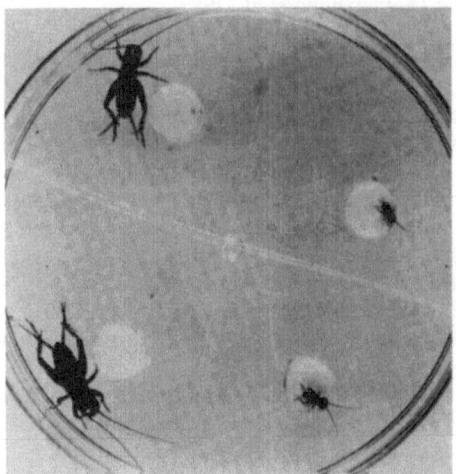

Figure 2. Effects of POT-II virus when fed to 6 week old black field crickets (*Teleogryllus commodus*). Left, crickets fed control diet; right, crickets fed POT-II at 0.33% (w/v). (Fig. courtesy of E.P.J. Burgess DSIR, Auckland, New Zealand.)

causes activation of gene expression (Pearce et al., 1991; Farmer & Ryan, 1992) and thus inhibitor production. Direct evidence for the role of protease inhibitors in providing protection in the 'field' was first provided by Gatehouse et al. (1979) who demonstrated that elevated levels of these inhibitors in one variety of cowpea were partly responsible for the observed resistance of the seeds to the major storage pest of this crop, the bruchid *Callosobruchus maculatus* F.. This particular trait was later exploited by conventional plant breeding whereby bruchid resistance was transferred to an agronomically improved background (Redden et al., 1983).

There have also been many examples of protease inhibitors being active against certain insect species both in *in vitro* assays against insect gut proteases (Birk et al., 1963; Applebaum, 1964) and in *in vivo* artificial diet bioassays, as illustrated in Fig. 2 (Steffens et al., 1978; Gatehouse et al., 1979; Broadway & Duffy, 1986; Gatehouse & Hilder, 1988; Burgess et al., 1991). A systematic study of a wide range of different inhibitors has been made in an attempt to find a kinetic parameter which would be useful in predicting the potential of any inhibitor to act as a resistance factor to the larvae of grass grubs (*Costelytra zealandica* (White)), a major insect pest in New Zealand (Christeller & Shaw, 1989); the dissociation constant of the inhibitor: protease complex was suggested as such a parameter. This strategy may well be a valuable tool in choosing the most appropriate enzyme inhibitors to control specific insect pests, though much work will be required to validate this, since food is consumed by organisms, not by enzymes (Hilder et al., 1991).

The mechanism of antimetabolic action of protease inhibitors is not yet fully elucidated; direct inhibition of digestive enzymes is unlikely to be the only effect and possibly a situation analogous to the effect of some protease inhibitors on mammals, where the major deleterious effect is loss of nutrients through pancreatic hypertrophy and overproduction of digestive enzymes (Liener, 1980), holds in insects. That they affect the nutritional biochemistry of C. *maculatus* has been clearly demonstrated since methionine supplementation has been shown to overcome the antinutritional effects of the cowpea trypsin inhibitor (CpTI) (Gatehouse & Boulter, 1983). Broadway suggests that they are only one part of the complex interaction between plant nutritional value and the insects' digestive physiology (Broadway et al., 1986); this may explain why there are some cowpea varieties with high levels of CpTI which bruchids are able to infest (Xavier-Filho et al., 1989).

Insect-resistant transgenic plants expressing protease inhibitors

The first gene of plant origin to be successfully transferred to another plant species resulting in enhanced insect resistance was that isolated from cowpea, encoding a trypsin/trypsin inhibitor (Hilder et al., 1987). This protein was considered to be a particularly suitable candidate for transfer to other plant species via genetic engineering for a number of reasons; it had been shown to be an effective antimetabolite against a range of field and storage pests including members of the Lepidoptera, Coleoptera and Orthoptera (Table 1); there was, however, no evidence that it had any deleterious effects upon mammals. It is a small polypeptide of about 80 amino acids, belonging to the Bowman-Birk inhibitor family (Gatehouse et al., 1980); homologous sequences are encoded by a moderately-repetitive gene family in the cowpea genome (Hilder et al., 1989).

A full-length cDNA clone encoding a trypsin/trypsin inhibitor from cowpea was produced and the coding sequence was placed under the control of a CaMV 35S promoter in the final construct produced for transfer to plants (Fig. 3). The construct employed the *Agrobacterium tumefaciens* Ti plasmid binary vector pROK2; a terminator from the nopaline synthetase gene was placed 3' to the coding sequence, and the construct also contained a *nos-aph(3')II* (usually referred to as *nos-neo*) gene to allow transformants to be selected on the basis of kanamycin resistance. The vector was mobilised into *Agrobacterium*, and the bacteria were used to infect

Table 1. Insect pests against which cowpea trypsin inhibitors (CpTI) are effective

Order	Insect pest	Primary crops attacked
Field pests		
Lepidoptera	*Heliothis virescens*[*]	tobacco, cotton
	Helicoverpa zea[*]	maize, cotton, beans, tobacco
	Helicoverpa armigera	cotton, beans, maize, sorghum
	Spodoptera littoralis[*]	maize, rice, cotton, tobacco
	Chilo partellus	maize, sorghum, sugarcane, rice
	Autographa gamma[*]	sugarbeet, lettuce, cabbage, beans, potato
	Manduca sexta[*]	tomato, tobacco, potato
Orthoptera	*Locusta migratoria*	polyphagous but preference for wild & cultivated grasses
Coleoptera	*Diabrotica undecimpunctata*	maize
	Costelytra zealandica	grasses, clover
	Anthonomus grandis	cotton
Storage pests		
Coleoptera	*Callosobruchus maculatus*	cowpea, soyabean
	Tribolium confusum	most flours

[*]Insects to which CpTI transgenic tobacco plants exhibit significantly enhanced levels of resistance.

tobacco leaf discs, by standard protocols; subsequent production of rooted plants after selection of regenerating shoots on kanamycin-containing media also followed normal procedures. By taking cuttings from the original transformants, and rooting them, numbers of clonal plants sufficient for insect bioassay could be produced from each of the original transformants. The transformed plants were shown to be expressing CpTI in the leaves at levels varying from undetectable to nearly 1% of total soluble protein by a dot blot immunoassay; this range of values has subsequently been found to be fairly typical for plant genes driven from the CaMV promoter. In control plants transformed with a construct where the coding sequence of CpTI had been inserted in the incorrect (*i.e.*, 3'-5') orientation relative to the CaMV promoter, no expression of the protein was detected. The expression of CpTI was confirmed in the 'correct' transformants by Western blotting, and by a direct *in vitro* assay for inhibition of bovine trypsin. The former technique showed that tobacco was capable of processing the precursor CpTI polypeptide encoded by the inserted coding sequence to a polypeptide resembling native CpTI on SDS-PAGE; other plant proteins have been shown to be correctly processed in transgenic tobacco plants (Ellis *et al.*, 1988; Cheung *et al.*, 1988). The latter technique showed that the CpTI synthesised in transgenic tobacco possessed its normal functional integrity. The lack of complication in obtaining relatively high levels of expression of functional CpTI in tobacco illustrates the advantage of expressing plant proteins in transgenic plants; there are no problems with codon usage, mRNA stability, protein processing, etc. that have been observed to occur if attempts to express in plants proteins derived from non-plant sources.

The critical test was to ascertain whether the CpTI producing tobacco plants exhibit enhanced levels of resistance/tolerance to insect infestation compared to the control plants. In the first instance bioassay of clones of selected transformants was carried out using first

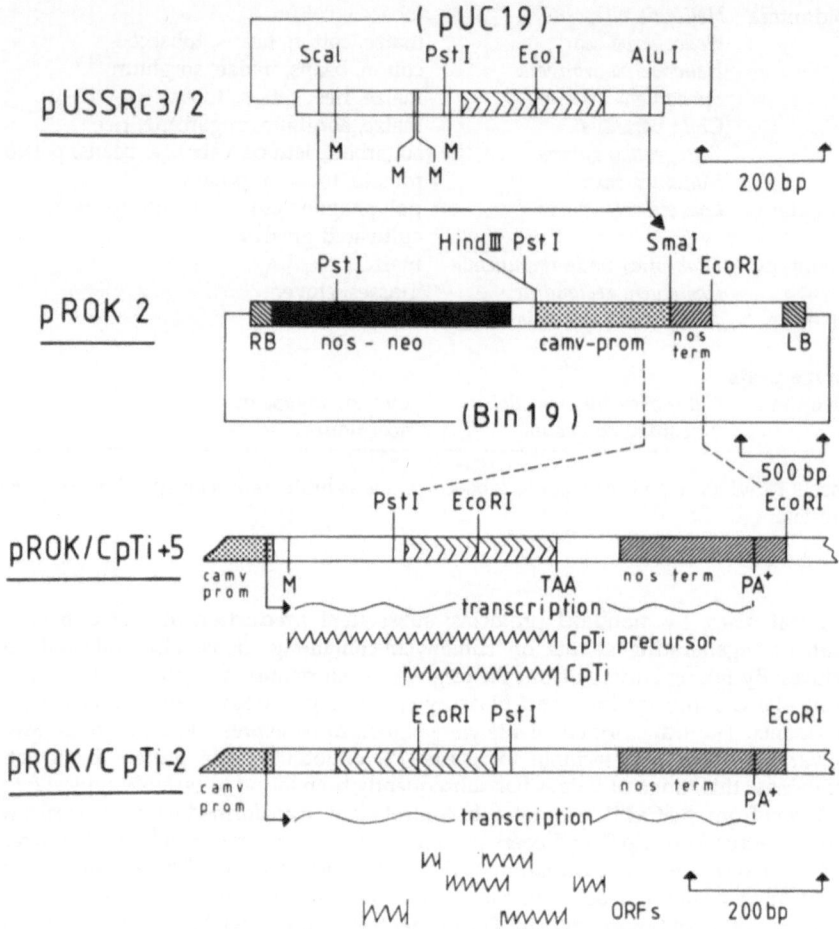

Figure 3. Construction of a CpTI expression vector for plant transformation (Hilder *et al.*, 1987). The CpTI cDNA pUSSRc3/2, containing a complete mature CpTI-coding sequence (indicated as >>>; in frame, initiator codons are marked M), was restricted with Alu I and Sca I and ligated into the Sma I site of the expression vector pROK 2. Clones with the coding sequence in the correct orientation relative to the promoter (pROK/CpTI + 5) and in the incorrect orientation (pROK/CpTI-2) were generated. Transcripts generated by the clone with the CpTI-coding sequence in the correct orientation will be translated to produce a CpTI precursor polypeptide; transcripts from the clone with the CpTI-coding sequence in the incorrect orientation contain six short open reading frames.

instar larvae of the tobacco budworm (*Heliothis virescens* F.); this insect was chosen as it is a serious pest of tobacco, cotton and maize and thus represents a pest of major economic importance. With these clonal plants, and subsequent generations derived from their self-set seed, the CpTI expressing plants showed only minor damage compared to the control plants (Fig. 4), which in some instances were reduced to stalks. Although the larvae begin to feed on the CpTI-expressing plants, causing some limited damage, they either die or fail to develop as they would on control plants. These observations are consistent with a mechanism

Figure 4. Bioassay of control and CpTI-expressing transgenic tobacco plants against larvae of of *Manduca sexta* (tobacco hornworm). Left, a control showing complete destruction; right a transgenic CpTI expressor, showing minimal damage.

of CpTI toxicity initially proposed by Gatehouse & Boulter (1983). This protection afforded by CpTI has subsequently been demonstrated for other lepidopteran pests including *H. zea*, *Spodoptera littoralis* (Boisd.), and *Manduca sexta* (Johanssen) (Fig. 4; Gatehouse et al., 1991a). Statistical analysis of the bioassay in terms of plant damage by leaf area, and insect survival and biomass, confirmed the highly significant protection afforded by CpTI. Recent trials carried out in California showed that expression of CpTI in tobacco afforded significant protection in the field against *Helicoverpa zea* (Boddie); results from these trials closely resembled those obtained previously in trials carried out under controlled environmental conditions in growth chambers (Hoffman et al., 1991). Unfortunately it is not possible to test the efficacy of CpTI against coleopteran pests in transgenic tobacco plants as most coleopterans of economic interest do not appear to attack tobacco. However, other species of CpTI expressing plants which are susceptible to coleopteran attack, including potato, oil seed rape and lettuce, have now become available.

Despite CpTI being an effective antimetabolite against a wide spectrum of insect pests (Table 1) recent mammalian feeding trials incorporating the purified protein at levels of 10% of the total protein have failed to demonstrate toxicity, at least in the short term (Pusztai et al., 1992). CpTI has been demonstrated to directly inhibit insect gut proteases, although this may not be its only site of action. It is not effective against the acidic proteases of the mammalian stomach. The differences in the organisation of the insect and mammalian digestive systems and the vast range of secondary compounds available from plants means that it should not be that difficult to find compounds within plants which are toxic to herbivorous insects but not to mammals (Hilder & Gatehouse, 1990; Hilder et al., 1991).

Not only have the genes encoding protease inhibitors isolated from cowpea been shown to confer resistance when expressed in tobacco (Hilder et al., 1987) but the tomato inhibitor II gene, when expressed in the same model plant, has also been shown to confer insect resistance (Johnson et al., 1989). The tomato, and potato inhibitor II gene encodes a trypsin inhibitor (with some chymotrypsin inhibitory activity) and expression of this gene in tobacco on the constitutive promoter CaMV 35S resulted in increased levels of protection against the larvae of the lepidopteran M. sexta. These workers showed that the decrease in larval weight was roughly proportional to the level of protease inhibitor II being expressed; at levels over 100 mg of the foreign protein per g of tissue, larval growth was severely retarded, whereas at lower levels (ca. 50 mg/g tissue) growth was retarded to a lesser degree. Several of the transgenic plants were shown to contain inhibitor levels over 200 mg/g tissue; these levels are within the range that is routinely induced by wounding leaves of either tomato or potato plants (Graham et al., 1986). However, tobacco plants expressing tomato inhibitor I (specific for chymotrypsin) at levels of 130 mg/g had no deleterious effects upon larval development.

Lectins

Many lectins belong to a homologous family of proteins based on an amino acid chain of approximately 220 residues, e.g., soyabean lectin (Lotan et al., 1975), although totally different sequence types have been shown to have similar functional properties. Toxicity of this type of protein in mammals (Evans et al., 1973; Liener, 1980) and birds (Jayne-Williams & Burgess, 1974) has been well documented, and detailed studies on their mechanism of toxicity have been carried out (Pusztai et al., 1979; King et al., 1980a, 1980b; De Oliveira 1986).

The first report of lectin toxicity towards insects was by Janzen et al. in 1976, who demonstrated that purified lectin from Phaseolus vulgaris L. (PHA) was toxic to C. maculatus when added to an artificial diet. Gatehouse et al. (1984) subsequently confirmed the toxicity of these seed lectins towards developing larvae and on the basis of indirect immunofluorescence investigations using monospecific antisera for globulin lectins showed that the molecules, when ingested, bound to the midgut epithelial cells. From these observations it was suggested that the mechanism of lectin toxicity is analogous to that believed to occur in rat, namely that the ingested lectin causes disruption of the midgut epithelial cells. This leads to a breakdown of nutrient transport, and also facilitates the absorption of potentially harmful substances. In pest species of P. vulgaris, such as Acanthoscelides obtectus (Say), the lectin molecules do not bind to these epithelial cells, and thereby, presumably, are unable to cause the harmful effects previously demonstrated (Gatehouse et al., 1989). More recent results have suggested that the toxicity of P. vulgaris lectin preparations towards bruchids is due to a contaminating α-amylase inhibitor which, while it is without lectin functionality, has sequence homology to the lectins (Altabella & Chrispeels, 1990); however, this observation appears to be specific to P. vulgaris, and is not the case with lectin preparations from other species. The lectins present in the mature seeds of the winged bean Psophocarpus tetragonolobus DC have been shown to be involved in seed resistance to non-pest species (Gatehouse et al., 1991b).

Not only have certain lectins been identified as being insecticidal compounds by demonstrating a protective role within the seed, as typified by the bean lectins and winged bean lectins (Janzen et al., 1976; Gatehouse et al., 1984, 1991b) but several studies have also been carried out involving the screening of purified lectins against insect pests in an attempt to identify insecticidal proteins (Murdoch et al., 1990) and hence isolate the genes encoding them for subsequent plant transformation. Shukle & Murdoch (1983) demonstrated that the lectin from soyabean was toxic to the developing larvae of the lepidopteran Manduca sexta when tested in bio-assay, and recently wheat germ agglutinin has been shown to have an inhibitory effect on development of the European corn borer (*Ostrinia nubilalis* (Hübner)) and the Southern corn root worm (*Diabrotica undecimpunctata* (Barber)), two important pests of maize (Czapla & Lang, 1992). The disadvantage, however, in using lectins from *P. vulgaris*, winged bean, soyabean or wheat germ is that they have also been shown to exhibit mammalian toxicity (Pusztai *et al.*, 1979; Higuchi *et al.*, 1984). This therefore will limit their potential use in the production of transgenic plants, particularly in crop plants.

The lectin isolated from pea, *Pisum sativum* L., on the other hand, has been shown to be innocuous to mammals as it is readily broken down in the gut (Begbie & King, 1985). However, it was shown to be insecticidal when the purified protein was incorporated into artificial seeds (Boulter & Gatehouse, 1986). A lectin purified from snowdrop (Van Damme *et al.*, 1987) has also been shown to be insecticidal, but does not exhibit mammalian toxicity (Pusztai *et al.*, 1991).

Not only have certain lectins been shown to be insecticidal to members of the orders Coleoptera and Lepidoptera, but recent results have demonstrated their toxicity to some homopteran pests. The snowdrop lectin (GNA) and wheatgerm agglutinin (WGA) were shown to be highly insecticidal to the brown plant hopper (*Nilaparvata lugens*), a major economic pest of rice, when incorporated into an artificial liquid diet medium at 0.1% (wt/vol) (Powell *et al.*, 1992); similar results were also obtained with the green leaf hopper (*Nephotettix*). In both cases survival on these lectin containing diets followed that of starved nymphs (Fig. 5). As with members from other insect orders, not all lectins tested against these sap-sucking insects were found to be entomocidal. This is the first direct evidence of proteins being effective against Homoptera.

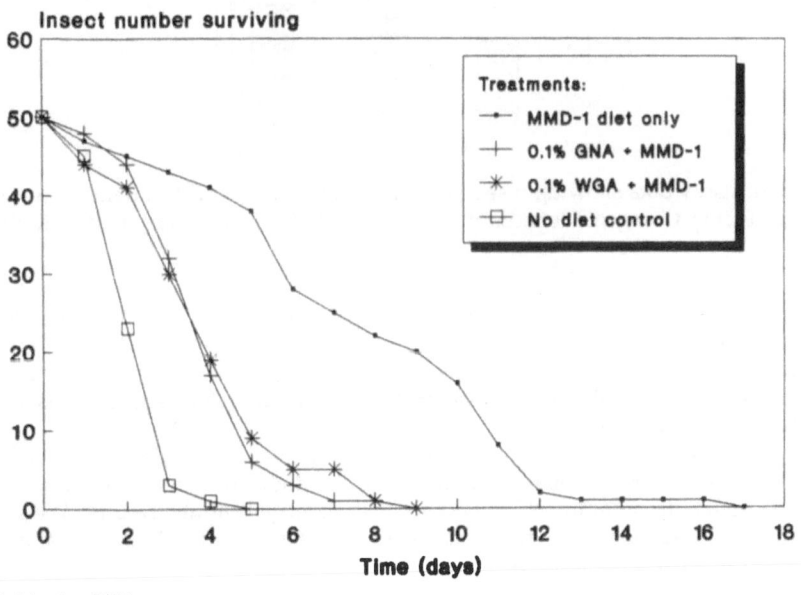

Third Instar BPH

Figure 5. Effects of wheat germ agglutinin (WGA) and *Galanthus nivalis* agglutinin (GNA) on survival of third instar brown plant hopper (*Nilapavarta lugens*). The lectins were each incorporated into an artificial diet at 0.1% (w/v).

Transgenic plants expressing lectin

A gene encoding the pea lectin (P-Lec) has been expressed at high levels in transgenic tobacco plants from the CaMV-35S promoter by *Agrobacterium* transformation (Edwards, 1988). P-Lec expressing plants were then tested in bio-assay for enhanced levels of resistance/tolerance to *H. virescens*. The results showed that not only was larval biomass significantly reduced on the transgenic plants, compared to those from control plants, but leaf damage, as determined by computer aided image analysis, was also reduced (Boulter *et al.*, 1990). Transgenic tobacco plants containing both the cowpea trypsin inhibitor gene (CpTI) and the P-Lec gene were obtained by cross-breeding plants derived from the two primary transformed lines. These plants expressing the two insecticidal genes, each at approximately 1% of the total soluble protein, were also screened for enhanced resistance to *H. virescens*. Although the insecticidal effects of the two genes was not synergistic they were additive (Fig. 6), with insect biomass on the double expressers being only 11% compared to those from

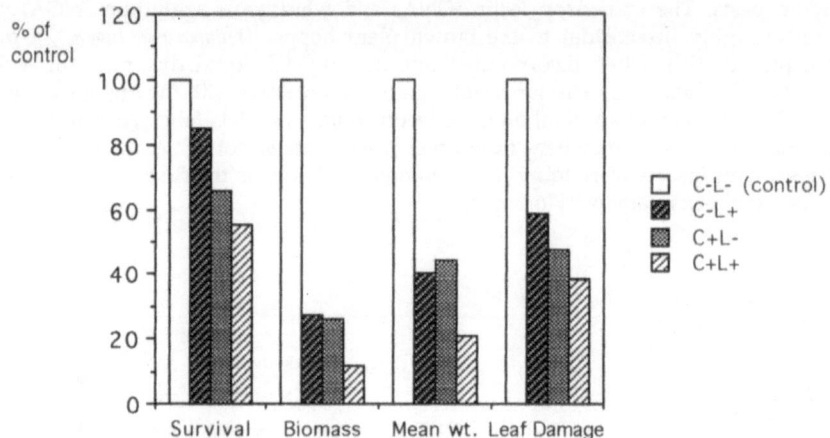

Figure 6. Bioassay data showing insect survival and biomass, and leaf damage of CpTI- and P-Lec-expressing transgenic tobacco plants against *Heliothis virescens*. (After Boulter *et al.*, 1990).

control plants and 50% of those from plants expressing either CpTI or P-Lec alone. Leaf damage was also the least on the double expressing plants. Not only is this the first example of a lectin gene being successfully transferred to another plant species resulting in enhanced insect resistance, but it is also the first demonstration of additive protective effects of different plant-derived insect resistance genes.

The gene encoding the snowdrop lectin has also been successfully transferred to other plants, including tobacco, where preliminary results indicate it too has conferred partial resistance, both in terms of insect biomass and leaf damage (unpubl. results). We are now in the process of expressing this gene in rice plants in order to control loss and damage of this crop caused by brown plant hopper.

Summary

The use of genetically engineered crops expressing plant-derived insect resistance genes appears to be a viable means of producing crops with significantly enhanced levels of resistance. The strategy employed by the authors is to produce gene packages whose products are targeted to different biochemical and physiological processes within the insect. In this way it is hoped to provide a multimechanistic form of resistance which can be tailored to the different crops and prevailing insect pests at a given time. The authors have recently demonstrated that tobacco plants can express plant-derived 'foreign' proteins to moderately high levels with no obvious deleterious effects, and thus this proposed strategy should be possible in other crops (Hilder et al., 1991; Hilder & Gatehouse, 1991).

References

Altabella, T. & M.J. Chrispeels (1990). Tobacco plants transformed with the bean αai gene express an inhibitor of insect α-amylase in their seeds. *Plant Physiol.* **93**: 805-810.

Applebaum, S.W. (1964). Physiological aspects of host specificity in the Bruchidae - I. General considerations of developmental compatibility. *J. Insect Physiol.* **10**: 783-788.

Begbie, R. & T.P. King (1985). The interaction of dietary lectin with porcine small intestine and the production of lectin-specific antibodies. In: T.C. Bog-Hansen & J. Breborowicz (eds), *Lectins*, Vol. IV, pp. 15-17. Berlin: Walter de Gruyter and Co.

Birk, Y., A. Gertler & S. Khalef (1963). Separation of a *Tribolium*-protease inhibitor from soybeans on a calcium phosphate column. *Biochim. Biophys. Acta* **67**: 326-328.

Boulter, D. & A.M.R. Gatehouse (1986). Isolation of genes involved in pest and disease resistance. In: E. Magnien (ed.), *Biomolecular Engineering in the European Community*, pp. 715-725. Dordrecht: Martinus Nijhoff.

Boulter, D., G.A. Edwards, A.M.R. Gatehouse, J.A. Gatehouse & V.A. Hilder (1990). Additive protective effects of incorporating two different higher plant derived insect resistance genes in transgenic tobacco plants. *Crop Protection* **9**: 351-354.

Broadway, R.M. & S.S. Duffy (1986). Plant proteinase inhibitors: mechanism of action and effect on the growth and digestive physiology of larval *Heliothis zea* and *Spodoptera exigua*. *J. Insect Physiol.* **32**: 827-833.

Broadway, R.M., S.S. Duffey, G. Pearce & C.A. Ryan (1986). Plant proteinase inhibitors: A defense against herbivorous insects? *Entomol. exp. appl.* **41**: 33-38.

Brown, W.E., K. Takio, K. Titani & C.A. Ryan (1985). Wound-induced trypsin inhibitor in alfalfa leaves: identity as a members of the Bowman-Birk inhibitor family. *Biochemistry* **24**: 2105-2108.

Burgess, E.P.J., P.S. Stevens, G.K. Keen, W.A. Laing & J.T. Christeller (1991). Effects of protease inhibitors and dietary protein level on the black field cricket *Teleogryllus commodus*. *Entomol. exp. appl.* **61**: 123-130.

Cheung, A.Y., L. Gogorad, M. van Montagu & J. Schell (1988). Relocating a gene for herbicide tolerance: a chloroplast gene is converted into a nuclear gene. *Proc. Natl Acad. Sci. USA* **85**: 391-395.

Christeller, J.T. & B.D. Shaw (1989). The interaction of a range of serine proteinase inhibitors with bovine trypsin and *Costelytra zealandica* trypsin. *Insect Biochem.* **19**: 233-239.

Czapla, T.H. & B.A. Lang (1992). Effect of plant lectins on the larval development of European corn borer (Lepidoptera: Pyralidae) and Southern corn rootworm (Coleoptera: Chrysomelidae). *J. Econ. Entomol.* (in press).

De Oliveira, J.T.A. (1986). Seed lectins; the effects of dietary *Phaseolus vulgaris* lectin on the general metabolism of monogastric animals. PhD dissertation, Aberdeen University.

Edwards, G.A. (1988). Plant transformation using an *Agrobacterium tumefaciens* Ti-plasmid vector system. PhD dissertation, Durham University.

Ellis, J.R., A.H. Shirsat, A. Hepher, J.N. Yarwood, J.A. Gatehouse, R.R.D. Croy & D. Boulter (1988). Tissue specific expression of a pea legumin gene in seeds of *Nicotiana plumbaginifolia*. *Plant Mol. Biol.* **10**: 203-214.

Evans, R.J., A. Pusztai, W.B. Watt & D.H. Bauer (1973). Isolation and properties of protein fractions from navy beans (*Phaseolus vulgaris*) which inhibit growth of rats. *Biochim. Biophys. Acta.* **303**: 175-184.

Farmer, E.E. & C.A. Ryan (1992). Octadecanoid precursors of jasmonic acid activate the synthesis of wound-inducible proteinase inhibitors. *Plant Cell* **4**: 129-134.

Feeny, P.P. (1976). Plant apparency and chemical defence. *Rec. Adv. Phytochem.* **10**: 1-40.

Gatehouse, A.M.R. & D. Boulter (1983). Assessment of the anti-metabolic effects of trypsin inhibitors from cowpea (*Vigna unguiculata*) and other legumes on development of the bruchid beetle *Callosobruchus maculatus*, *J. Sci. Food Agric.* **34**: 345-350.

Gatehouse, A.M.R. & V.A. Hilder (1988). Introduction of genes conferring insect resistance. In: *Proceedings of Brighton Crop Protection Conference*, Vol. 3, pp. 1234-1254. Suffolk UK: Lavenham Press Ltd.

Gatehouse, A.M.R., J.A. Gatehouse, P. Dobie, A.M. Kilminster & D. Boulter (1979). Biochemical basis of insect resistance in *Vigna unguiculata*, *J. Sci. Food Agric.* **30**: 948-958.

Gatehouse, A.M.R., J.A. Gatehouse & D. Boulter (1980). Isolation and characterisation of trypsin inhibitors from cowpea. *Phytochemistry* **19**: 751-756.

Gatehouse, A.M.R., F.M. Dewey, J. Dove, K.A. Fenton & A. Pusztai (1984). Effect of seed lectin from *Phaeolus vulgaris* on the development of larvae of *Callosobruchus maculatus*; mechanism of toxicity, *J. Sci. Food Agric.* **35**: 373-380.

Gatehouse, A.M.R., S.J. Shackley, K.A. Fenton, J. Bryden & A. Pusztai (1989). Mechanism of seed lectin tolerance by a major insect storage pest of *Phaseolus vulgaris*, *Acanthoscelides obtectus*, *J. Sci. Food Agric.* **47**: 269-280.

Gatehouse, A.M.R., H.B. Minney, P. Dobie & V.A. Hilder (1990). Biochemical resistance to bruchid attack in legumes; investigation and exploitation. In: K. Fujii, A.M.R. Gatehouse, C.D. Johnson, R. Mitchel & T. Yoshida (eds), *Bruchids and Legumes: Economics, Ecology and Coevolution*, pp. 241-256. Dordrecht: Kluwer Academic Publishers.

Gatehouse, J.A., V.A. Hilder & A.M.R. Gatehouse (1991a). Genetic engineering of plants for insect resistance. In: D. Grierson (ed.), *Plant Genetic Engineering (Plant Biotechnology Series*, Vol. 1), pp. 105-135. London: Blackie & Son; New York: Chapman & Hall.

Gatehouse, A.M.R., D.S. Howe, J.E. Flemming, V.A. Hilder & J.A. Gatehouse (1991b). Biochemical basis of insect resistance in winged bean seeds (*Psophocarpus tetragonolobus*) seeds. *J. Sci. Food Agric.* **55**: 63-74.

Graham, J.S., G. Hall, G. Pearce & C.A. Ryan (1986). Regulation of synthesis of proteinase inhibitors I and II mRNAs in leaves of wounded tomato plants. *Planta* **169**: 399-405.

Green, T.R. & C.A. Ryan (1972). Wound-induced proteinase inhibitor in plant leaves: a possible defense mechanism asgainst insects. *Science* **175**: 776-777.

Heinrichs, E.A. & O. Mochida (1983). From secondary to major pest status: The case of insecticide-induced rice brown planthopper, *Nilaparvata lugens* resurgence. *Proc. XV Pacific Science Congress, New Zealand.*

Higuchi, M., I. Tsuchiga & K. Iwai (1984). Growth inhibition and small intestinal lessions in rats after feeding with isolated winged bean lectin. *Agric. Biol. Chem.* **48**: 695-701.

Hilder, V.A. & A.M.R. Gatehouse (1990). In: G.W. Lycett & D. Grierson (eds), *Genetic Engineering of Crop Plants*, pp. 51-66. London: Butterworths.

Hilder, V.A. A.M.R. Gatehouse (1991). Phenotypic cost to plants of an extra gene. *Transgenic Res.* **1**: 54-60.

Hilder, V.A., A.M.R. Gatehouse, S.E. Sheerman, R.F. Barker & D. Boulter (1987). A novel mechanism of insect resistance engineered into tobacco. *Nature* **330**: 160-163.

Hilder, V.A., R.F. Barker, R.A. Samour, A.M.R. Gatehouse, J.A. Gatehouse & D. Boulter (1989). Protein and cDNA sequences of Bowman-Birk protease inhibitors from the cowpea (*Vigna unguiculata* Walp), *Plant Mol. Biol.* **13**: 701-710.

Hilder, V.A., A.M.R. Gatehouse & D. Boulter (1991). Transgenic plants for conferring insect tolerance-protease inhibitor approach. In: S. King & R. Win (eds), *Transgenic Plants*, Vol. I. London: Butterworths (in press).

Hoffman, M.P., F.G. Zalom, J.M. Smilanick, L.D. Malyj, J. Kiser, L.T. Wilson, V.A. Hilder & W.M. Barnes (1991). Field evaluation of transgenic tobacco containing genes encoding *Bacillus thuringiensis* d-endotoxin or Cowpea trypsin inhibitor: efficacy against *Helicoverpa zea* (submitted).

Janzen, D.H., H.B. Juster & I.E. Liener (1976). Insecticidal action of the phytohemagglutinin in black bean on a bruchid beetle. *Science* 192: 795-796.

Jayne-Williams, D.J. & C.D. Burgess (1974). Further observations on the toxicity of navy beans (*Phaseolus vulgaris*) for Japanese quail (*Coturnix coturnix japonica*). *J. Appl. Bacteriol.* 37: 149-169.

Johnson, R., J. Narvaez, G. An & C.A. Ryan (1989). Expression of proteinase inhibitors I and II in transgenic tobacco plants: effects on natural defense against *Manduca sexta* larvae. *Proc. Natl Acad. Sci. U.S.A.* 86: 9871-9875.

King, T.P., A. Pusztai & E.M.W. Clark (1980a). Immunocytochemical localisation of injested kidney bean (*Phaseolus vulgaris*) lectins in rat gut. *Histochem. J.* 12: 201-208.

King, T.P., A. Pusztai & E.M.W. Clarke (1980b). Kidney bean (*Phaseolus vulgaris*) lectin-induced lesions in rat small intestine. 3. Ultrastructural studies. *J. Comp. Pathol.* 92: 357.

Liener, I.E. (1980). *Toxic constituents of Plant Foodstuffs*, 2nd edition, New York: Academic Press.

Lotan, R., R. Cacan, M. Cacan, H. Debray, W.C. Carter & N. Sharon (1975). On the presence of two types of subunit in soybean agglutinin. *FEBS Lett.* 57: 100.

Meiners, J.P. & T.C. Elden (1978). Resistance to insects and diseases in *Phaseolus*. In: R.S. Summerfield & A.H. Buting (eds), *Advances in Legume Science, International Legume Conference, Kew*, pp. 359-364.

Metcalf, R.L. (1986). The ecology of insecticides and the chemical control of insects. In: M. Kogan (ed.), *Ecological Theory and Integrated Pest Management*. New York: John Wiley.

Murdoch, L.L., J.E. Huesing, S.S. Nielsen, R.C. Pratt & R.E. Shade (1990). Biological effects of plant lectins on the cowpea weevil. *Phytochem.* 29: 85-89.

Pearce, G., D. Strydom, S. Johnson & C.A. Ryan (1991). A polypeptide from tomato induces wound-inducible proteinase inhibitor proteins. *Science* 253: 895-898.

Powell, K.S., A.M.R. Gatehouse, V.A. Hilder & J.A. Gatehouse (1992). Antimetabolic effects of plant lectins and plant and fungal enzymes on the nymphal stages of two important rice pests, *Nilaparvata lugens* and *Nephotettix nigropictus*. *Entomol. exp. appl.* (in press).

Pusztai, A., E.M.W. Clarke & T.P. King (1979). The nutritional toxicity of *Phaseolus vulgaris*. *Nutr. Soc.* 38: 115-120.

Pusztai, A., G. Grant, S. Bardocz, D.J. Brown, J.C. Stewart, S.W.B. Ewen, A.M.R. Gatehouse & V.A. Hilder (1992). Nutritional evaluation of the trypsin inhibitor from cowpea. *British J. Nutr.* (in press).

Redden, R.J., P. Dobie & A.M.R. Gatehouse (1983). The inheritance of seed resistance to *Callosobruchus maculatus* F. in cowpea (*Vigna unguiculata* L. Walp.). I. Analysis of parental, F_1, F_2, F_3 and backcross seed generations. *Aust. J. Agric. Res.* 34: 681.

Shukle, R.H. & L.L. Murdock (1983). Lipoxygenase, trypsin inhibitor, and lectin from soybeans: effects on larval growth of *Manduca sexta* (Lepidoptera: Sphingidae). *Environ. Entomol.* 12: 787-791.

Shumway, L.K., V.V. Yang & C.A. Ryan (1976). Evidence for the presence of proteinase inhibitor I in vacuolar protein bodies of plant cells. *Planta* 129: 161-165.

Steffens, R., F.R. Fox & B. Kassel (1978). Effect of trypsin inhibitors on growth and metamorphosis of corn borer larvae *Ostrinia nubilalis* (Hübner). *J. Agric. Food Chem.* **26**: 170-174.

Van Damme, E.J.M., A.K. Allen & W.J. Peumans (1987). Isolation and characterization of a lectin with exclusive specificity towards mannose from snowdrop (*Galanthus nivalis*) bulbs. *FEBS Lett.* **215**: 140-144.

Walker-Simmons, M. & C.A. Ryan (1977). Immunological identification of proteinase inhibitors I and II in isolated tomato leaf vacuoles. *Plant Physiol.* **60**: 61-63.

Xavier-Filho, J., F.A.P. Campos, M.B. Ary, C.P. Silva, M.M.M. Carvalho, M.L.R. Macedo, F.J.A. Lemos & G. Grant (1989). Poor correlation between levels of proteinase inhibitors found in seeds of different cultivars of cowpea (*Vigna unguiculata*) and the resistance/ susceptibility to predation by *Callosobruchus maculatus*. *J. Agric. Food Chem.* **37**: 1139-1143.

Proc. 8th Int. Symp. Insect-Plant Relationships, Dordrecht: Kluwer Acad. Publ.
S.B.J. Menken, J.H. Visser & P. Harrewijn (eds), 1992

The development of host-plant resistance to insect pests: outlook for the tropics

N.A. Bosque-Pérez[1] and I.W. Buddenhagen[2]
[1] *International Institute of Tropical Agriculture, Ibadan, Nigeria*
[2] *University of California, Davis, California, USA*

Key words: Durable resistance, International Agricultural Research Centers, tropical insect pests

Abstract

Host-plant resistance (HPR) to insect pests is considered one of the key tactics for insect control, particularly in developing countries where utilization of other control methods such as pesticides is often difficult or unwise. The search for more sustainable methods of pest control, make host-plant resistance more attractive every day. Additionally, the opportunity for utilizing new insect resistance genes and the ability to move these across plant species, through plant biotechnology, open new doors to the field of HPR. Pest/host-plant complexes are dynamic systems influenced by numerous factors and the development of pest control strategies has to take these into consideration. Major changes in the pest complex of various tropical crops have occurred in recent years as a result of among others: the accidental introduction of new pests or the emergence of new biotypes that make previously resistant varieties ineffective, the introduction of new agronomic practices, and the deployment of new varieties with hidden susceptibility to previously minor pests. Analysis of the origin of crops and their pests is, thus, essential for devising pest control strategies and it is important to have flexible breeding programs that allow for the incorporation of new objectives as new pests come along. There is also a need to breed for varieties with durable resistance, particularly in countries where farmers cannot afford failure of HPR.

Introduction

In discussing plant resistance to insects we should keep in mind that many insects evolved utilizing plants or their products as their major food source. Myriad species and large fluctuating insect populations are tied to our plant biota and always will be. We are concerned here with the level of depredation of major tropical food crops and how this may be lessened through appropiate plant breeding strategies and methods.

We recognize that other means exist to reduce insect depredations: biological control, cultural practices, insecticides and their more rational use through integrated pest management. But the extent to which all these are needed is based on the level of susceptibility of the variety the farmer grows, and this exists because it was selected by a plant breeder when he/she chose the plant or plants that were to be the base of the new variety. For some crops in the tropics, where modern-bred varieties are not yet grown, the susceptibility levels are the product of natural and farmers' selection. How can we as

235

entomologists influence positively the development of higher and more durable levels of resistance in the new varieties of the future?

Host-plant resistance

Host-plant resistance (HPR) has been defined as "the property that enables *a plant* to avoid, tolerate, or recover from injury by insect populations that would cause greater damage to other plants of the same species under similar environmental conditions" (Kogan, 1975). A farmer is, however, concerned with *varietal* performance over a season, or seasons, and total yield and quality are the ultimate objectives. A breeder has similar objectives but he/she works with many different individual plants or lines in a non-farm environment to establish a "founder" of a new variety. How much an individual plant's reaction to insect pests or pathogens in a breeders' plot will reflect its performance as a cultivar in vasts fields is often not clear. Moreover, although many cultivars are genetically uniform, outbreeding crops are composed of populations of genetically different individuals. A definition, possibly more restrictive, but more relevant to plant breeding is: "Insect resistance of a plant or cultivar is a genetic characteristic as a result of which less depredation occurs from insects than on other plants or cultivars challenged equally by the appropiate insect pest".

Although identifying sources and studying the complexity of mechanisms of resistance to insects receives much attention, one should remember that the ultimate goal is less insect damage in the varieties of *the future*. These are still to be developed and it is in the developmental process (selection and breeding) where resistance must be incorporated, judged accurately, and selected among other criteria of the variety-to-be. Unfortunately, often little effort is made during breeding to have insect resistance qualities in the new variety.

HPR is an ideal means of insect control. This is particularly true in developing countries (mostly tropical countries) where utilization of other control methods such as pesticides is often difficult or unwise. HPR requires minimum input and action by the farmer (*i.e.*, purchase of seed of the right variety only), often has a cumulative effect, is compatible with other control methods and there are no negative environmental effects. In fact, the level of HPR is the base on which all other methods of control must be built! If sufficiently high, nothing else will be needed for economic crop production.

While the numbers of insect-resistant cultivars are mounting (Tingey & Steffens, 1990) the figures on pesticide use reveal the enormous gap between these two control methods. According to Pimentel (1990) by 1989 over 2.5 billion kg of pesticides were being applied annually in the world, with 34% of the total used in North America, 45% in western and eastern Europe and 21% in the rest of the world. A large proportion of this total is insecticide and acaricide use. Estimates are that these figures will increase in the coming years. This reveals how important it is to invest more on breeding for greater insect resistance and thereby have a more sustainable and less environmentally negative means of pest control. Also, the opportunity for utilizing new insect resistance genes and to move these into many plant species through plant biotechnology open new doors to the field of HPR.

HPR in the tropics

Emphasis on HPR in the tropics is more recent than in North America or Europe. Although some breeding programs existed in tropical countries prior to 1940, it was not until the 1950's that major programs were initiated (by the Rockefeller Foundation) to improve tropical food crops such as maize (*Zea mays* L.), potatoes (*Solanum tuberosum* L.), and rice (*Oryza sativa* L.). These efforts eventually lead to the formation of the International Agricultural Research Centers (IARC's) (Plucknett & Smith, 1982). The IARC's and their influence on National Agricultural Research Systems (NARS) have contributed greatly to new efforts on HPR to insects as an integral part of varietal improvement programs.

Improving the major food crops of the tropics is a major objective of the IARC's. The crops of major emphasis of those IARC's involved in HPR research in the tropics and institute locations are presented in Table 1. Many pests occur on each crop, but we present a summary of only major ones for which active screening is carried out (Table 2). Examples of some of these activities follow.

Table 1. International Agricultural Research Centers conducting host-plant resistance research on major tropical food crops

Center	Headquarters	Tropical crops of emphasis
International Center of Tropical Agriculture (CIAT)	Cali, Colombia	Cassava, common beans, rice, tropical pastures
International Crops Research Institute for the Semi-Arid Tropics (ICRISAT)	Hyderabad, India	Sorghum, millet, groundnut, pigeon pea
International Maize and Wheat Improvement Center (CIMMYT)	El Batan, Mexico	Maize
International Institute of Tropical Agriculture (IITA)	Ibadan, Nigeria	Cowpea, soybean, maize, cassava, plantain
International Potato Center (CIP)	Lima, Peru	Potato, sweet potato
International Rice Research Institute (IRRI)	Los Banos, Philippines	Rice

CIAT. Resistance has been identified to major pests of cassava (*Manihot esculenta* Crantz) including mites, thrips, whiteflies and mealybugs (Belloti *et al.*, 1987). Cassava varieties with some resistance to the first three groups of pests have been released (Lapointe *et al.*, 1990).

On common beans (*Phaseolus vulgaris* L.), bruchids have received major emphasis. Resistance to the Mexican bean weevil, (MBW) *Zabrotes subfasciatus* (Bohemen) is conferred by arcelin, a protein found in wild dry bean accessions and transferred to bean cultivars through conventional breeding methods (Cardona *et al.*, 1990). Lines with resistance to this pest were released for testing in 1990 (Lapointe *et al.*, 1990). Tolerance to the leafhopper *Empoasca kraemeri* Ross and Moore has been found and significant progress made by using yield under leafhopper attack as the major selection criterion (Kornegay & Cardona, 1990).

For the rice delphacid [*Tagosodes orizicolus* (=*Sogatodes oryzicola)* (Muir)] emphasis is on developing tolerance of mechanical damage and resistance to the rice hoja blanca virus of which this insect is a vector (A. Pantoja, pers. comm.; Lapointe *et al.*, 1990). Work on tropical forages includes efforts to find stable resistance to spittle bugs in *Brachiaria* spp. (Lapointe *et al.*, 1990).

Table 2. Status of screening and breeding for varietal resistance to major tropical insect pests[a]

Crop and insect common name	Insect scientific name	Resistance sources identified	Resistant breeding lines available	Resistant varieties released	Mechanisms of resistance studied[b]
Rice					
Brown planthopper	*Nilaparvata lugens*	+	+	+[c]	+
Green leafhopper	*Nephotettix virescens*	+	+	+[c]	+
Whitebacked planthopper	*Sogatella furcifera*	+	+	+	+
Rice delphacid	*Tagosodes orizicolus*	+	+	+	-
Rice striped borer	*Chilo suppressalis*	+	+	+	+
Gall midge	*Orseolia oryzae*	+	+	+[c]	+
Rice weevil	*Sitophilus oryzae*	+	-	-	-
Sorghum					
Spotted stem borer	*Chilo partellus*	+	+	-	+
Sorghum midge	*Contarinia sorghicola*	+	+	+	+
Sorghum shoot fly	*Atherigona soccata*	+	+	-	+
Sorghum head bug	*Calocoris angustatus*	+	+	-	+
Millet					
Earhead caterpillar	*Heliochielus albipunctella*	+	-	-	-
Stemborer	*Acigona ignefusalis*	+	-	-	-
Maize					
Fall armyworm	*Spodoptera frugiperda*	+	+	-	-
Southwestern corn borer	*Diatraea grandiosella*	+	+	-	-
Sugarcane borer	*D. saccharalis*	+	+	-	-
Pink stalk borer	*Sesamia calamistis*	+	+	-	-
African sugarcane borer	*Eldana saccharina*	+	+	-	-
Maize storage weevil	*Sitophilus zeamais*	+	-	-	+
Larger grain borer	*Prostephanus truncatus*	-	-	-	-
Common bean					
Mexican bean weevil	*Zabrotes subfasciatus*	+	+	-	+
Leafhopper	*Empoasca kraemeri*	+	+	-	+
Pigeon pea					
Pod borer	*Helicoverpa armigera*	+	+	-	-
Groundnut					
Cowpea aphid	*Aphis craccivora*	+	+	-	+

238

Cowpea

Cowpea aphid	*Aphis craccivora*	+	+	+	-
Cowpea storage bruchid	*Callosobruchus maculatus*	+	+	+	+
Flower thrips	*Megalurothrips sjostedti*	+	+	+	+
Legume pod borer	*Maruca testulalis*	+	-	-	-

Potato

Potato tuber moth	*Phthorimaea operculella*	+	+	-	-

Sweet potato

Sweet potato weevil	*Cylas* spp.	+	-	-	-

Cassava

Thrips	*Frankliniella williamsi*	+	+	+	+
Whitefly	*Aleurotrachelus socialis*	+	+	+	-
Cassava mealybugs	*Phenacoccus* spp.	+	-	-	-
Cassava green mite	*Mononychellus tanajoa*	+	-	-	+

[a] Adapted from various sources, see text and references for details. Screening techniques have been developed for all of the above pests, except for millet insects for which development is in progress.

[b] Refers to in-depth studies identifying the physical or biochemical factor(s).

[c] Biotypes have been a major constraint for resistance breeding (Heinrichs, 1992).

CIMMYT. HPR efforts include the development of tropical maize materials with resistance to the fall armyworm (FAW), *Spodoptera frugiperda* (Smith), and other maize pests (Mihm et al., 1988). A Multiple Borer Resistance (MBR) population has been formed and improved over the years. This population combines high levels of resistance to FAW and moderate resistance to various maize borers, including the southwestern corn borer (SWCB), *Diatraea grandiosella* (Dyar), and the sugarcane borer (SCB), *D. saccharalis* Fabricius. Additionally, a maize population with tropical adaptation and resistance to FAW and SCB has been formed [Multiple Insect Resistance, Tropical (MIRT)] and together with the MBR made available to national programs and international collaborators (Mihm et al., 1988).

CIP. Major emphasis has been on the potato tuber moth (PTM), *Phthorimaea operculella* (Zeller), one of the most damaging potato pests worldwide. Recent work has focused on utilizing wild potato species as sources of resistance to PTM and progenies of crosses between wild and cultivated potatoes which show some resistance to PTM have been obtained (Chavez et al., 1988).

CIP has recently initiated work on sweet potato (*Ipomoea batatas* (L) Lam.). On this crop the major pests of concern are the sweet potato weevils *Cylas* spp. Low levels of resistance to weevils have been reported; potential exists for the development of cultivars with higher levels of resistance and wild species of *Ipomoea* are being evaluted as possible sources of resistance to *Cylas* spp. (Jansson et al., 1991).

ICRISAT. HPR activities on millet (*Pennisetum americanum* (L.) Schum.) concentrate on the earhead caterpillar *Heliochielus* (= *Raghuva*) *albipunctella* (de Joannis), and the stemborer *Acigona ignefusalis* Hampson (Nwanze, 1985). Major efforts are on the development of screening techniques. Plant characters common in millet landraces which confer some degree of resistance/tolerance to insect pests have been identified but incorporation of these characters into germplasm with desirable agronomic characters has been difficult (Nwanze, 1985).

For sorghum (*Sorghum bicolor* (L.) Moench), genotypes with some degree of resistance to the spotted stem borer, *Chilo partellus* Swinhoe, have been identified, the mechanisms of resistance determined, and borer resistant lines with moderate yield and proper grain quality developed (Agrawal *et al.*, 1990). High yielding varieties resistant to the sorghum midge, *Contarinia sorghicola* Coq., are available. Moderate levels of resistance to the shoot fly, *Atherigona soccata* Rond., have been found. Genotypes with resistance to the sorghum head bug, *Calocoris angustatus* Lethiery, have been identified (Sharma & Lopez, 1991). Evaluation of germplasm for multiple insect resistance has resulted in identification of genotypes with moderate resistance to more than one pest (Nwanze *et al.*, 1991).

For pigeon pea (*Cajanus cajan* (L.) Millsp.), lines with tolerance/moderate resistance to the pod borer, *Helicoverpa armigera* (Hübner) have been identified and made available to farmers in India for on-farm testing (Lateef & Pimbert, 1990). On groundnut (*Arachis hypogaea* L.) resistance to *Aphis craccivora* Koch, vector of groundnut rosette, has been reported and mechanisms of resistance studied (Wightman *et al.*, 1990). The use of wild species of groundnuts as sources of resistance to insects is now receiving attention.

IITA. HPR research has been carried out on several cowpea (*Vigna unguiculata* Walp.) pests. High levels of resistance to the cowpea aphid, *A. craccivora*, and the leafhopper *Empoasca dolichi* Paoli have been identified and resistant varieties developed; additionally, varieties with moderate levels of resistance to the cowpea storage bruchid, *Callosobruchus maculatus* (Fabricius) have been developed (Singh *et al.*, 1990). Moderate to low levels of resistance to the flower thrips, *Megalurothrips sjostedti* (Trybom) have been incorporated into improved varieties. Resistance to post-flowering pests like the legume pod borer, *Maruca testulalis* (Geyer) and a complex of pod sucking bugs including *Clavigralla tomentosicollis* Stål., has been difficult to identify; moderate or low levels of resistance have been found, sometimes in agronomically poor genotypes (Singh *et al.*, 1990). Wild *Vigna* species are being screened for resistance to these pests (Singh *et al.*, 1990). Collaboration with other laboratories to obtain interspecific crosses, and on biotechnology is underway.

For maize, the stem borers *Sesamia calamistis* Hampson and *Eldana saccharina* Walker and the storage weevils *Sitophilus* spp. have received major emphasis. Sources of resistance to the two stem borers have been found and maize populations with moderate resistance to either *S. calamistis* or *E. saccharina* have been formed; these are undergoing selection to increase the levels of resistance (Bosque-Pérez & Mareck, 1990). These maize populations have been made available to various NARS in Africa for utilization in their breeding programs. Further screening to identify new sources of resistance is being conducted. Work on *Sitophilus* weevils includes studies on comparative methods to assess for resistance and screening of diverse germplasm. A collaborative project with the Natural Resources Institute (NRI) in the U.K. has been established to identify potential biochemical sources of resistance to post-harvest storage pests.

IRRI. Work on insect resistance at IRRI began around the mid 1960's and detailed HPR studies on many pests have been carried out since then. Thousands of rice accessions have been screened for resistance to insect pests and numerous sources of resistance identified (Heinrichs, 1992). In addition, wide crosses have been used to incorporate insect resistance genes from wild into cultivated rice. Insects of major emphasis are the green leafhoppers, *Nephotettix* spp., the brown planthopper (BPH) *Nilaparvata lugens* Stål, the whitebacked planthopper (WBPH), *Sogatella furcifera* (Horváth), and the gall midge (GM), *Orseolia oryzae* Wood-Mason. Varieties with resistance to green leafhoppers, BPH, WBPH, GM, and the stem borer *Chilo suppressalis* Walker have been released and are being grown in several countries in Asia (Heinrichs, 1992). However, the emergence of biotypes of BPH and GLH has been a major constraint to breeding (Heinrichs, 1992). We discuss this issue in more detail later in the paper.

Other crops

Although this paper emphasizes HPR on the major food crops handled by the IARC's, one should recognize the importance of crop production in the tropics of crops other than the few handled by the IARC's. There are many perennial fruit, nut, oil, and industrial crops of the tropics where breeding new varieties is either absent or at a level not involving entomologists: *i.e.*, citrus, bananas, plantains, mangoes, oil palm, coffee, cocoa, tea, rubber. Levels of insect resistance/susceptibility in these and many similar crops can be expected to remain as they now are for the foreseeable future (discounting potential biotechnology interventions which *might* be applied). Diverse vegetables in the tropics have many important insect pests, and although new varieties are constantly being bred, this group has unique facets. First, cosmetic market requirements reinforce pesticide use and have reduced breeders' concern for insect resistance. Second, resistance factors in the plant parts we eat must not be toxic to humans. And third, private sector breeding is paramount. Seed companies release and promote varieties for the tropics. Although they may include insect resistance as a minor target along with major ones of quality and yield, most actual breeding (as compared to *testing*) is done outside the tropics with the exception of the work done at the Asian Vegetable Research and Development Center (AVRDC) and through its collaborators in Asia. The possibility of introducing anti-insect genes into insect-susceptible, but otherwise good varieties of vegetables, via biotechnology, is a real one, given the private sector involvement with the vegetable seed industry.

The changing pest complex in the tropics

Pest/host-plant complexes cannot be viewed as static, but rather as dynamic systems influenced by numerous factors, including many that are human-driven. The development of control methods, including HPR, has to take these factors into consideration. The key to improved balance between production and depredation involves applying the best possible strategy and tactics within a framework of understanding of evolution, environment and population interactions. We discuss three types of activities which have resulted in major changes in the pest complex of various tropical crops.

New encounters/re-encounters and coevolved systems. Many pests have evolved with the crop species they attack (*e.g.*, the brown planthopper and rice in Asia; the sorghum stem borers and sorghum in Africa); these are considered coevolved systems (Buddenhagen, 1977). On the other hand, many pest problems are new-encounter situations, in which the crop has been moved to a new area and pests that have evolved with other, often related, plant species come to attack the introduced crop (Buddenhagen & De Ponti, 1983); (*e.g.*, stem borers and maize in Africa; the delphacid *T. orizicolus* and rice in South America).

Re-encounter situations may occur after historical separations such as those of the cassava mealybug (CMB), *Phenacoccus manihoti* Matile-Ferrero and the cassava green mite (CGM), *Mononychellus tanajoa* (Bondar), introduced accidentally from South America to Africa in the early 1970's (Herren & Neuenschwander, 1991) after being "separated" from their cassava host (of American origin) for approximately 400 years. Another example is the recent accidental introduction into Africa of the larger grain borer (LGB), *Prostephanus truncatus* Horn, a pest of stored maize in Central America and Mexico. In both cases, and in the absence of natural enemies and any degree of resistance in these two staple crops in Africa, pest spread and major outbreaks ensued, resulting in significant losses and in some cases food shortages.

Analysis of the origin of the crops and their pests has been essential in devising strategies for the control of these pests (*i.e.*, the use of natural enemies for the control of the CMB and CGM (Herren & Neuenschwander, 1991) and a combination of tactics for the control of LGB). For the development of HPR it is important to determine the degree of coevolution of host plant/pest prior to embarking in a breeding program, as this will influence decisions on the selection of parental sources, the kind of resistance to be selected and the breeding strategy (Buddenhagen & De Ponti, 1983) as well as the relative emphasis on breeding vis-à-vis other control strategies.

Since major continental movement of crops has occurred during the last 400 years, and many of the coevolved pests of several staple foods in tropical countries have been "left behind" it is to be expected that re-encounter situations will continue to occur in the future. Thus, it is important to have flexible breeding programs that allow for the incorporation of new objectives as new pests come along. In some cases when the arrival of a new pest is imminent, preventive breeding might be required to avoid or reduce the devastating losses that otherwise might occur.

Coevolved systems need special consideration. Where a crop was domesticated locally from wild ancestors, it receives predation from many pests which coevolved with its ancestors. The domestication of the crop sometimes resulted in "domestication" of its pests and pathogens as well. They remain in the modified environment of the primitive agriculture of the local farming system. As the system changes with time and innovation the pests/pathogens also change in their evolutionary drive for survival. Both major vertical and minor modifying genes for virulence/avirulence, high aggressiveness/low aggressiveness will have evolved. Relationships beyond the cultivated fields, into the wild reservoir species and local flora may continue as part of the time/space survival mechanisms. The coevolved systems present a much more intractable system for the plant breeder/entomologist/pathologist wishing better varieties with greater HPR. This should be recognized and new tactics devised, with this appreciation.

For example, cowpeas evolved in Africa, are affected there by many important coevolved pests, and have been very difficult to improve in terms of insect resistance in spite of much effort to do so. Without application of insecticides, even moderately resistant cowpea cultivars cannot be *intensified* successfully in modern agriculture in their homeland. This leads us to a second major factor: changed agronomy.

Changed agronomy. The changed agronomy that accompanies the intensification of cropping in modern agriculture changes opportunities for pests and pathogens. A classical example in the tropics is that of brown planthopper (BPH) and rice in Asia. BPH only became a major pest of rice in tropical Asia during the 1960's (IRRI, 1979). With the introduction of new high yielding varieties, rice cultivation in Asia was intensified, fertilizer use increased and planting two to three crops per year became the norm over large areas. These factors resulted in better conditions for the reproduction and perpetuation (in space and time) of pests like the brown planthopper and green leafhoppers (Heinrichs, 1988). Outbreaks followed. Attempts to control BPH by using heavy doses of pesticides backfired, as the natural enemy complex was killed. BPH-resistant varieties were developed but these proved to be non-durable as biotypes which could attack the previously resistant varieties, emerged (Heinrichs, 1988). An integrated pest management program which combines several control tactics is being promoted at present. The need to breed for varieties with durable insect resistance remains.

The very tactics used to increase yields as "peasant" agriculture is pushed towards "modern" agriculture tend to destabilize a balanced system. Fertilizer application, especially nitrogen, increases biomass production, canopy density, and succulence. Dwarf varieties, developed to respond to higher nitrogen without lodging, create denser canopies. In much of Africa and Latin America the long term rotational effects inherent in slash and burn agriculture are lost as cropping is intensified in time and space. Formerly scattered plots in the bush become contiguous fields. In the Colombian Llanos, for example, heavy disease and arthropod pest outbreaks occurred after susceptible cassava varieties and intensive cultivation practices were introduced to a region of previously scattered cassava plantings (Hershey et al., 1988). In other cases, farmers formerly growing landraces, which are mixtures of genotypes, grow a uniform variety, removing any dampening effect of varietal mixture on pest and pathogen buildup. Polycropping tends to be replaced by monocropping, often changing conditions to favor pest/pathogen buildup. Greater yields require more storage facilities and changes in storage structures or practices may well enhance insect depredation in the store.

Breeding new varieties. Few people realize that while breeding new varieties attempts to solve existing problems, this same process often results in the creation of "new" ones. Two common problems are: 1) Previously unknown or minor pests suddenly become major ones when new varieties with hidden susceptibility are released; 2) New pest biotypes emerge which render the previously resistant varieties susceptible.

For maize, the adoption of improved, disease resistant varieties in some countries of West Africa has been hindered by their poor storability under farmer's conditions. In Benin Republic, farmers report these new varieties to be more susceptible to *Sitophilus* weevils, than "local" ones (Anonymous, 1989). This is to a great extent due to poor husk cover of the improved varieties. In response to this we at IITA have initiated a comprehensive research program to address the problem. This includes breeding for better husk cover in maize and detailed studies on the biology and behavior of *Sitophilus* weevils as influenced by maize storage form. We have found that the oviposition behavior of *Sitophilus* weevils is drastically different on shelled maize compared to maize on the cob (Kossou et al., in press). Traditionally, screening for resistance to weevils has been done using small samples of shelled maize. Since most farmers in Africa store their maize on the cob with the husks on, resistance identified using the traditional screening method misses important resistance factors associated with the cob/husk structure. Additionally, it might result in selecting as resistant, varieties that do not perform as such under farmers storage practices. We now use a combination of the traditional screening method and techniques which involve the screening of maize on the cob.

In cases like this, resistance breeding would be successful only after: 1) having a clear understanding of farmer's practices, 2) studying the biology of the pest in relation to these practices and, 3) the development of innovative screening techniques.

This calls for on-site selection and breeding so that varieties are adapted to local stresses and fit into local farmer's practices. Although to some, this might appear to be self-evident, enormous efforts are devoted to test and select varieties in ecosystems quite disimilar from those in which they were bred (Buddenhagen, 1983a). Unfortunately, the IARC's in general promote this, and the very nature of their existence and funding often depends on proving that their varieties have wide adaptation and are widely adopted.

Much has been said about the second type of problem, the emergence of new biotypes. The classical example for the tropics is that of the rice brown planthopper in Asia where the development of biotypes has been a major constraint in breeding for resistance (Heinrichs, 1992), but also it has been a major *result* of the rice breeding and selection approaches used. Resistance was sought under severe challenge of seedlings and only vertical resistance (VR) ,i.e., biotype-specific resistance, was selected as was proven by its transient nature soon after release. The deployment of IR 26, a rice variety with a high level of resistance to BPH, resulted in rapid suppression of the pest in the Philippines, Indonesia and other Southeast Asian countries, but after two to three years outbreaks of

BPH on IR 26 were reported (Heinrichs, 1988). The monophagous nature of this pest coupled with the strong selection pressure which highly resistant varieties exert on it are believed to explain the emergence of biotypes. The heart of the matter of "resistant biotypes" rests in how VR genes affect pest evolution as compared with horizontal resistance (HR) genes. The subject of HR has recently been reviewed by Simmonds (1991). Hopefully, entomologists and breeders working on insect resistance will not have to repeat the mistakes made in breeding against pathogens. The key question is: can we devise breeding strategies that will minimize biotype selection and enhance the durability of resistance? (Buddenhagen, 1983b; Gould, 1988). The answer is yes, but it will take a clear understanding of the difference between VR and HR and of what specific steps need to be taken to obtain varieties with HR and not transitory VR. Much learning, analytical thinking, innovative research and long term committment will be required. When host-plant resistance programs are under pressure to produce quick solutions, as is often the case in the tropics, it is difficult to spend sufficient time analyzing the ecosystem and getting a clear picture of the intricacies of the host-plant/pest complex. However, it is only by doing so that long lasting solutions can be obtained.

Strategies that have been devised for dealing with new insect biotypes and pathogen races in temperate zones, such as sequential release of resistant varieties, use of multilines, pyramiding genes and gene rotation (in time and space) have been recommended for use in Asia (Heinrichs, 1988) and have been suggested as a possible way to deal with biotypes in other tropical regions. In the case of the Mexican bean weevil (MBW), plans at CIAT are for incorporating resistance to this pest into "all commercial grain classes of dry beans with tropical adaptation and multiple disease resistance" (Cardona et al., 1990). It is planned that high yielding MBW resistant lines will be distributed to national programs in Africa and Latin America. Studies are being conducted to determine the possible impact of MBW biotype emergence (Cardona et al., 1990). The development of multilines individually containing the arcelin alelles 1, 2 and 5 is one of CIAT's future strategies for addressing the potential threat of biotypes. Whether or not this strategy might work, however, is completely unknown. Storage pests may behave quite differently from wheat rust pathogens for which "multilines" were first proposed.

We believe gene deployment strategies are not practical in a continent like Africa where extension services are scarce and adoption of new varieties often takes years. Many African countries often face critical food shortages, and it is precisely in these countries where farmers can least afford failure of HPR. Thus, the need to breed for varieties with durable resistance cannot be overstated. The review on horizontal resistance by Simmonds (1991) summarizes all documented instances of HR to diseases and pests. It is interesting to note that while 168 cases of HR are reported for plant pathogens (fungi, bacteria and viruses) only 26 are cited for insects and mites (21 arthropod species, 14 crops). Of these, only one is from the tropics, that of the sorghum shoot fly, A. soccata (Borikar & Chopde, 1982, cited by Simmonds, 1991). Indeed, Simmonds calls attention to this relatively low number of HR to arthropod pests. The question as to whether breeding for insect resistance is more difficult than for disease resistance has never been properly answered.

It is possible that there are more cases of HR to insects (especially in the tropics) but these have not been documented. There is evidence that minor genes play a role in determining resistance to N. virescens in rice (Heinrichs & Rapusas, 1985). Due to the development of brown planthopper biotypes and ineffectiveness of resistant cultivars, attempts have been made at IRRI to incorporate "field" (i.e., mature plant) resistance into these cultivars (Velusamy et al., 1987). Methods have been developed at IRRI to evaluate rice for tolerance to BPH (Heinrichs, 1992). It is our view, however, that not enough efforts are being made in the tropics to ensure durability of resistance and as such this remains a major challenge for those of us engaged in host-plant resistance efforts.

When is HPR practical?

Careful analysis of the agroecosystem is required for answering this question. What exactly is the nature of the problem? Is it solvable through means other than HPR? If breeding for HPR is attempted what are the possibilities of success? These are useful questions to ask prior to embarking into a breeding program. We have already mentioned evolutionary considerations. Aditionally, one should consider if the insect pest is important because of the direct damage it causes or as a vector of viruses and other plant pathogens. Viruses such as rice tungro in Asia, maize streak virus and African cassava mosaic in Africa, and hoja blanca of rice in South America, are major production constraints. In Asia, the problem of rice tungro viruses was originally tackled by attempting to control the vectors, green leafhoppers in the genus *Nephotettix*, with insecticides and with vector-resistant varieties. IR 8 and 26 other IR varieties released subsequently have moderate or high levels of vector-resistance (Heinrichs, 1986). While these varieties have played a role in decreasing losses due to tungro, this approach appears insufficient as disease epidemics still occur.

Experiments have demonstrated that adult survival and female fecundity of *N. virescens* significantly increased on the moderately resistant variety IR 8 after only 15 generations of selection for "virulence" on this variety; tungro infection went from 35% infected plants to 85% after only four generations of selection of the vector (Heinrichs, 1988; Heinrichs & Rapusas, 1990). Although response to selective pressures from resistant varieties in the field is not expected to occur at a rate as rapid as in the laboratory (Heinrichs & Rapusas, 1990), there is evidence for increased "virulence" of *N. virescens* on the rice cultivars IR 36 and IR 42 in the field (Rapusas & Heinrichs, 1986). Additionally, irrespective of the development of biotypes, under heavy pressure from the vectors, the varieties with moderate insect resistance have high incidence of tungro virus (Heinrichs & Rapusas, 1983). Thus, in spite of the breeding efforts the problem of tungro remained more or less unsolved (Buddenhagen, 1983a). More recent attempts are directed at breeding for resistance to the virus, in addition to the vector (Heinrichs & Rapusas, 1990).

The strategy followed at IITA for a similarly serious virus disease with a complex of insect vectors was different. Maize streak virus (MSV), transmitted by leafhoppers of the genus *Cicadulina*, is the most widespread and economically important virus of maize in Africa. Damage to maize due to MSV can be insignificant in some years but epidemics of the disease can devastate crops with yield losses up to 100%. A program was initiated at IITA in 1975 to breed for resistance to the virus. Techniques were developed to mass rear the vector and to carry out infestations in the screenhouse and the field (Soto *et al.*, 1982). Sources of resistance to MSV were rapidly found and resistance is now routinely incorporated into all maize varieties bred at IITA. Additionally, resistance sources have been made available to National Programs across Africa. Results from field trials have shown that resistance to MSV is stable across locations in Africa; when tested under natural streak epidemics resistant varieties have always outyielded susceptible ones (Efron *et al.*, 1989), proving the effectiveness of IITA's approach. Breeding efforts have been complemented with studies on virus ecology and epidemiology in relation to vector biology, as well as with studies on the nature and geographical distibution of virus isolates/strains. If instead of breeding for resistance to the virus, efforts would have been devoted to developing vector resistance we believe little progress would have been made.

The role of biotechnology

Interest is growing in the potential use of molecular biology techniques for the development of insect resistant varieties for the tropics. While biotechnology activities are far more advanced in temperate countries, many seed companies that operate in the tropics as well as international institutions such as the IARC's, are involved in efforts that should lead to the utilization of genetically engineered crops in the tropics in the future. We think this is a healthy development as biotechnology offers a greater gene pool from where to obtain resistance and also offers the possibility of combining multiple resistance factors in a single variety, a task that might take many years for a conventional breeding program (Gould, 1988). However, biotechnology is not a panacea and cannot be viewed as a substitute for a conventional breeding program.

When developing insect resistance using molecular techniques the evolutionary considerations which apply to a classical breeding program, are relevant. Gould (1988; 1991) has recently reviewed this subject and encourages molecular biologists to pay attention to the potential evolutionary responses of target pests to genetically engineered crops. Emphasis should be on plant resistance factors to which pests have low potential for genetic adaptation (Gould, 1988). Both physiological and behavioral mechanisms by which pests can adapt should be studied. If the single resistance genes to be manipulated through biotechnology act as vertical genes their use will be negative in most situations. They may work where more than one gene can be combined and where the crop is invaded each season by insects propagating in alternate host plants, as may be the case of some new-encounter pests. Much research is needed to clarify these issues.

We think that detailed studies of the agroecosystem where a resistant variety is to be deployed are also essential. Pest problems in the tropics can only be solved through careful research and analysis, there are no magic solutions for them. Host-plant resistance, whether conventional or molecular, cannot operate in a vacuum. Tropical agroecosystems are usually far more complex than those in temperate countries, both biologically and socioeconomically. Basic and applied research on tropical agroecosystems requires greater funding than ever as human populations in the tropics continue to increase.

The gap between HPR research and less crop loss

Breeders try to develop "better" varieties. "Better" means more yield, higher quality, greater adaptability for a target area, and higher levels of disease and insect resistance. Insect resistance is just one of the many objectives. But the more objectives, the greater investment needed, the slower the progress, and less likelihood of success.

In spite of much research on HPR, fewer hectares are grown of varieties that suffer less from insect pests, than one would expect. Why is this? There are several diverse reasons:

1) Entomologists can screen lines from a germplasm bank and come up with a list of "resistant" lines; publications can be written, rewarding the investigator. No collaboration with a breeder (often difficult) is required.

2) Entomologists and/or biochemists can work on resistance mechanisms and come up with new findings and publications. Often this is where major interest and investment in HPR goes. Again, no collaboration with a breeder is needed.

3) Breeders control the germplasm, the crosses, progeny selection and the eventual release of a new variety. Even if an entomologist is involved, it is often peripheral. Unless a more resistant, incipient variety is also the highest yielding it will be very difficult for it to be released and promoted. Who is to make the subjective judgement that a variety with a lower yield potential but more resistance should be released instead? Unfortunately, a new variety may be so vulnerable that farmers, after an initial try, will revert to their old varieties that are much lower yielding (than even a moderately resistant variety that they never had access to). All this emphasizes the need for teamwork.

Entomologists often feel frustrated that after much effort to identify sources of resistance plant breeders make no use of these sources. Breeders often complain that entomologists sometimes fail to understand basic breeding principles and make unrealistic demands. Unfortunately, the way we tend to be educated by our universities enhances these problems rather than the contrary. Students and professionals tend to be rewarded for working on their own and team spirit is not fomented. Host-plant resistance can only be effective as a practical control method in the field if breeders, entomologists and plant pathologists work together. Every effort should be made to understand each others' disciplines. The role of the entomologist does not end when he/she identifies resistance sources and the breeder should not work in isolation. Institutions must facilitate and encourage regular interaction amongst scientists of different disciplines. We wonder if the trend for multidisciplinary crop improvement teams in the IARC's has not passed its zenith. Certainly, there are strong pressures to move in other directions. Unless regular, real interaction between crop protection scientists and plant breeders can be strengthened, the varieties of the future may end up being more susceptible to insects and diseases than ever.

Acknowledgements. We wish to thank colleagues from various IARC's and Universities who kindly provided literature for the preparation of this paper, especially E.A. Heinrichs, A.C. Belloti, A. Pantoja, S.L. Lapointe, C. Cardona, K.T. Raman, J.H. Mihm, K.F Nwanze, K. Leuschner, S.S. Lateef, H.C. Sharma, G.V. Ranga Rao, W.M. Tingey and F. Gould.
The opinions expressed in this paper are those of its authors and do not necessarily represent the views of the institutions for which they work.

References

Agrawal, B.L., S.L. Taneja, L.R. House & K. Leuschner (1990). Breeding for resistance to *Chilo partellus* Swinhoe in sorghum. *Insect Sci. Applic.* **11**: 671-682.

Anonymous (1989). Développement rural integre de la province du Mono. Institut de Recherches et d'Applications des Méthodes de Développement, Ministere du Développement Rural et de l'Action Cooperative. Republique du Benin.

Belloti, A.C., C. Hershey & O. Vargas (1987). Recent advances in resistance to insect and mite pests of cassava. In: C. Hershey (ed.), *Cassava Breeding: A Multidisciplinary Review*, pp. 441-470. Cali, Colombia: Centro Internacional de Agricultura Tropical.

Bosque-Pérez, N.A. & J.H. Mareck (1990). Screening and breeding for resistance to the maize stem borers *Eldana saccharina* and *Sesamia calamistis*. *Pl. Res. Ins. Newsl.* **16**: 119-120.

Buddenhagen, I.W. (1977). Resistance and vulnerability of tropical crops in relation to their evolution and breeding. In: P.R. Day (ed.), *The Genetic Basis of Epidemics in Agriculture*, pp. 309-326. New York: NY Acad. Sci.

Buddenhagen, I.W. (1983a). Agroecosystems, disease resistance, and crop improvement. In: T. Kommedahl & P.H. Williams (eds), *Challenging Problems in Plant Health*, pp. 450-460. St. Paul, Minnesota: American Phytopathological Society.

Buddenhagen, I. W. (1983b). Breeding strategies for stress and disease resistance in developing countries. *Annu. Rev. Phytopathol.* **21**: 385-409.

Buddenhagen, I. W. & O.M.B. de Ponti (1983). Crop improvement to minimize future losses to diseases and pests in the tropics. *FAO Plant Prot. Bull.* **31**: 11-30.

Cardona, C., J. Kornegay, C.E. Posso, F. Morales & H. Ramirez (1990). Comparative value of four arcelin variants in the development of dry bean lines resistant to the Mexican bean weevil. *Entomol. exp. appl.* **56**: 197-206.

Chavez, R., P.E. Schmiediche, M.T. Jackson & K.T. Raman (1988). The breeding potential of wild potato species resistant to the potato tuber moth, *Phthorimaea operculella* (Zeller). *Euphytica* **39**: 123-132.

Efron, Y., S.K. Kim, J.M. Fajemisin, J.H. Mareck, C.Y. Tang, Z.T. Dabrowski, H.W. Rossel & G. Thottappilly (1989). Breeding for resistance to maize streak virus: a multi-disciplinary team approach. *Plant Breeding* 103: 1-36.

Gould, F. (1988). Evolutionary biology and genetically engineered crops. *BioScience* 38: 26-33.

Gould, F. (1991). The evolutionary potential of crop pests. *Am. Scient.* 79: 496-507.

Heinrichs, E.A. (1986). Perspectives and directions for the continued development of insect-resistant rice varieties. *Agric. Ecos. Environ.* 18: 9-36.

Heinrichs, E.A. (1988). Variable resistance to Homopterans in rice cultivars. *ISI Atlas of Sci: Plants & Animals* 1: 213-220.

Heinrichs, E.A. (1992). Host-Plant Resistance. In: E.A. Heinrichs (ed.), *Biology and Management of Rice Insects*, pp. 510-540. New Delhi: Wiley.

Heinrichs, E.A. & H.R. Rapusas (1983). Correlation of resistance to the green leafhopper, *Nephotettix virescens* (Homoptera: Cicadellidae) with tungro virus infection in rice varieties having different genes for resistance. *Environ. Entomol.* 12: 201-205.

Heinrichs, E.A. & H.R. Rapusas (1985). Cross-virulence of *Nephotettix virescens* (Homoptera: Cicadellidae) biotypes among some rice cultivars with the same major-resistance gene. *Environ. Entomol.* 14: 696-700.

Heinrichs, E.A. & H.R. Rapusas (1990). Response to selection for virulence of *Nephotettix virescens* (Homoptera: Cicadellidae) on resistant rice cultivars. *Environ. Entomol.* 19: 167-175.

Herren, H.R. & P. Neuenschwander (1991). Biological control of cassava pests in Africa. *Annu. Rev. Entomol.* 36: 257-283.

Hershey, C., K. Kawano & J.C. Lozano (1988). Breeding cassava for adaptation to a new ecosystem: a case study from the Colombian Llanos. VIIth Symposium International Society for Tropical Root Crops, pp. 525-540. Paris: INRA.

International Rice Research Institute (IRRI) (1979). Brown planthopper: threat to rice production in Asia. Los Baños, Philippines: IRRI.

Jansson, R.K., K.V. Raman & O.S. Malamud (1991). Sweet potato pest management: future outlook. In: R.K. Jansson & K.V. Raman (eds), *Sweet Potato Pest Management: a Global Perspective*, pp. 429-437. Boulder: Westview Press.

Kogan, M. (1975). Plant resistance in pest management. In: R.L. Metcalf & W.H. Luckmann (eds), *Introduction to Insect Pest Management*, pp. 103-146. New York: Wiley & Sons.

Kornegay, J.L. & C. Cardona (1990). Development of an appropriate breeding scheme for tolerance to *Empoasca kraemeri* in common bean. *Euphytica* 47: 223-231.

Kossou, D.K., N.A. Bosque-Pérez & J.H. Mareck (1992). Effects of shelling maize cobs on the oviposition and development of *Sitophilus zeamais* Motschulsky. *J. Stored Prod. Res.* (in press).

Lapointe, S.L., A.C. Belloti, C. Cardona & A. Pantoja (1990). The role of plant resistance in CIAT's commodity programs. *Pl. Res. Ins. Newsl.* 16: 20-21.

Lateef, S.S. & M.P. Pimbert (1990). The search for host-plant resistance to *Helicoverpa armigera* in chickpea and pigeonpea at ICRISAT. In: Summ. Proc. First Consultative Group Meeting on the Host Selection Behavior of *Helicoverpa armigera*, pp. 14-18. Patancheru, A.P., India: ICRISAT.

Mihm, J.H., M.E. Smith & J.A. Deutsch (1988). Development of open-pollinated varieties, non-conventional hybrids and inbred lines of tropical maize with resistance to fall armyworm, *Spodoptera frugiperda* (Lepidoptera: Noctuidae) at CIMMYT. *Florida Entomol.* 71: 262-268.

Nwanze, K.F. (1985). Some aspects of pest management and host plant resistance in pearl millet in the Sahel. *Insect Sci. Applic.* 6: 461-465.

Nwanze, K.F., Y.V.R. Reddy, S.L. Taneja, H.C. Sharma & B.L. Agrawal (1991). Evaluating sorghum genotypes for multiple insect resistance. *Insect Sci. Applic.* **12**: 183-188.

Pimentel, D. (1990). Introduction. In: D. Pimentel (ed.), *Handbook of Pest Management in Agriculture*, pp. 3-11. Boca Raton, Florida: CRC Press.

Plucknett, D.L. & N.J.H. Smith (1982). Agricultural research and third world food production. *Science* **217**: 215-220.

Rapusas, H.R. & E.A. Heinrichs (1986). Virulence of green leafhopper (GLH) colonies from Luzon, Philippines, on IR36 and IR42. *Int. Rice Res. Newsl.* **11**: 15.

Sharma, H.C. & V.F. Lopez (1991). Stability of resistance in sorghum to *Calocoris angustatus* (Hemiptera: Miridae). *J. Econ. Entomol.* **84**: 1088-1094.

Simmonds, N.W. (1991). Genetics of horizontal resistance to diseases of crops. *Biol. Rev.* **66**: 189-241.

Singh, S.R., L.E.N. Jackai, J.H.R. Dos Santos & C.B. Adalla (1990). Insect pests of cowpea. In: S.R. Singh (ed.), *Insect Pests of Tropical Food Legumes*, pp. 43-89. Chichester, England: Wiley & Sons.

Soto, P.E., I.W. Buddenhagen & V.L. Asnani (1982). Development of streak virus-resistant maize populations through improved challenge and selection methods. *Ann. Appl. Biol.* **100**: 539-546.

Tingey, W.M. & J.C. Steffens (1990). The environmental control of insects using plant resistance. In: D. Pimentel (ed.), *Handbook of Pest Management in Agriculture*, pp. 131-155. Boca Raton, Florida: CRC Press.

Velusamy, R., E.A. Heinrichs & G.S. Khush (1987). Genetics of field resistance of rice to brown planthopper. *Crop Sci.* **27**: 199-200.

Wightman, J.A., K.M. Dick, G.V. Ranga Rao, T.G. Shanower & C.G. Gold (1990). Pests of groundnut in the semi-arid tropics. In: S.R. Singh (ed.), *Insect Pests of Tropical Food Legumes*, pp. 243-322. Chichester, England: Wiley & Sons.

Proc. 8th Int. Symp. Insect-Plant Relationships, Dordrecht: Kluwer Acad. Publ.
S.B.J. Menken, J.H. Visser & P. Harrewijn (eds), 1992

Endogenous proteinase inhibitors induced in tobacco in response to herbivory: effects on the interpretation of insect resistance of transgenic tobacco plants

Maarten A. Jongsma, Bert Visser and Willem J. Stiekema
DLO-Centre for Plant Breeding and Reproduction Research, Wageningen, The Netherlands

Key words: Tomato, wound-induction, inhibitor assay

Relatively little is known about the endogenous proteinase inhibitors induced in tobacco in response to insect herbivory, despite several reports on insect resistance based on transgenic tobacco, constitutively expressing trypsin inhibitors (Hilder et al., 1987). No reference has been made to potential changes in the inhibitor level during the 5-7 days of infestation with insects. A report by Walker-Simmons & Ryan (1977) surveyed wound-induced accumulation of trypsin inhibitor activities in several plant genera. They described that tobacco (*Nicotiana tabacum* L.) induces inhibitors to higher levels than tomato (*Lycopersicon esculentum* Miller) in response to crude elicitors from plants. In our experiments we found that after an insect attack of seven days very high levels of endogenous trypsin, chymotrypsin and subtilisin inhibitors were induced in tobacco (*N. tabacum* cv. Petit Havana SR1). These endogenous levels exceeded the reported transgenic inhibitor levels. This apparent major endogenous inhibitor contribution should be taken into account when considering the total inhibitor levels required *in planta* to reduce insect growth. We decided to investigate how these inhibitors are regulated in response to a simulated insect attack in tobacco (*N. tabacum* cvs Samsun NN and Petit Havana SR1) and tomato (*L. esculentum* cv. Moneymaker). For studies of insect resistance in plants it is necessary to know the differential as well as the overall inhibitor level at which insect larval growth is affected. As far as we know no one has followed overall inhibitor activity levels for any length of time. We developed a quantitative assay to monitor overall inhibitor activity levels against trypsin, chymotrypsin and subtilisin, which allows rapid estimation of inhibitor levels in large numbers of samples.

Method

We mimicked continuous insect herbivory by removing a thin leaf strip across the midvein of leaf number 5 (counting from the cotyledons from tobacco) each day, using a pair of scissors. In the case of tomato, a leaf strip was removed from every major leaflet of leaf 5. An identical set of control plants was not wounded. To study both local and systemic effects, samples were collected daily from leaves number 3, 5 and 7. The sample from leaf number 5 was a leaf strip parallel to the wound, and the leaf tips from leaves 3 and 7 were removed. Sampled plants were discarded. The details of the analysis of the samples cannot be presented here. The method was a modification of the method of Kourteva et al. (1987).

Results

We obtained the following results:

1. Tobacco induces subtilisin inhibitors to a quarter of the trypsin and chymotrypsin inhibitor levels. Tomato induces equivalent levels for all three serine proteinases in the test.
2. Local and systemic responses are 10 times as strong in tomato as in tobacco. At the site of wounding, tomato induces levels of up to 800 IU/mg (1 IU/mg = one microgram of inhibited proteinase per milligram of total soluble leaf protein) after 8-9 days, whereas tobacco obtains levels of only 100 IU/mg.
3. The systemic response of tomato and tobacco (when present) is 10-fold weaker than the local response.
4. The systemic response in the higher leaves of tomato increases over time. In tobacco cv. Petit Havana it decreases over time, while in Samsun NN it is entirely absent.
5. The systemic response in the lower leaves of tomato and cv. Samsun NN is very late. It takes 7-8 days of wounding before induction takes place. In cv. Petit Havana we did not observe any response in the lower leaves.

Discussion

The consequences of the interpretation of insect resistance of transgenic tobacco plants are two-fold. First, we can conclude that the endogenous levels of proteinase inhibitors present after an initial period of induction are equal or higher than the transgenic level. Second, we have observed that the diversity of inhibitor specificity is greater than the specificity of the transgenic inhibitor. We suggest that these two factors, intrinsic to tobacco, may have enhanced the reported growth limiting effects of the transgenic inhibitor on lepidopteran insects. This implies that one should avoid generalizing the effects reported for tobacco as being applicable to other plant species as well. The reason why transgenic tobacco plants are resistant and control plants are not, most likely hinges on the constitutive mode of expression of the transgenic inhibitor. It is puzzling, therefore, why plants haven't evolved constitutive inhibitor levels in their leaves to protect themselves more effectively against herbivory.

References

Hilder, V.A., A.M.R. Gatehouse, S.E. Sheerman, R.F. Barker & D. Boulter (1987). A novel mechanism of insect resistance engineered into tobacco. *Nature* 330: 160-163.

Kourteva, I., R.W. Sleigh & S. Hjerten (1987). Assay for enzyme inhibition: detection of natural inhibitors of trypsin and chymotrypsin. *Anal. Biochem.* 162: 345-349.

Walker-Simmons, M. & C.A. Ryan (1977). Wound-induced accumulation of trypsin inhibitor activities in plant leaves. *Plant Physiol.* 59: 437-439.

Proc. 8th Int. Symp. Insect-Plant Relationships, Dordrecht: Kluwer Acad. Publ.
S.B.J. Menken, J.H. Visser & P. Harrewijn (eds), 1992

Effect of induced resistance mechanisms in potato and tomato plants on the Colorado potato beetle

Caroline J. Bolter
Agriculture Canada, London, Ontario, Canada

Key words: Cysteine protease inhibitor, methyl jasmonate, pathogens

Attack by herbivores causes biochemical changes in plant tissue that lead to the induction of resistance mechanisms. It has been suggested by various authors that a relationship exists between plant resistance to pathogens, such as viruses, bacteria and fungi, and to insects (Krischik *et al.*, 1991). Protection against insects conferred by compounds that render the plant resistant to pathogens has been demonstrated by many authors (Bronner *et al.*, 1991). While others have shown that plants inoculated with pathogens were at least partially protected form a subsequent attack by insects (McIntyre *et al.*, 1981). It was with this background that I decided to investigate the possibility that the development of the Colorado potato beetle, *Leptinotarsa decemlineata* (Say), (CPB), a major pest of potatoes and tomatoes in Ontario, is affected when insects are provided with a food source treated with materials that induce resistance to pathogens.

Treatments

Immediately after emerging, first-instar larvae were placed on treated plants or their appropriate controls. Insects were reared until adult emergence. Tomato plants (Bonny Best) were treated with (1) linoleic acid, 20 mg/ml applied to the upper surface of the lower three leaves of the plant (Cohen *et al.*, 1991), (2) a low dose, soil treatment with the herbicide trifluralin (2ppm) (Cohen *et al.*, 1992), or (3) a hypersensitive-inducing dose of the bacterial pathogen *Pseudomonas syringae* (2.5-5 x10^7 CFU/ml) applied to the leaves. Potato tubers (*cv*. Chieftain) were treated with the growth-promoting rhizobacteria, *Pseudomonas* spp. (Frommel *et al.*, 1991), and both tomato and potato plants were treated with methyl jasmonate (MJ) for 24 hours at 30°C (Farmer & Ryan, 1990).

Results and discussion

No significant difference was observed between duration of CPB development from larva to adult in insects fed on tomato plants treated with linoleic acid or *P. syringae* and controls, or with potatoes treated with the growth-promoting rhizobacteria. However, although duration of development was the same, there was 45% reduction in the number of adults emerging from pupation, compared to controls, when insects were reared on tomatoes treated with trifluralin. When larvae were provided with MJ-treated tomato leaves, 40% fewer adults emerged, five days later than controls. No significant difference was observed with MJ-treated potatoes. In addition, it was observed that there was, on average, 20% more cysteine protease inhibitor (CPI) activity, measured spectrophoto-metrically (McLaughlin & Flaubert, 1977), in tomato leaves treated with MJ than controls.

CPI levels in MJ-treated potato leaves were not significantly different from controls.

These preliminary experiments suggest that some of the defence-inducing treatments used do not have any effect on CPB development, while others, such as trifluralin and MJ, do. MJ-treatment of tomatoes results in the induction of compounds that decrease the rate of development of CPB and increase mortality. Farmer & Ryan (1990) showed that serine protease inhibitors (PI I & II) are induced in tomatoes after MJ treatment. However, unlike the majority of insects, Coleoptera including CPB rely on cysteine proteases for digestion, not on serine proteases (Wolfson & Murdock, 1987). Tomato leaves accumulate cysteine protease inhibitors under stressful conditions (Akers & Hoff, 1980) and CPB larvae fed with inhibitors of cysteine proteases have high mortality rates compared to controls (Wolfson & Murdock, 1987). Preliminary studies suggest that cysteine protease inhibitor(s) activity is induced in tomato plants after exposure to MJ.

These data encourage further investigation, and screening experiments using variations of the above mentioned treatments and others are in progress. Longevity, sex ratio, fecundity and overwintering potential of adults will also be investigated.

References

Akers, C.P. & J.E. Hoff (1980). Simultaneous formation of chymopapain inhibitor activity and cubical crystals in tomato. *Can. J. Bot.* **58**: 1000-1004.

Bronner, R., E. Westphal & F. Dreger (1991). Pathogenesis-related proteins in *Solanum dulcamara* L. resistant to the gall mite *Aceria cladophthirus* (Nalepa) (syn. *Eriophyes cladophthirus*). *Physiol. Mol. Plant Pathol.* **38**: 93-104.

Cohen, R., D.A. Cuppels, R.A. Brammall & G. Lazarovits (1992). Induction of resistance towards bacterial pathogens of tomato by exposure of the host to dinitroanaline herbicides. *Phytopathology* (in press).

Cohen, Y., G. Ulrich & E. Mosinger (1991). Systemic resistance of potato plants against *Phytophthora infestans* induced by unsaturated fatty acids. *Physiol. Mol. Plant Pathol.* **38**: 255-263.

Farmer, E.E. & C.A. Ryan (1990). Interplant communication: Airborne methyl jasmonate induces synthesis of proteinase inhibitors in plant leaves. *Proc. Natl. Acad. Sci. USA.* **87**: 7713-7716.

Frommel, M.I., J. Nowak & G. Lazarovits (1991). Growth enhancement and developmental modifications of *in vitro* grown potato (*Solanum tuberosum* ssp. *tuberosum*) as affected by nonfluorescent *Pseudomonas* spp. *Plant Physiol.* **96**: 928-936.

Krischik, V.A., R.W. Goth & P. Barbosa (1991). Generalized plant defense: effects on multiple species. *Oecologia* **85**: 562-571.

McIntyre, J.L., J.A. Dodds & J.D. Hare (1981). Effects of localized infections of *Nicotiana tabacum* by tobacco mosaic virus on systemic resistance against diverse pathogens and an insect. *Phytopathology* **71**: 297-301.

McLaughlin, J. & G. Flaubert (1977). Partial purification and some properties of a neutral sulfhydryl and an acid protease from *Entamoeba histolytica. Can J. Microbiol.* **23**: 420-425.

Wolfson, J.L. & L.L. Murdock (1987). Suppression of larval Colorado potato beetle growth and development by digestive proteinase inhibitors. *Entomol. exp. appl.* **44**: 235-240.

Proc. 8th Int. Symp. Insect-Plant Relationships, Dordrecht: Kluwer Acad. Publ.
S.B.J. Menken, J.H. Visser & P. Harrewijn (eds), 1992

Influence of phenolic compounds on the relationship between the cassava mealybug and its host plants

P.A. Calatayud, M. Tertuliano and B. Le Rü
Laboratory of Entomology, Orstom, Brazzaville, Congo

Key words: Antibiosis, flavonoids, *Manihot esculenta*, *Phenacoccus manihoti*, phenolic acids

The cassava mealybug, *Phenacoccus manihoti* Matile-Ferrero (Homoptera, Pseudococcideae) is a phloemophagous insect feeding on cassava *Manihot esculenta* Crantz (Euphorbiaceae; Calatayud (unpubl.). No correlation could be found between primary nutrients (sucrose and free amino acids in sieve and intercellular fluid of cassava) and antibiotic resistance, (Tertuliano & Le Rü, 1992). Analysis of secondary compounds in the phloem sap of cassava showed us only phenolic compounds (Calatayud, unpubl.). Using liquid chromatography, we assayed phenolic acid and flavonoid contents of phloem sap and intercellular fluid of leaves (sampling by the modified centrifugation method of Rohringer *et al.*, 1983) from cassava mealybug host plants, to see whether the presence of these compounds is related to plant resistance (antibiosis).

Table 1. Intrinsic rates of increase (Rc) of *P. manihoti* on eight host plants (Tertuliano & Le Rü, 1992). Means (± S D) with different letters are significantly different (P < 0.05; n = 417 to 574)

Host plants	Rc
Poinsettia	0.038 ± 0.003 c
Incoza	0.133 ± 0.003 b
Faux caoutchouc	0.141 ± 0.003 b
Zanaga	0.155 ± 0.009 a
3 M 8	0.141 ± 0.005 b
Talinum	0.150 ± 0.001 ab
Ganfo	0.160 ± 0.003 a
30 M 7	0.150 ± 0.011 ab

We tested eight plants differing in host resistance (antibiosis) to cassava mealybug (Table 1): five varieties of cassava (*M. esculenta*), "Faux caoutchouc" (hybrid between *M. esculenta* x *M. glaziovii* Mull. Arg.), Poinsettia (*Euphorbia pulcherrina* Willd.) and Talinum (*Talinum triangularae* Jacq.).

Mean concentrations of total phenols from two substitute plant species (Poinsettia and Talinum) were found to be higher in infested plants than in unifested ones (Fig. 1). This

255

response to mealybug attack was significantly higher (P < 0.05) in Poinsettia, considered to be the most resistant plant (antibiosis) among the host plants tested. After discriminant analysis, major phenolic compounds were found to be higher in infested cassavas than in unifested ones. Only one compound (glycosyl flavonoid) was correlated ($r^2 = 0.7$; $P < 0.05$) to cassava resistance (Fig. 2). This compound seems to reduce the development of *P. manihoti*.

Mean values (mg/ g dry weight of extracts)

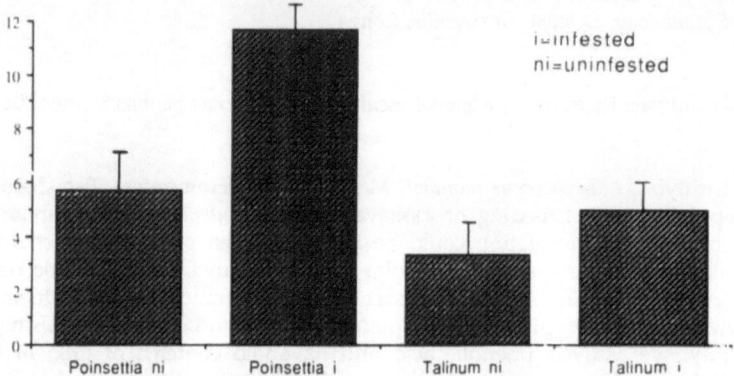

Figure 1. Mean concentrations of total phenols in phloem sap and intercellular fluid of leaves from Poinsettia and Talinum.

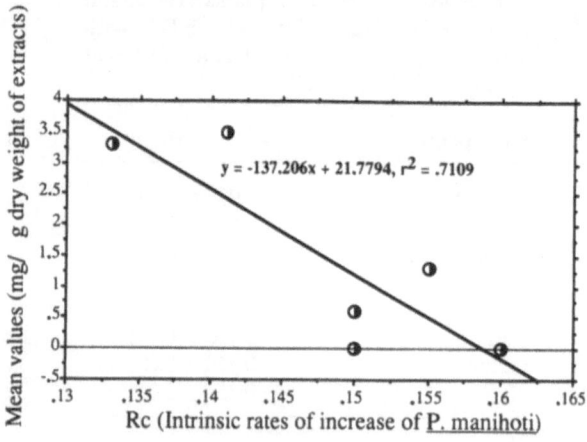

Figure 2. Linear relationship existing between mean concentrations of one glycosyl flavonoid (*) and Rc values of tested cassavas.

These results show that phenolic compounds, and in particular glycosyl flavonoid, present in phloem sap and intercellular fluid of leaves are correlated with plant resistance (antibiosis) to cassava mealybug. Since the correlations observed alone are insufficient to prove causal relationships between phloem composition and mealybug performance, we must complement the analyses by experiments using artificial diets simulating the sap of cassava to test some of the factors presumed to be active in the mealybug response.

Acknowledgements. We thank S. Dossou-Gbete for statistical analysis and A. Kiakouama for technical assistance.

References

Rohringer, R., F. Ebrahim-Nesbat & G. Wolf (1983). Protein in intercellular washing fluids from leaves of barley (*Hordeum vulgare*). *J. exp. Bot.* **34**: 1589-1605.

Tertuliano, M. & B. Le Rü (1992). Relationship between cassava mealybug and its host plants: Influence of various host plants on the expression of mealybug growth potential. *Entomol. exp. appl.* **64**: 1-9.

Proc. 8th Int. Symp. Insect-Plant Relationships, Dordrecht: Kluwer Acad. Publ.
S.B.J. Menken, J.H. Visser & P. Harrewijn (eds), 1992

Measurements of host-plant resistance in *Chrysanthemum* to *Frankliniella occidentalis*

F.R. van Dijken
Centre for Plant Breeding and Reproduction Research, Wageningen, The Netherlands

Key words: Damage, flowers, image-analysing-system

The most important and difficult-to-control pest in ornamental crops grown in greenhouses both in Europe and in the U.S.A., is the Western Flower Thrips, *Frankliniella occidentalis* (Pergande) (Robb, 1989). This insect can go unnoticed and builds up a large population before it is detected. Therefore the growers treat the plants preventively with insecticides. Over 100 treatments are carried out every year in some crops, which results in a significant pollution of the environment.

In the Netherlands, chrysanthemum flowers are grown in greenhouses, covering an area of approximately 735 ha. The flower production amounts to more than 300 million dollars each year. There is no biological control and 15 kg/ha/year of pesticides are sprayed to prevent the plants becoming infested with thrips. The Dutch government demands a 20% reduction in the amount of pesticides applied over the next five years. Host-plant resistance to thrips will help to realise this goal. After 1992 new strong reductions are foreseen.

Several crops (*e.g.*, melon, cotton, cucumber, rose and carnation) exhibit variation in the susceptibility to *F. occidentalis* (Mantel, 1989). Variation in host-plant resistance of chrysanthemum to some insects and mites has been found (*e.g.*, to *Liriomyza trifolii* (Burgess), *Myzus persicae* (Sulz.) and *Tetranychus urticae* Koch). In 1987, Mollema started an investigation on host-plant resistance to the Western Flower Thrips in chrysanthemum. His pilot experiments showed considerable variation in resistance and therefore a large project on this subject was started.

Survey of host-plant resistance

An inquiry among Dutch breeders resulted in a sample of 13 different chrysanthemum cultivars, which were thought to vary considerably in their reaction to *F. occidentalis*. These cultivars were grown in a greenhouse under normal conditions (temp. 21-35°C; 40-80% RH). To induce flowers, the light-dark cycle was 8-16 h. Several experiments were performed with young plants without flowers, and with flowering plants. The plants were kept in cages, one plant per cage, and infested with 20 adult females. Measurements of insect damage to plants were made four weeks after infestation. Plant length, number of leaves, total leaf area and the area of the leaves that had been damaged were measured. At harvest, a sample of the plant was taken to count the thrips.

Damage assessment. The leaf area damaged was measured using the Technical-Command-Language-Image analyzing system developed by the Technical University of Delft. The software had to be adapted to recognize the different types of damage. The damage depended on the stage of the plant attacked and also on the cultivar. Only the damage on the upperside of the green leaves was measured, because it turned out that in chrysanthemums the thrips mainly feed here.

Plants without flowers. The classification on young plants without flowers was as follows: 1) resistant plants with less than 10 mm^2 damage; 2) partial resistant plants with 10-100 mm^2 damage and 3) susceptible plants with more than 100 mm^2 damage per plant. Table 1 shows the results of three cultivars selected to represent each of the three categories (91003 susceptible, 91010 partly resistant and 91013 resistant).

Table 1. Measurements of the effects of *F. occidentalis* on three chrysanthemum cultivars (n = 5). (Number of plants per category, n = 5)

Cultivar	91003		91010		91013	
Character	Infested	Control	Infested	Control	Infested	Control
Plant length cm	53.4	59.4	40.0	43.6	42.2	43.6
Total leaf area (cm^2)	260.7	298.4	486.0	602.4	425.7	436.5
Area of damaged leaf by thrips (mm^2)	111.6	1.2	20.6	0	7.4	0
No. of thrips in sample	1.8	0.2	1.0	0	0.2	0

Analysis of variance showed that there was no significant interaction between genotype and treatment for plant length ($F_{2,24}$ = 0.817) and total leaf area ($F_{2,24}$ = 3.008).

The number of thrips counted was extremely low and this contributed to the opinion that chrysanthemums are not suitable host plants for *F. occidentalis*.

Plants with flowers. The number of thrips (ñ) found on flowering plants was very high which led to the conclusion that waiting four weeks in my experiment after infestation is too long to get an amenable number of thrips. (*cv.* 91003: ñ = 762; *cv.* 91010: ñ = 698; *cv.* 91013: ñ = 350).

The rank order of the number of thrips in flowering plants is the same as in the damaged leaf area in non-flowering plants. When compared to the damaged leaf area in flowering plants (x) however, the order is reversed: *cv.* 91003: x = 21.8 mm^2; *cv.* 91010: x = 58.6 mm^2 and in *cv.* 91013: x = 73.6 mm^2.

In these three cultivars there was no correlation at all between the damaged leaf area of the flowering plants and the number of thrips, neither within nor between the cultivars. None of the cultivars of chrysanthemum tested had an absolute resistance to *F. occidentalis*.

Recommendations for further research

There is a great potential for continued research, such as:
1. The level of resistance is age/stage dependent so there may be resistance in other stages;
2. the growers' chrysanthemum, *Dendranthema grandiflorum* (Ramat) S. Kitamura has a huge genetic variability and resistance may be found in other cultivars;

3. there are several uncultivated related species, which can possibly be used as a source of resistance;
4. and last but not least it may be possible to introduce resistance genes from completely different organisms by means of transformation.

In all these cases tests for host-plant resistance will depend on both the characters of the insect and of the plant.

References

Mantel, W.P. (1989). Bibliography of the western flower thrips *Frankliniella occidentalis* (Pergande) (Tripanoptera: Thripidae). *Bull. SROP/WPRA* 12: 29-66.

Robb, K.L. (1989). Analysis of *Frankliniella occidentalis* (Pergande) as a pest of floricultural crops in California greenhouses. Ph.D. Dissertation, Univ. Calif. Riverside: 135 pp.

Proc. 8th Int. Symp. Insect-Plant Relationships, Dordrecht: Kluwer Acad. Publ.
S.B.J. Menken, J.H. Visser & P. Harrewijn (eds), 1992

An automatic and accurate evaluation of thrips-damage. Image-Analysis: a new tool in breeding for resistance

Chris Mollema, Folchert van Dijken, Kees Reinink and Ritsert Jansen
DLO-Centre for Plant Breeding and Reproduction Research, Wageningen, The Netherlands

Key words: Damage-rating, grey level values, partial resistance

The Western Flower Thrips (WFT) *Frankliniella occidentalis* (Pergande) is a major pest world-wide (Brødsgaard, 1989). Besides causing direct feeding damage, this insect transmits tomato spotted wilt virus. As chemical and biological control of WFT is difficult, the development of resistant crops is vital. At CPRO-DLO, research is aimed at obtaining resistance in cucumber and chrysanthemum.

The application of partial resistance in plant breeding requires precise methods of measurement. Visual damage-rating of infested plants is time-consuming or inaccurate. Therefore, an automatic, fast and accurate method was developed using Image-Analysis techniques (Groen *et al.*, 1988; McClatchy, 1988).

Equipment

Image-Analysis consists of a high resolution CCD video camera, a colour monitor, a personal computer and a framegrabber. The software operating this system is "TCL-Image*". We developed a special software application "DAMEASURE" for automatic discrimination between a damaged and an undamaged leaf area.

Procedure

Image-Analysis uses images of 512 x 512 pixels as input. Each pixel has a grey level value between 0 (= black) and 255 (= white). When a damaged leaf (showing yellowish spots) is exposed to the camera, the system automatically separates the pixels corresponding to the damaged area from the pixels corresponding to the undamaged area. The result is visible by presentation of the spots in contrasting colours. Comparison of this presentation with the original image allows the operator to decide whether the discrimination between the damaged and undamaged area is correct or not. If it is correct, the following output is shown: plant number; number of spots; number of small spots; number of large spots; mean spot size (mm^2); total area damaged (mm^2); total leaf area (mm^2); total damaged area (%). Conversion to mm^2 is according to previous calibration.

* TCL stands for Technical Command Language. See TCL-Image user's manual (1990). TNO Institute of Applied Physics, Delft, The Netherlands.

If the automatic discrimination is not correct, a number of other options (manual corrections) are available:
1. Automatic separation within area of interest (mouse-operated).

2. Manual separation of damaged and undamaged area (mouse-operated). The appropriate threshold can be selected by visual comparison between the original image and the image after thresholding. Pixels in the image with grey level values exceeding the threshold value are indicated in red.
3. Rub out of objects which that are NOT related to damage, *e.g.*, veins or chlorotic spots with grey values similar to the thrips-damage (mouse-operated).
4. Indication of objects that are related to damage. In case most objects are NOT related to damage, it is quicker to disregard all objects except those indicated (mouse-operated).
5. Saving the image for later processing.
6. Combination of options.

Results

Speed. One automatic measurement takes approximately 30 s. If manual corrections are necessary, the total time may increase to 100 s. Another advantage is that the data are fed directly into the computer and can be transferred to MS-DOS compatible programs (*e.g.*, Lotus) without delay.

Accuracy. The automatic measurement was compared with a detailed visual method (counting spots in three categories for size). The highly significant correlation ($R^2 = 0.80$; $N = 23$) between the visual method and the automatic method proves its accuracy and reliability. The visual method, however, over-estimated individual spot sizes.

Conclusion

The automatic method is unbiased, independent of the operator's condition and more than 10 times quicker.

References

Brødsgaard, H.F. (1989). *Frankliniella occidentalis* (Thysanoptera: Thripidae) - a new pest in Danish glasshouses. A review. *Tidsskr. Planteavl* **93**: 83-91.
Groen, F.C.A., R.J. Ekkers & R.H. de Vries (1988). Image processing with personal computers. *Signal Processing* **15**: 279-291.
McClatchy, W. (1988). Extracting data from images. *MacWeek* Vol. **2**: 45.

Proc. 8th Int. Symp. Insect-Plant Relationships, Dordrecht: Kluwer Acad. Publ.
S.B.J. Menken, J.H. Visser & P. Harrewijn (eds), 1992

Susceptibility of *Chrysanthemum* cultivars to thrips (*Frankliniella occidentalis*) infestation and the role of some physical plant characters

C.M. de Jager and R.P.T. Butôt
University of Leiden, Leiden, The Netherlands

Key words: Hairiness, leaf toughness, leaf thickness, resistance, leaf hairines

The use of pesticides in crop protection is a great threat to the environment. The use of pest-resistant crops could possibly help to reduce this risk. The aim of this project is to develop a biotechnological test-kit to scan young chrysanthemum plants for resistance to pests like *Frankliniella occidentalis* (Pergande) and *Liriomyza trifolii* (Burgess). Chrysanthemum is an ornamental crop of value to the dutch economy and the thrips species *F. occidentalis* is one of the most important pests on chrysanthemum and many other crops. Before developing a test-kit we need to know more about the mechanisms which cause resistance for *F. occidentalis* in chrysanthemum.

Methods

Differences in susceptibility. Two methods were developed to measure differences in susceptibility of five chrysanthemum cultivars.
Method 1: five first instar larvae were put into a leaf cage. The leaf cage was placed between leaf number five and leaf number ten. Ten plants per cultivar were used. The mortality of the larvae was determined and the amount of leaf damage was estimated every other day for three weeks.
Method 2: whole plants without (A) and with (B) flowers were put into tubes with twenty adult female thrips. Ten plants per cultivar were used. After four weeks, the amount of damage was estimated and the thrips were counted after being dislodged from the plants by washing with 50% aethanol.

All plants were kept under growth room conditions: 18°C day and 19°C night temperature, 8 h light and 70% RH. Once a week the plants received *ca* 75 ml of a 2 EC nutrient solution. For the first method, the plants were five weeks old; for the second method they were three weeks (2A) and nine weeks (2B) old. The plants that were used for measuring the physical leaf characters were five weeks old. The thrips for the experiments were reared on flowers of a susceptible chrysanthemum cultivar at 20°C both day and night, 12 h light and 70% RH. The experimental conditions were similar.
Physical characters. Toughness, hairiness and thickness were measured on the eighth leaf of each plant. Toughness was measured with a penetrometer. Hairiness was determined by counting the number of T-hairs per square centimetre. Leaf area and the water content measurements were done on leaf number seven and nine. The water content of the leaves was calculated from the fresh and dry weight of the leaves. Physical leaf characters were correlated with the susceptibility found in leaf cages.

Results and discussion

Differences in susceptibility. There was a significant difference in damage and number of thrips between the five cultivars (ANOVA, $P < 0.0001$) in leaf cages. In tubes a twenty- to sixty-fold increase in thrips numbers was found when plants without flowers were compared to plants with flowers (Fig. 1a and 1b). Both experiments, with and without flowers, showed a significant difference in thrips numbers ($P < 0.001$), and damage (2A: $P = 0.020$. 2B: $P = 0.0014$) between cultivars. In the experiment with flowers, the number of flowers explained 44% ($P < 0.0001$) of the differences in thrips numbers (Fig. 1a and 2a). When corrected for the number of flowers, there was a striking resemblance with the differences in susceptibility found in the experiment with plants without flowers (Fig. 1b and 2b).

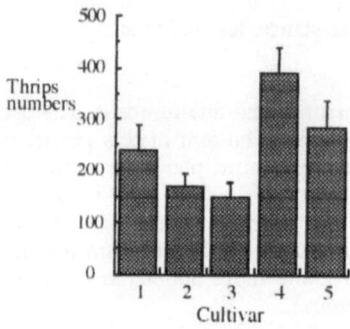

Figure 1a. Thrips numbers on plants with flowers (2B).

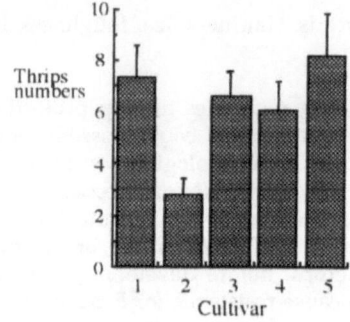

Figure 1b. Thrips numbers on plants without flowers (2A).

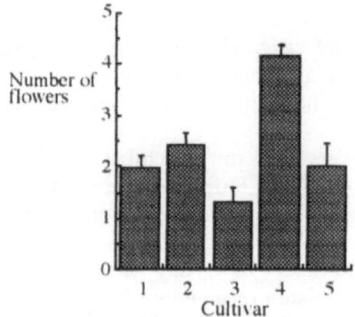

Figure 2a. Mean number of flowers per plant (2B).

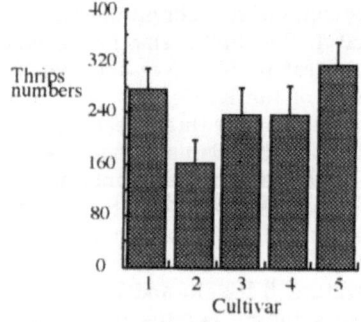

Figure 2b. Thrips numbers corrected for number of flowers (2B).

Physical characters. There were significant differences in all physical characters between the five cultivars (ANOVA, $P < 0.0001$). Toughness was positively correlated ($R = 0.88$, $P = 0.047$) and the leaf area was negatively correlated ($R = -0.89$, $P = 0.041$) with survival of thrips larvae in the leaf cages (method 1). The mechanism of these relationships is still unclear.

Acknowledgements. The authors wish to thank Dr T.J. de Jong, Dr P.G.L. Klinkhamer and Prof. Dr E. van der Meijden for their useful comments on the experimental design and help with the data evaluation.

Proc. 8th Int. Symp. Insect-Plant Relationships, Dordrecht: Kluwer Acad. Publ.
S.B.J. Menken, J.H. Visser & P. Harrewijn (eds), 1992

Effects of resistance in cucumber upon life-history components of *Frankliniella occidentalis*

Carmen Soria and Chris Mollema
DLO-Centre for Plant Breeding and Reproduction Research, Wageningen, The Netherlands

Key words: Antibiosis, antixenosis, reproduction, survival, thrips

The Western Flower Thrips (WFT), *Frankliniella occidentalis* (Pergande) is one of the most serious insect pests in greenhouses in the Netherlands. Chemical and biological control of this insect is difficult, therefore breeding for resistant varieties is a solution to this problem. Although cucumber genotypes with low levels of damage after standard infestation with WFT have recently been found by Mollema *et al.* (unpubl. results), nothing is known about the effects of these genotypes upon the life-history components of WFT. Therefore, we decided to study three parameters (reproduction, generation time and survival). The importance of this approach in describing effects of host-plant resistance has already been stated by Trichilo & Leigh (1985).

Material and methods

The life-history components of WFT were determined on leaf discs (1.5 cm ø) laid upside down on a little tap water inside wells of tissue culture plates and covered with transparent plastic film (Mollema *et al.*, 1990). The experiments were carried out in a climate room at 27 ± 1°C, 50% RH and 16:8 (L:D) photoperiod.

Reproduction. One-day-old females together with one male were isolated on 24 leaf discs of each genotype. After one day, the males were removed and females were transferred to new leaf discs daily for four days, and the number of hatched larvae per disc was recorded.

Generation time and survival. Four WFT females per leaf disc were allowed to lay eggs for a period of five hours. Newly hatched larvae were isolated on fresh leaf discs. Two plants of each genotype were used and 48 leaf discs were taken from each plant. The transition to later stages and survival was recorded twice daily and the leaf discs were refreshed twice weekly. The experiment was continued until adult emergence.

Results

Reproduction. The reproduction rates of WFT females were lower on the resistant genotypes than on the susceptible genotype used as control (G6). Peak reproduction of the WFT females was observed at the age of two days (Table 1).

Table 1. Average number of hatched larvae per WFT female on seven cucumber genotypes

Age females (days)	Genotypes						
	G6	9143	9153	9130	9140	9104	9127
1	0.68	0.58	1.00	0.86	0.43	0.53	0.26
2	3.00	1.42	1.59	1.14	1.43	2.00	2.00
3	1.82	0.37	0.41	0.24	1.05	0.47	0.58
4	1.27	0.00	0.12	0.05	0.28	0.05	0.00

Generation time and survival. The total period from egg to adult was shortest on the susceptible control genotype G6 (12.6 days). On most resistant genotypes this period was ± 13.7 days, except on genotype 9140 where it was 20.1 days. The generation time is estimated two days more for all genotypes.

The survival of all WFT developmental stages on the susceptible control genotype (G6) was higher than on the resistant genotypes. The second larval stage was the most critical. On the resistant genotypes, survival in this stage was 10% or less while on the susceptible control it was more than 60% (Table 2).

Table 2. Survival (%) of *Frankliniella occidentalis* developmental stages on seven cucumber genotypes

Developmental stage	Genotypes						
	G6	9143	9153	9130	9140	9104	9127
Until L2	92.70	82.70	78.20	79.15	71.85	69.80	80.20
Until Prepupa	61.42	10.40	8.78	0.00	3.30	6.25	0.00
Until Pupa	61.42	6.85	5.55	0.00	3.10	5.20	0.00
Until Adult	56.25	5.60	4.50	0.00	3.10	4.20	0.00

Conclusions

- Antixenotic resistance is expressed by reduced reproduction of WFT.
- Developmental period is prolonged on the resistant genotypes.
- For all resistant genotypes, antibiotic resistance is expressed by high larval mortality, particularly at the second larval stage.

References

Mollema, C., G. Steenhuis & P. van Rijn (1990). Development of a method to test resistance to western flower thrips (*Frankliniella occidentalis*) in cucumber. *IOBC/WPRS Bulletin* **13(6)**: 113-116.

Trichilo, P.J. & T.F. Leigh (1985). The use of life tables to assess varietal resistance of cotton to spider mites. *Entomol. exp. appl.* **39**: 27-33.

Proc. 8th Int. Symp. Insect-Plant Relationships, Dordrecht: Kluwer Acad. Publ.
S.B.J. Menken, J.H. Visser & P. Harrewijn (eds), 1992

The impact of environmental conditions on survival of the leaf miner *Liriomyza trifolii* on *Chrysanthemum* cultivars

Marinke J. van Dijk, Cilke Hermans, Jan de Jong and Ed van der Meijden
Dept of Population Biology, University of Leiden, Leiden, The Netherlands

Key words: Agromyzidae, Diptera, host plant resistance, light intensity, nitrogen gift, performance

Is resistance to leaf miners in chrysanthemum affected by culture practices? Probably the most important quality factor for insects is leaf nitrogen content. We decided to manipulate plants by varying the nitrogen gift. As we noticed that plants raised in greenhouses in the winter showed much less susceptibility than plants raised in the summer, we also decided to study the effects of light intensity. A third factor that we varied in the experiment was the day length. We expected this to lead to nutrient uptake stress, and therefore a more pronounced effect of the decrease in nitrogen gift.

Materials and methods

The experiment was set up on rockwool plugs (9.5 cm diam. x 6 cm high) in two series: the first with continuous light, the second with 16 hours day length. We used a susceptible and an intermediately susceptible cultivar for each series. The high light intensity chamber was set at 64 W/m², and the half light intensity chamber at 32 W/m². The temperature was maintained at 20°C, 70% RH. Each chamber was given three different nitrogen treatments (blocks): normal nutrient mixture, nitrogen reduced to a half, and nitrogen reduced to a quarter of the normal gift.

We assessed oviposition differences, larval development, and measured plant dry weight, nitrogen content, total phenolics (chlorogenic acid equivalents), leaf toughness (penetrometer), leaf area, fresh weight, number of leaves and stem length. For the oviposition experiments, pairs of flies were placed in leaf cages on three seperate leaves of each plant. After one day the flies were removed and after the eggs had hatched, the leaves were boiled for three min after which we counted the eggs per fly pair as well as the number of first stage larvae (L1). For the larval survival experiments, plants were individually placed in perspex cylinders covered with gauze lids with two pairs of flies. The flies were removed after one day and the larvae were allowed to develop. The number of L1 mines, survival from L1 to pupa and number of pupae were recorded.

Results and conclusion

Influence on leaf-miner performance. Leaf-miner performance was influenced by the various environmental conditions as shown in Table 1. The effects of reducing N-gift were strongest on larval survival. Continuous day length also had the strongest impact on larval survival. Reducing the light intensity did not unequivocally reduce larval survival, although it did reduce oviposition. Number of pupae, as cumulative measure of success,

Table 1. Effects of the different treatment variations on leaf-miner performance measures and plant characteristics. S = susceptible cultivar, IS = intermediately susceptible cultivar. + = significant positive effect, - = significant negative effect

Leaf-miner performance	Cultivar	Reducing N-gift	Reducing light intensity	Day length 16 → 24 h
Oviposition	no effect	-	-	
Egg → L1 survival	no effect	+	no effect	+
L1 → pupa survival	S > IS	--	S ++, IS -	--
No. of pupae	S > IS	-	no effect	-
Plant characteristics				
% dry weight	no effect	+	-	+
% N in leaf	no effect	-	+	-
Phenol content	S < IS	+	-	+
Leaf toughness	S < IS	+	-	+
Plant growth		-	-	-

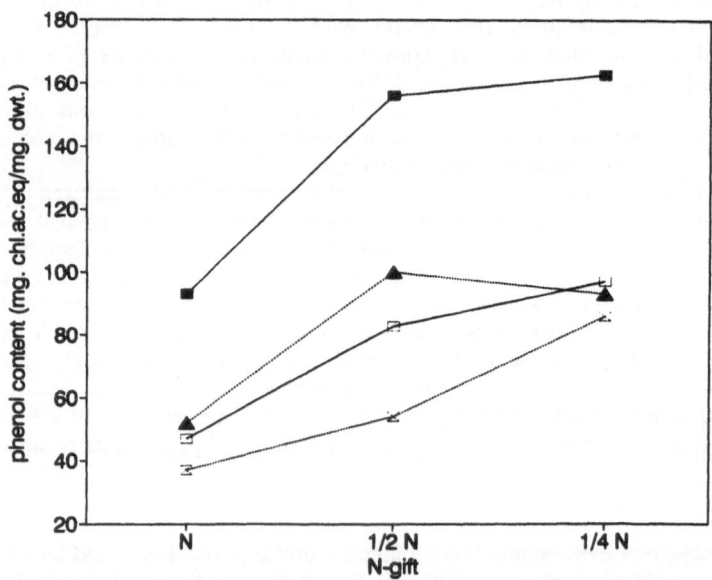

Figure 1. Interaction between the effects of nitrogen gift, day length and plant genotype (cultivar) on phenol content of the leaves. Intermediately susceptible genotype at 24 h (■) and at 16 h day length (□); susceptible genotype at 24 h (▲) and at 16 h day length (⊠).

was only negatively influenced by reducing N-gift and elongating day length to 24 h.

Reducing N-gift had a much more pronounced effect on survival under continuous light conditions than it has under normal long-day conditions.

Influence on plant characteristics. We expected the following traits to have a negative impact on fly performance: % dry weight, lower % N, phenol content and leaf toughness. The effects of the different treatment levels on those traits are shown in Table 1, together with the mean influence on plant growth.

As shown, reducing the light intensity had exactly the opposite effect on the resistance components measured as the other treatment factors. This partially explains why no clear negative effect of light intensity reduction was found on leaf-miner performance. Elongating the day length had the same effect on plant traits as reducing N-gift, and probably by the same underlying mechanism: continuous light induces nutrient uptake stress and therefore effectively means the same as reducing N-gift. The interaction between cultivar, N-gift and day length for phenol content is shown in Fig. 1. In the continuous light treatment both cultivars seemed to have reached their maximum phenol production at the ½ N treatment.

Unfortunately, all factors clearly reduce plant growth, a phenomenon that is unacceptable for growers, who sell their cut chrysanthemums in weight classes. The reduction in plant fresh weight caused by reducing N-gift is important; it ranges from 20% at ½ N to 40-50% at ¼ N.

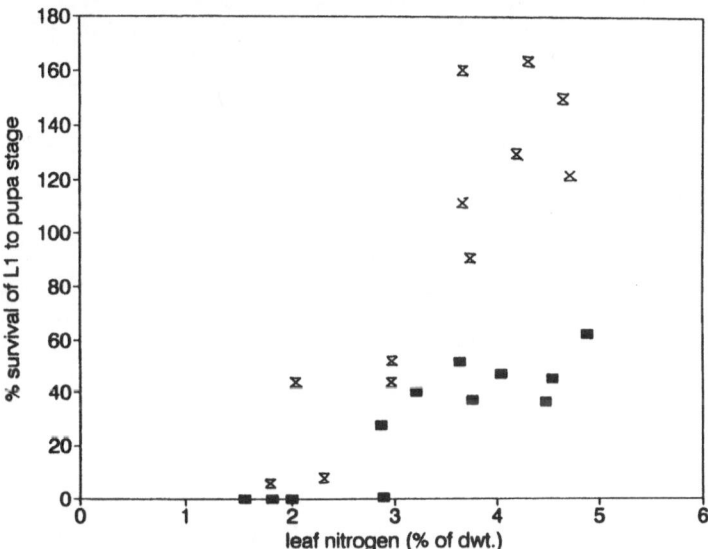

Figure 2. Relationship between leaf nitrogen content (% of dry weight) and leaf-miner survival from first instar to pupa for the two plant genotypes. Percentages higher than 100% are due to errors in estimation of first instar miner (■ = intermediately susceptible genotype, ⵝ = susceptible genotype).

Relationship between plant traits and leaf-miner performance. Using the cultivars as a factor, we determined the relative influence of % nitrogen in the leaf, phenol content and toughness on larval survival. The difference in survival between the two cultivars used could not be explained by any of the earlier mentioned resistance traits, implying that there must be a more important trait for between cultivar differences. For the within cultivar differences in survival, only % nitrogen in the leaf offers a satisfactory explanation. Phenol content and toughness are strongly negatively correlated to % nitrogen, but do not significantly contribute to the explained variance. Fig. 2 shows the relationship between % N in the leaf and larval survival for both cultivars.

In conclusion, the culture measures that negatively influenced leaf nitrogen content automatically reduce leaf-miner larval survival. However, it must be realized that plant growth will also be influenced. In high light conditions it is unwise to supply more nitrogen than is absolutely necessary to maintain normal growth.

Proc. 8th Int. Symp. Insect-Plant Relationships, Dordrecht: Kluwer Acad. Publ.
S.B.J. Menken, J.H. Visser & P. Harrewijn (eds), 1992

Selection for resistance to insects causing ramification in *Salix viminalis* plantations

Inger Åhman
Resistance Breeding Department, Svalöf AB, Svalöv, Sweden

Key words: Cecidomyiidae, gall midge, Lepidoptera, plant breeding, willow

During the mid 1980ies a breeding program was started in Sweden to improve yield, resistance to pests and diseases, and frost tolerance in a new type of crop, *Salix viminalis* L. grown to produce biomass for fuel. There are several species of insects that kill the apical meristem of shoots in *Salix* (willows), thereby stopping terminal growth. Subsequently side shoots develop. Such ramification causes prolems at the production of stem cuttings for new plantations. Ideally, the cuttings are produced from one-year-old rods that are long, unbranched and 1-2 cm in diameter. In the breeding program, the plantations of clones with one-year-old shoots are scored for the proportions of shoots with apical damage by the end of the season. A sample of five plants per replication is

Proportion of shoots without apical damage

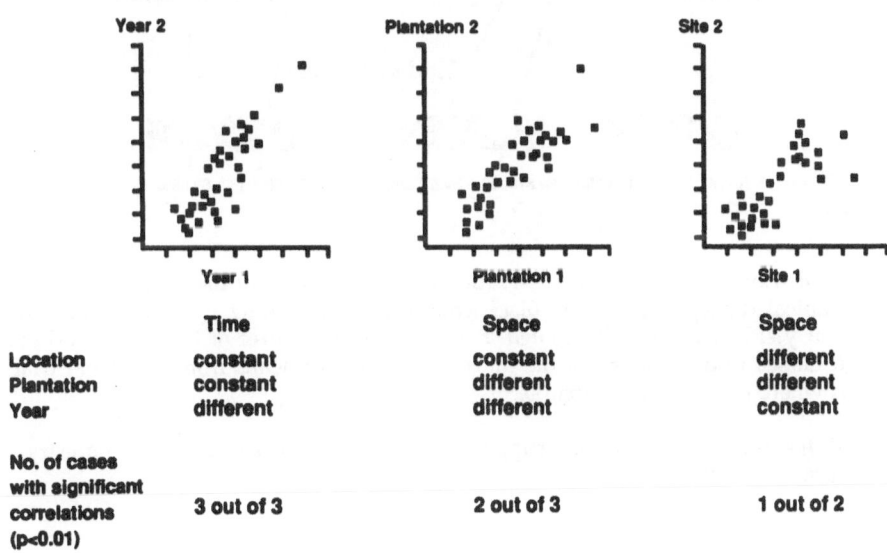

Figure 1. Consistency between scores of insect attack of clones scored in different years, plantations and locations.

scored. Damage by chewing larvae of Lepidoptera can be distinguished from that made by an unidentified gall midge. Damage due to this gall midge, *Dasineura (Rabdophaga) sp.n.* predominates in plantations where the gall midge occurs.

Results

For resistance breeding to be successful, there needs to be accordance in time and space concerning how seriously the various clones are attacked. The comparisons of the damage scores between years, plantations and locations were in accordance in 6 out of 8 comparisons (Fig. 1).

The clones to be released on the market have got a better or an equally good resistance to insects causing apical damage to shoots than the clones common in plantations today, even though yield and leaf rust scores are given a higher priority than insect attack when clones are selected in the breeding program (Fig. 2).

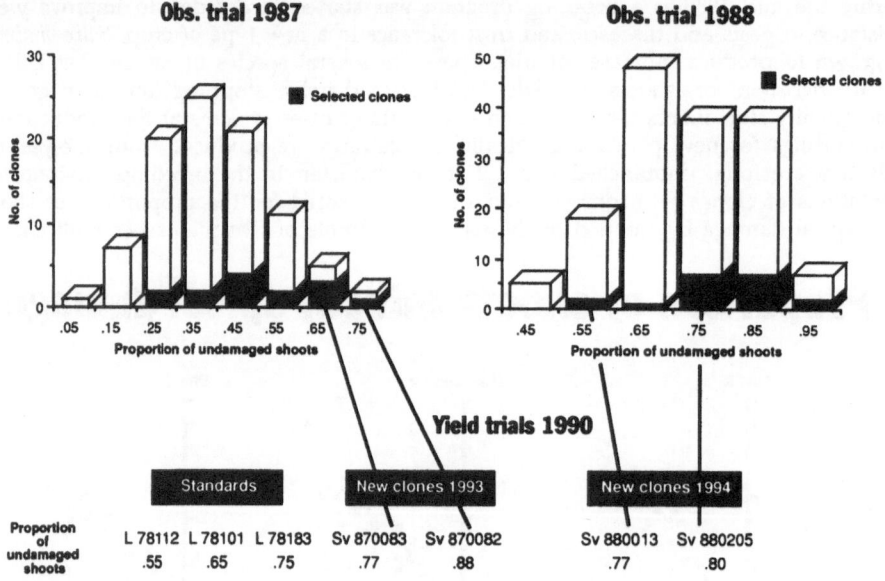

Figure 2. Distribution of clones in observation trials on classes for the proportion of shoots without apical damage by insects. Black columns indicate how the clones selected for subsequent yield trials were distributed. Below, the mean scores of three standard clones and four clones to be released on the market are shown (means from two yield trials in locations where the gall midge occurs).

Acknowledgement. This work was supported by the Swedish Council for Forestry and Agricultural Research.

Proc. 8th Int. Symp. Insect-Plant Relationships, Dordrecht: Kluwer Acad. Publ.
S.B.J. Menken, J.H. Visser & P. Harrewijn (eds), 1992

Development of tolerance to plant allelochemicals by the Colorado potato beetle

William W. Cantelo
USDA, ARS, Beltsville Agricultural Research Center, Beltsville, Maryland, USA

Key words: Host-plant resistance, *Solanum berthaultii*, *Solanum chacoense*

Efforts to develop potato breeding lines resistant to the Colorado potato beetle (CPB), *Leptinotarsa decemlineata* (Say), in the United States have concentrated on the use of *Solanum chacoense* Bitt. and *Solanum berthaultii* Hawkes as parental material. *S. chacoense* is unique in that a few accessions contain leptine, the only glycoalkaloid known that is toxic to the CPB but is not present in the tubers (Sinden *et al.*, 1991). *S. berthaultii* contains glandular trichomes on the foliage that entraps small insects, retards CPB development and deters CPB oviposition (Wright *et al.*, 1985).

Development of tolerance. The genetic lability that enables the CPB to develop tolerance to insecticides might be expected to enable it to develop tolerance to plant allelochemicals. To test this hypothesis, CPB neonates were placed on a highly resistant line selected from *S. chacoense* (P.I. 320287) and kept on this foliage until they moulted to the 2nd instar. These were then transferred to tomato foliage (*cv.* Manapal) for rearing to adulthood. Tomato is a good alternate host for the CPB in the Mid-Atlantic states where the CPB is a major pest of tomatoes. This selection process was repeated for four generations. Because of the unexpected rapidity of tolerance increase, the test was repeated. The same procedure was followed using samples from a highly resistant line of *S. berthaultii* (P.I. 265858). After moulting, the 2nd instars were transferred to potato foliage (*cv.* Kennebec) and reared through to adulthood.

Results and discussion

The percentage of the neonates that survived to the 2nd instar after having fed on the *S. chacoense* clone in the first test was 11, 37, 36, and 56 for generations 1 through 4, respectively. In the second test the percentages were 12, 24, 31, and 39 for the four generations. For comparison, the percentage that survived after having been reared only on tomato was 86. Neonates placed on *S. berthaultii* had a per cent survival of 40, 42, 45, 34, and 37 for generations 1 through 5, respectively. Those reared on 'Kennebec' had 84% moult to the 2nd instar. The number of days in the first stadium was: potato 2.9, tomato 3.5, *S. chacoense* first test 5.3, *S. chacoense* second test 5.2, *S. berthaultii* 5.3.

Both tests with *S. chacoense* indicated that the CPB can rapidly develop tolerance to a plant allelochemical. Development of insecticide resistance is considered to be the result of insecticide applications killing all but those few individuals that contain the genes for the resistance mechanism, perhaps one pair in a million. The results in this study suggest that *ca.* 10% of the population contained genes for resistance to the toxins of this *S. chacoense* clone, presumably leptine. It should be noted that the CPB is not found in the Andes mountains, where *S. chacoense* is indigenous, and thus had no need for developing a means for overcoming the specific defenses of this plant. During the five generations of observing *S. berthaultii*, the percentage moulting did not change significantly. The ability to overcome the physical defenses of glandular trichomes may be less common in the CPB. Despite the difference between the *Solanum* spp. in their effect on moulting percentages, they both extended the time required for the 1st stadium beyond that on 'Kennebec' by 2.4 days.

Applying current theory for the prevention of insecticide resistance, host-plant resistance could best be maintained by mixed plantings of genotypes, such as hybrids of *S. chacoense* and *S. berthaultii* interplanted with susceptible *cvs*. If the rapid development of tolerance by the CPB to *S. chacoense* is genetically based and occurs in the evolutionary process, evolutionary changes may be much more rapid than generally believed.

References

Sinden, S.L., W.W. Cantelo, L.L. Sanford & K.L. Deahl (1991). Allelochemically mediated host resistance to the Colorado potato beetle, *Leptinotarsa decemlineata* (Say) (Coleoptera: Chrysomelidae). *Mem. Entomol. Soc. Can.* **159**: 19-28.
Wright, J.W., M.B.Dimock, W.M. Tingey & R.L. Plaisted (1985). Colorado potato beetle (Coleoptera: Chrysomelidae) expression of resistance in *Solanum berthaultii* and interspecific hybrids. *J. Econ. Entomol.* **78**: 576-582.

Proc. 8th Int. Symp. Insect-Plant Relationships, Dordrecht: Kluwer Acad. Publ.
S.B.J. Menken, J.H. Visser & P. Harrewijn (eds), 1992

The use of restriction fragment length polymorphism analysis for identifying biotypes of the virus vector aphid *Amphorophora idaei*

A.N.E. Birch[1], G. Malloch[1], B. Harrower[1], M.A. Catley[2] and A.T. Jones[1]
[1]   *Scottish Crop Research Institute, Invergowrie, Dundee, UK*
[2]   *John Innes Centre, Plant Science Research, Colney Lane, Norwich, UK*

Key words: DNA fingerprint, molecular probes, raspberry, virulence

The large raspberry aphid, *Amphorophora idaei* (Born.) is the main vector of four viruses commonly found infecting red raspberry (*Rubus idaeus* L.) in Europe. Resistance to the aphid, based on several major and minor genes, has been incorporated into a range of cultivars and greatly reduces the numbers of *A. idaei* and spread of viruses transmitted by this aphid (Jones, 1979). However, three virulent biotypes of *A. idaei* able to overcome different resistant genes were identified by Briggs (1965).

The major resistance gene A_1, which confers resistance to *A. idaei* biotypes 1 and 3, was first made available to growers in the UK in 1947 with the release of the red raspberry *cv.* Malling Landmark. Subsequently, gene A_1 has been incorporated into other raspberry cultivars that are widely grown in the UK. More recently gene A_{10}, which confers resistance to all four known aphid biotypes, has been introduced into a few raspberry *cvs.* Almost all raspberry *cvs* newly released in the UK contain *A. idaei* resistance genes and at present 80% of the raspberry hectarage is planted with cultivars containing such genes. Not surprisingly, this is imposing a strong selection pressure on *A. idaei* and, in recent years, field populations of biotype 2, which is able to overcome gene A_1, have increased in abundance. Information on changes in prevalence of biotypes in field populations is lacking but would enable different *A. idaei* resistance genes to be deployed with maximum effectiveness.

Existing tests to identify *A. idaei* biotypes, based on differential colonisation of a range of resistant raspberry genotypes (Birch & Jones, 1988), are too laborious for monitoring large numbers of aphid samples from the field. We are evaluating molecular probes for more rapid diagnosis of genetic diversity in field populations of *A. idaei*, with a view to monitoring changes in biotype prevalence in the field. We present preliminary results of tests with molecular probes on two *Rubus*-colonising *Amphorophora* species and on single clones of three *A. idaei* biotypes.

Methods

Genomic DNA was extracted from laboratory cultured clonal populations of *A. idaei* biotypes 1 and 2 and of biotype X, a possible new biotype collected in 1986 from *cv.* Glen Prosen in Scotland. Biotype X was differentiated from biotype 2 by its greater reproductive rate on Scottish *cvs* Glen Prosen and Glen Moy. Both these cultivars were presumed to contain the A_1 resistance gene, but their resistance is much less than other cultivars containing this gene (Birch & Jones, 1988). For comparison, genomic DNA was also extracted from a clonal population of the blackberry aphid, *Amphorophora rubi* (Kalt.).

The DNA was digested using the restriction enzymes *EcoRI, EcoRV, HindIII, ApaI, BamHI* or *BglII* and electrophoresed in agarose gels. The gels were Southern blotted and probed with a ^{32}P-end-labelled oligonucleotide probe from M13 protein III gene (15 bp) or a randomly primed flax ribosomal DNA probe (8.6 kbp).

Results and discussion

In Southern blots, the M13 probe differentiated all three *A. idaei* biotypes tested. Thus, the probe distinguished biotype X from biotypes 1 and 2 in DNA digests made with *EcoRI, EcoRV, HindIII* and *BamHI*, and distinguished biotype 1 from biotype 2 in digests made with *BglII*.

Two years after the first restriction fragment length polymorphism (RFLP) analysis, further DNA was extracted from the same three *A. idaei* biotype clones. Tests with the M13 probe and the same set of restriction enzymes used previously, gave RFLP patterns identical to those obtained in 1989, indicating genetic stability through *c.* 50 parthenogenetic generations. In later tests the flax ribosomal DNA probe distinguished between *A. rubi* and single clones of each of the three *A. idaei* biotypes in digests with only one enzyme, *BglII*.

These preliminary results are encouraging and further work in progress on field populations of *A. idaei* collected from a range of raspberry *cvs* in different regions of the UK over two years indicates that virulent biotypes 2 and X are now widespread, due to cultivation of raspberry *cvs* containing *A. idaei*-resistance genes for up to 40 years.

References

Birch, A.N.E. & A.T. Jones (1988). Levels and components of resistance to *Amphorophora idaei* in raspberry cultivars containing different resistance genes. *Ann. appl. Biol.* **113**: 567-578.

Briggs, J.B. (1965). The distribution, abundance and genetic relationships of four strains of the Rubus aphid (*Amphorophora rubi* (Kalt.)) in relation to raspberry breeding. *J. Hort. Sci.* **40**: 109-117.

Jones, A.T. (1979). Further studies on the effect of resistance to *Amphorophora idaei* in raspberry (*Rubus idaeus*) on the spread of aphid-borne viruses. *Ann. appl. Biol.* **92**: 119-123.

Proc. 8th Int. Symp. Insect-Plant Relationships, Dordrecht: Kluwer Acad. Publ.
S.B.J. Menken, J.H. Visser & P. Harrewijn (eds), 1992

Biochemical adaptations of cereal aphids to host plants

A. Urbanska and B. Leszczynski
Agricultural and Pedagogic University, Dept of Biochemistry, Siedlce, Poland

Key words: Aphid enzymes, bird cherry-oat aphid, chemical interactions, phenolic compounds, winter wheat

Phenolic compounds are an important group of allelochemicals in cereal - aphid interactions (Leszczynski *et al.*, 1985). Phytophagous insects have developed various behavioural, physiological and biochemical adaptations to their host plants, one of them being enzymes that metabolize allelochemicals (Brattsten, 1988). Little is known about these enzymes in aphids. Polyphenol oxidase and peroxidase are important enzymes in the metabolism of phenols. Polyphenol oxidase (E.C. 1.10.3.1) catalyzes oxidation and hydroxylation of phenols converting them to melanine pigments, which are nondetrimental to phytophagous insects (Miles, 1968). Peroxidase (E.C. 1.11.1.7) oxidizes phenolic compounds to quinones that polymerize to insoluble pigments.

Experimental

The following aspects of research conducted by us are reported here: (1) determining of the enzyme activity in saliva and in whole bird cherry-oat aphids (*Rhopalosiphum padi* L.), (2) measuring the oxidation of phenolic compounds to quinones by the aphid saliva enzymes, (3) assessing the activity of the enzymes in aphids raised on wheat with different levels of phenols, (4) measuring activity of the enzymes in the aphids fed on a diet with the phenolic compounds studied.

Table 1. Enzyme activity in the saliva and homogenates of *R. padi*

	Polyphenol oxidase[a]	Peroxidase[b]
Saliva - larvae; apterae; alatae	A_{460}/24 h/100 aphids 0.05; 0.03; 0.05	A_{430}/24 h/100 aphids 0.05; 0.07; 0.10
Homogenates - apterae; alatae	A_{460}/30 min/1 mg protein 0.08; 0.09[x]	A_{430}/30 min/1 mg protein 0.10; 0.11[x]
Optimum pH	8.2 - 9.4	7.0 - 7.4

[x] Value significantly different at $P < 0.05$
[a] Dihydroxyphenylalanine was used as substrate
[b] Pyrogallol was used as substrate

We found polyphenol oxidase and peroxidase in the saliva of larvae, apterae, alatae and in the homogenates of the whole apterae and alatae of the bird cherry-oat aphid (Table 1). Among the phenolic compounds studied: quercetin, chlorogenic, gentisic and gallic acids were oxidized the most by the aphid salivary enzymes, whereas o-coumaric, p-coumaric, caffeic, syringic and vanillic acid were less oxidized and m-coumaric, p-hydroxybenzoic, ferulic and gallic acids as well as scopoletin and catechol, least (Table 2). The higher activity of the polyphenol oxidase and peroxidase was observed in the aphids raised on the susceptible wheat variety Liwilla, which had a lower phenol content than the moderately resistant varieties Saga and Grana (Table 3).

Table 2. Content of quinones in 4×10^{-4} M solutions of the studied phenolic compounds after 24 h exposure to feeding of 100 aphids, compared with unexposed controls

Compound	Optical density at 480[a] nm, compared with control
Benzoic acid derivatives	
p-Hydroxybenzoic; genetisic;	0.03; 0.12
protocatechuic[b]; gallic[b]; tannic;	0.04; 0.12; 0.02
syringic[b]; vanillic[b]	0.07; 0.05
Trans-Cinnamic acid derivatives	
o-coumaric; m-coumaric; p-coumaric[b];	0.09; 0.04; 0.09
caffeic[b]; ferulic[b]; chlorogenic[b]; scopoletin	0.07; 0.03; 0.15; 0.03
Catechol	0.03
Quercetin[b]	0.14

[a] Specific reaction with dimethylaniline was used
[b] Phenolic compounds found in wheat tissues

Table 3. The polyphenol oxidase and peroxidase activity of *R. padi* raised on winter wheat varieties containing different levels of total phenols

Factor / variety	Moderately resistant		Susceptible
	Saga	Grana	Liwilla
Total phenols mg/g dry weight	7.84[*]	6.94[*]	4.65
polyphenol oxidase A_{460}/30 min/1 mg protein	0.05[*]	0.08[*]	0.16
Peroxidase A_{430}/30 min/1 mg protein	0.05[*]	0.08[*]	0.16

[*] Value significantly different at $P < 0.05$

In vitro most of the tested phenolic compounds decreased the aphid's polyphenol oxidase and peroxidase activity, with the exception of quercetin, chlorogenic, gentisic and gallic acids (Table 4).

The results suggest that the defence mechanism of the bird cherry-oat aphid toward the phenolic compounds dependent on the polyphenol oxidase and peroxidase, is limited and connected with the quality and quantity of these allelochemicals in the host plant tissues. The higher concentration of these substances in the moderately resistant wheat varieties inhibited the activity of the aphid enzymes. Thus the wheat resistance mechanism based on phenols might be very effective against the bird cherry-oat aphid.

Table 4. Changes in the enzyme activity in whole aphids fed with the studied phenolic compounds

Enzyme	Phenolic compound	Increasing conc. of phenols		
		1.0×10^{-4}M	2.0×10^{-4}M	4×10^{-4}M
Polyphenol oxidase	gentisic, gallic, chlorogenic acids, quercetin	activity increases		
	catechol, protocatechuic, tannic, caffeic acids	activity decreases		
Peroxidase	gentisic, gallic, chlorogenic acids, quercetin	activity increases		
	catechol, p-hydroxybenzoic protocatechuic, tannic, syringic, vanillic, o-, m-, p-coumaric, caffeic, ferulic acids, scopoletin	activity decreases		

References

Brattsten, L.B. (1988). Enzymic adaptations in leaf-feeding insects to host-plant allelochemicals. *J. Chem. Ecol.* **14**: 1919-1939.

Leszczynski, B., J. Warchol & S. Niraz (1985). The influence of phenolic compounds on the preference of winter wheat cultivars by cereal aphids. *Insect Sci. Applic.* **6**: 157-158.

Miles, P.W. (1968). Insect secretions in plants. *Annu. Rev. Phytopath.* **6**: 137-164.

Proc. 8th Int. Symp. Insect-Plant Relationships, Dordrecht: Kluwer Acad. Publ.
S.B.J. Menken, J.H. Visser & P. Harrewijn (eds), 1992

Criteria for host-plant acceptance by aphids

W.F. Tjallingii[1] and A. Mayoral[2]
[1] *Dept of Entomology, Wageningen Agricultural University, Wageningen, The Netherlands*
[2] *Centro Investigationes Biológicas (CSIC), Madrid, Spain*

Key words: Electrical penetration graph, EPG, phloem sap, TEM

Aphid probing or stylet penetration behaviour is needed (1) to get appropriate chemical information by internal gustation (Wensler, 1977) and (2) to reach the sieve elements in the vascular bundle for actual feeding. Transmission electron microscopy (TEM) showed that *Aphis fabae* (Scop.) on its host *Vicia faba* (L.) punctured many, if not all plant cells along the stylet track (Hogen Esch & Tjallingii, 1990; Tjallingii, 1990b). The stylet path, however, remained completely extracellular and the cells generally survived. It was intriguing that 11 sieve elements were reached and punctured before sap feeding started from the last one, which had evidently been punctured earlier, as we inferred on basis of saliva tracks. These data suggest that once a sieve element is reached its acceptability is not nessecarily self-evident. Presumably, its suitability might depend on the plant's reactions to aphid activity. Whether sap ingestion from a sieve element had taken place could not be concluded from the histology, but from an electrical penetration graph (EPG), recorded from this aphid previous to TEM processing. EPGs can record aphid activity during plant penetration (Tjallingii, 1988). Moreover, mechanical work, punctures of plant cells, saliva secretion, active and passive ingestion and other activities can also be derived from EPGs. So, the EPG from this aphid was able to show a number of sieve element punctures without ingestion preceding the final puncture with sustained sap ingestion.

The aim of the present study was to investigate if the situation described above is representative of this and other aphid-plant combinations. Fortunately, repeating the (very time consuming) histological TEM procedures was not necessary, because once the sieve element punctures with and without sap ingestion can be recognized in EPGs, these electrical recordings can provide adequate information.

Materials and methods

A. fabae and *Sitobion avenae* (F.) were reared on broad beans (*cv.* Drie maal wit) and barley (*cv.* Agrar) respectively, (16 h light per day, 20°C). Eight apterous adults were wired and connected simultaneously to the EPG amplifier. For 8 h, EPGs from about 15 aphids per treatment, on bean stems and leaves (*A. fabae*) or leaves of wheat (*cv.* Chinese Spring) and barley (*S. avenae*) were stored on a computer hard disk. The STYLET 2.0 computer program in ASYST (TM) was developed and used for successive analysis. Starting time and duration of wave-form patterns were retrieved. Only separate penetrations (PEN) and sieve element puncturing (E) are presented here. Within the sieve element punctures, distinction was made between the first (1st E) and later punctures. Furthermore, the first sieve element penetrations (E > 10 min) and those longer than one hour (E > 1 h) were distinguished. The number and durations of separate penetrations was also recorded.

Results

Penetration behaviour appeared to be composed of a series of successive penetrations (PEN) of increasing duration separated by decreasing non-penetration periods (Table 1). As time progressed, the duration of sieve element puncture within penetrations increased (E/PEN) whereas the number of new punctures (new E) increased up until to the 4th h and then decreased gradually. This implies that from then on a puncture could last several hours. The time to reach the sieve elements (1st E), to start sustained ingestion (E > 10 min) or long duration ingestion (E > 1 h) differed greatly between aphid-plant combinations (Table 2).

Table 1. Mean numbers (#), durations (h) or relative time spent by *A. fabae* on bean leaves

hour		1	2	3	4	5	6	7	8
PEN	(#)	4.6	2.3	2.4	1.2	1.5	0.6	0.3	0.5
PEN	(h)	0.75	0.87	0.87	0.91	0.93	0.95	0.97	0.96
E/PEN	(%)	0.5	1.9	4.5	18.2	43.5	65.1	64.6	78.5
new E	(#)	0.1	0.3	0.2	0.8	0.5	0.5	0.5	0.5

There is no reason to consider the stem as a less (1st E) or more (E > 10 min) suitable feeding site, or barley as being less suitable as a food plant than wheat on basis of other (*e.g.*, population dynamic) criteria. *A. fabae* takes somewhat longer to reach the sieve elements on bean leaves and stems than *S. avenae* does on barley and wheat (Table 3).

Table 2. Mean time needed to reach sieve elements or to accept them (h)

	1st E	E>10min	E>1h
leaf	3.4	5.2	5.3
stem	3.6	4.2	5.4
wheat	2.9	3.5	3.8
barley	5.8	6.3	6.5

Table 3. Mean time needed to reach the sieve element within a probe (h)

	1st E	E>10min
leaf	0.7	1.0
stem	0.8	0.9
wheat	0.4	0.5
barley	0.6	0.6

Discussion and conclusions

It appears that the time needed for aphids to accept a plant for feeding takes many hours. It is unlikely that wiring or other artefacts could have biased these results as was tested by control experiments recording honeydew secretion from free aphids. Many probes are needed and within these probes a number of sieve element punctures generally precede long sustained phloem sap feeding and thus acceptance. Most, but not all, E > 10 min appeared to take longer than 1 h as well. The question whether the 1st E really reflects the first sieve element puncture is doubtful (Tjallingii & Hogen Esch, this vol., pp. 283-285): earlier brief punctures may have occurred. The conclusion, however, remains the same because the delay between the first real sieve element puncture and sustained sap ingestion will be even greater. Therefore, the reconstructed stylet path with its large number of sieve element punctures before actual sap feeding earlier described would seem to be rather normal.

Furthermore, the situation in *S. avenae*-cereal combinations is comparable to that of *A. fabae-Vicia* with respect to this point. Why these earlier punctures did not lead to sap feeding remains unclear. A rejected sieve element in early punctures, can apparently be accepted later, and thus supports the hypothesis that the plant reaction may provide important cues for the real acceptance of the phloem sap. Further research is planned to test this hypothesis.

References

Hogen Esch, Th. & W.F. Tjallingii (1990). Fine structure of aphid stylets in plant tissue. *Symp. Biol. Hung.* **39**: 475-476.

Tjallingii, W.F. (1988). In: A.K. Minks & P. Harrewijn (eds), *Aphids, their Biology, Natural Enemies and Control* Vol. 2B, pp. 95-107. Amsterdam: Elsevier.

Tjallingii, W.F. (1990a). Continuous recording of stylet penetration activities by aphids. In: R.K. Campbell & R.D. Eikenbary (eds), *Aphid-plant Genotype Interactions*, pp. 89-99. Amsterdam: Elsevier.

Tjallingii, W.F. (1990b). Stylet penetration parameters from aphids in relation to host-plant resistance. *Symp. Biol. Hung.* **39**: 411-419.

Wensler, R.J.D. (1977). The fine structure of distal receptors on the labium of the aphid *Brevicoryne brassicae* L. (Homoptera). *Cell Tiss. Res.* **181**: 409-421.

Proc. 8th Int. Symp. Insect-Plant Relationships, Dordrecht: Kluwer Acad. Publ.
S.B.J. Menken, J.H. Visser & P. Harrewijn (eds), 1992

Ultrastructure and electrical recording of sieve element punctures by aphid stylets

Th. Hogen Esch and W.F. Tjallingii
Wageningen Agricultural University, Dept of Entomology, Wageningen, The Netherlands

Key words: Intracellular activities, membrane potential, phloem ingestion

Though stylet tracks have been intensely studied using light microscopy (Pollard, 1973), only transmission electron microscopy (TEM) can provide any realistic information (Kimmins, 1986). TEM of stylet tracks, whose formation was recorded by electrical penetration graphs (EPGs) revealed correlations between the recorded wave-forms and ultrastructure (Tjallingii, 1988). Cell wall damage due to intracellular punctures of plant cells could clearly be observed in micrographs (Hogen Esch & Tjallingii, 1990) and the membrane punctures are distinct in EPGs. The E pattern during these sieve element punctures have two distinct variants: E1 and E2 (Tjallingii, 1990). So far, clear correlations have only been demonstrated for E2, *i.e.*, passive feeding for the wave parts and continuous saliva secretion for the peak parts.

The aim of the present study was to correlate the E1 variant with the stylet tip position in the plant tissue.

Materials and methods

An adult apterous virginopara of *Aphis fabae* (Scop.) was wired and connected to an EPG amplifier. Signals were stored on analog tape (FM); if the E1 pattern occurred during recording, stylectomy was performed. The plant tissue with the stylet stump was then cut out of the leaf and processed for TEM. About 1200 sections containing the complete stylet track were collected serially. Micrographs were made of relevant parts, magnified between 700x (overviews) and 63000x. Durations of wave-forms and numbers of potential drops were retrieved from the EPG analysis.

Results

The stylet penetration lasting 75.5 min consisted of 67 min pathway activities (pattern C including A) and 8.5 min of pattern E1, after which it was terminated by stylectomy. The EPG (Fig. 1) showed 59 distinct potential drops preceding E1, each representing a brief intracellular puncture. Micrographs showed up a main track going from the epidermis towards the vascular bundle and reaching into the third sieve element encountered. A small side branch was made in the mesophyll. In total 35 cells (Table 1) were 'touched', *i.e.*, their walls made contact with the salivary sheath enveloping the stylets. Many cells showed breaks in their walls due to stylet penetration. The epidermal and mesophyll cells had about one break per cell whereas the phloem cells had about four times as many breaks. The total number of breaks was 69, *i.e.*, 10 more than the number of potential drops revealed in the EPG.

Discussion and conclusions

Not all breaks in the cell walls can be considered as having being caused by an intracellular puncture; *i.e.*, a definite passage of the stylet tips through the plasmalemma. Breaks showed little or no saliva deposition inside the cell wall and no traces were left behind within the plasmalemma. The secondary wall material was disrupted and filled with salivary sheath material. Often some callose formation was seen inferring an apparent wound reaction. Protoplasts, however, remained unaltered or were restored. So, we can conclude that although most of the breaks were related to cell punctures, the breaks were not completely unambiguous morphological indicators. Therefore, potential drops cannot be used as land marks for histology which makes linkage to the EPG more difficult.

Since 14 breaks were found in the sieve elements, E1 cannot be considered as the first sieve element puncture. *Vice versa*, however, there is no reason to doubt that E1 is indicative of a sieve element puncture, as was suggested earlier (Hogen Esch & Tjallingii, 1990). Other sieve element punctures very likely to occur, are reflected as short potential drops and are indistinguishable as such from punctures of other cells. So, E1 seems to represent a long intracellular sieve element puncture with some special aphid activity, the nature of which is unknown. More experiments are planned to study E1 and its role in sieve element feeding.

Table 1. TEM data.
For abbreviations see legend to Fig. 1

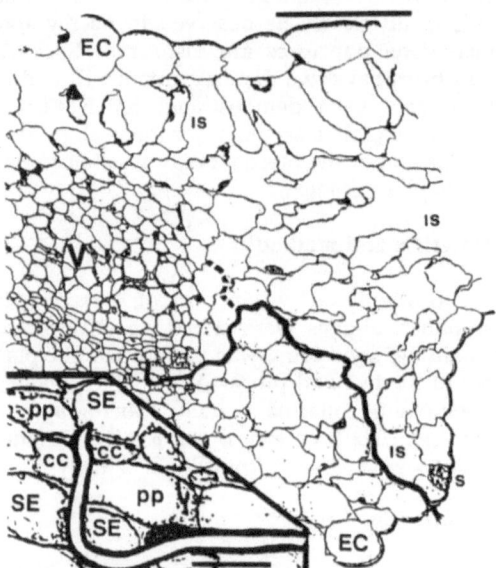

| | Numbers | |
| | Cells | Breaks |
Tissue	Touched	Touched + breaks	
EC	2	1	1
MC	21	9	19
PP	6	6	24
CC	3	3	11
SE	3	3	14
total	35	22	69

Figure 1. Overview of reconstructed stylet path to sieve element. Black track with stylets. Dashed area, empty side branch; S, stoma; IS, intercellular air space; V, vascular bundle; EC, epidermal; MC, mesophyllic; cell bar 100 μm. *Inset*: CC, companion cell; PP, phloem parenchyma; SE, sieve element; bar 10 μm.

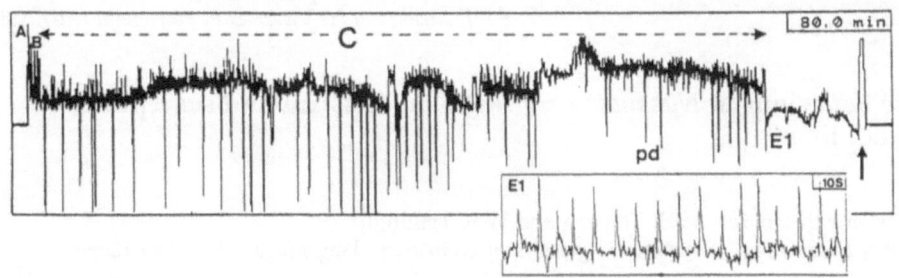

Figure 2. EPG from stylet penetration in Fig. 1. Downward peaks in C are potential drops (pd). C, from A to E1, tissue penetration; E1, sieve element puncture; arrow, stylectomy.

References

Hogen Esch, Th. & W.F. Tjallingii (1990). Fine structure of aphid stylets in plant tissue. *Symp. Biol. Hung.* **39**: 475-476.

Kimmins, F.M. (1986). Ultrastructure of the stylets pathway in host plant tissue. *Entomol. exp. appl.* **41**: 283-290.

Pollard, D.G. (1973). Plant penetration by feeding aphids (Hemiptera, Aphidoidea): a review. *Bul. Entomol. Res.* **62**: 631-714.

Tjallingii, W.F. (1988). In: P. Harrewijn & A.K. Minks (eds), *Aphids, their Biology, Natural Enemies and Control*, Vol. 2B, pp. 95-107. Amsterdam: Elsevier.

Tjallingii, W.F. (1990). Continuous recording of stylet penetration activities by aphids. In: R.K. Campbell & R.D. Eikenbary (eds), *Aphid-plant Genotype Interactions*, pp. 89-99. Amsterdam: Elsevier.

Proc. 8th Int. Symp. Insect-Plant Relationships, Dordrecht: Kluwer Acad. Publ.
S.B.J. Menken, J.H. Visser & P. Harrewijn (eds), 1992

The behaviour of *Nasonovia ribisnigri* on resistant and susceptible lettuce lines

Maarten van Helden, M.H. Thijssen and W.F. Tjallingii
Dept of Entomology, Wageningen Agricultural University, Wageningen, The Netherlands

Key words: Aphididae, behaviour, EPG, Homoptera, plant acceptance

Absolute resistance to *Nasonovia ribisnigri* (Mosley) was transferred from *Lactuca virosa* to *L. sativa*. The resistance is based on a single dominant gene (Nr-gene, Eenink *et al.*, 1982a, b). An earlier study compared Direct Current Electrical Penetration Graphs (DC-EPGs) from *N. ribisnigri* on resistant and susceptible lettuce (*L. sativa*) (Van Helden & Tjallingii, 1990). It showed that the EPGs differed only after the aphid had reached the phloem. During the EPG the aphid is attached to a gold wire which hampers its movements (Tjallingii, 1985). In this study, we tried to reveal possible effects of tethering on the behaviour of the aphid, and possible differences in behaviour on resistant and susceptible lines.

Materials and methods

Aphids, plants and experiments. Mass rearing of *N. ribisnigri* and culture of plants was done as described by Van Helden & Tjallingii (1990). We used two (near) isogenic lettuce lines which differed only in the Nr-gene (Van Helden & Tjallingii, 1990). The plants were about four weeks old (6-8 fully expanded leaves). The aphids were recently moulted alate virginoparae. Experiments were conducted in the laboratory at 22 ± 1°C under artificial illumination (HF fluorescent tubes *ca* 6000 Lux/72 $\mu E.m^{-2}.s^{-1}$).

Behavioural observations

Pretreatment of the aphids. Teneral alate virginoparae were brushed from their food plant (lettuce *cv.* "Taiwan") into a petri dish between 8.30 and 9.00 a.m. They were marked with water soluble colour paint in different colours and placed on a Taiwan plant for at least 3 h.

Half hour continuous observations. An aphid was carefully picked up with a brush and transferred to an adhesive label (8 mm diameter) on the fourth leaf of the plant (seen from the growing point). Its behaviour was recorded for 30 min, the parameters recorded being: activity, antennal movements, location on the plant and penetration site on the leaf.

48 Hour interval observations. In an identical set up, aphids were observed for 10 s at 0.5, 1, 2, 4, 8, 16, 24, 28, 32, 40, and 48 h. To prevent the aphids walking off the plant and later returning to it, tape with Fluon was attached over the pot (making the surface slippery); aphids walking over the Fluon dropped down and were recorded as having left the plant.

Results and discussion

Differences between 'free' and tethered aphids. The results of the half hour continuous observations showed a difference between free and tethered aphids for several of the parameters. Differences were especially great for the duration of the first non-penetration period (1 *vs* 3 min). This could have been due to a slight difference in pretreatment of the EPG treatment (Van Helden & Tjallingii, 1990), but a direct effect of the tethering was more likely. There was also a small, but statistically significant reduction of the total time spent penetrating. This result was also seen in the 48 h observations (Table 1).

Table 1. "Free" observations compared to EPG recordings (tethered) on resistant and susceptible lettuce. NP = Non penetration, SP = Stilet penetration, * = Significant difference between FREE and EPG (α = 0.05), # = Significant difference between resistant and susceptible (α = 0.05). Time in seconds ± SE

	Susceptible		Resistant	
	Free	EPG	Free	EPG
Half hour observations				
No. of observations	33	15	34	20
Total time NP	458 ± 34 *	777 ± 105 *	504 ± 41 *	746 ± 83 *
Number of NP	4.2 ± 0.2	5.1 ± 0.4	4.9 ± 0.3	4.9 ± 0.4
Mean duration NP	118 ± 10	149 ± 17	116 ± 11 *	164 ± 24 *
Total time SP	1342 ± 34 *	1023 ± 105 *	1296 ± 41 *	1054 ± 83 *
Number of SP	4.1 ± 0.2	4.8 ± 0.3	4.8 ± 0.3	4.8 ± 0.4
Mean duration SP	307 ± 22 #	252 ± 34	365 ± 23 #	248 ± 40
Duration first NP	55.5 ± 6.8 *	179.9 ± 32.2 *	68.5 ± 12.7 *	205.2 ± 49.1 *
Duration first SP	23.9 ± 1.8 #	115.6 ± 54.8	43.1 ± 7.2 #	47.4 ± 10.1
48 h observations and 16 h EPGs				
No. of observations	36	15	33	20
% of time penetrating	92.7	77.0	81.1	74.2
% aphids leaving	33.3	0	91.4	0

Differences between resistant and susceptible lines. The EPGs did not show any significant differences between resistant and susceptible lines during the first 30 min. The "free" aphids, however, spent a longer time on the first penetration on the resistant line and the total time spent on penetration was apparently longer (the long first penetration time on susceptible plants during EPG was due to a single exceptionally long penetration of 884 s; without this value the mean would have been 61 s). Possibly, any recordable difference between resistant and susceptible lines which influenced the aphids was masked by the EPG treatment. The interval observations, like the EPGs, showed that more time was spent in penetration on the susceptible line.

Time until leaving the plant compared to EPGs. On the resistant plants almost all aphids left the plant between hour 4 and hour 24. This fraction was lower on the susceptible lines but the time of leaving was similar (Fig. 1). The EPGs showed that on both lines nearly all aphids had reached the phloem after 16 h and that no differences in time to first E pattern were observed (Van Helden & Tjallingii, 1990) (see Fig. 1 for change in time, results of resistant and susceptible lines for EPG combined). Therefore, the aphids had ample time to reach the phloem before they left the plant, confirming the results of the EPG experiments, which showed no clear differences in EPGs before the phloem was reached (Van Helden & Tjallingii, 1990).

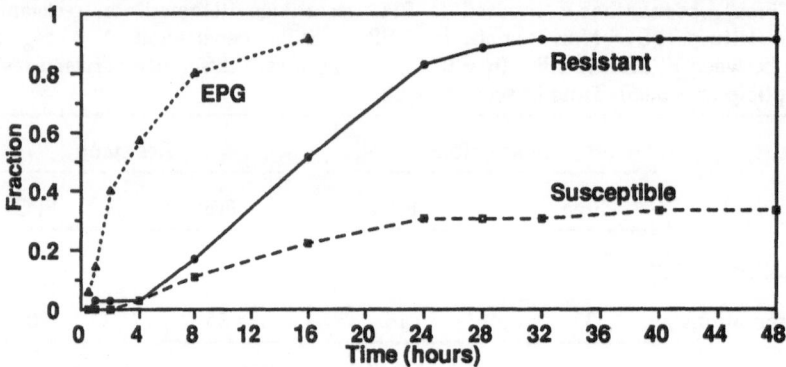

Figure 1. The fraction of aphids leaving the plant in the 48 h interval observations on resistant an susceptible lettuce (no aphids left after 30 h) compared to the fraction reaching an E pattern during EPG recordings (both lines are combined since no difference was observed).

Conclusions

The absolute resistance to *N. ribisnigri* in lettuce causes a severe disruption of the feeding behaviour of the aphid, eventually causing the aphid to reject the plant as a food source. Although the resistance influences the aphid during early short penetrations final rejection occurs only after the phloem is reached. The reduced freedom of movement due to tethering does not seriously disturb the aphids' behaviour.

References

Eenink, A.H., R. Groenwold & F.L. Dieleman (1982a). Resistance of lettuce (*Lactuca*) to the leaf aphid *Nasonovia ribisnigri* 1: Transfer of resistance from *L. virosa* to *L. sativa* by interspecific crosses and selection of resistant breeding lines. *Euphytica* **31**: 291-300.

Eenink, E.H., R. Groenwold & F.L. Dieleman (1982b). Resistance of lettuce (*Lactuca*) to the leaf aphid *Nasonovia ribisnigri* 2: Inheritance of the resistance. *Euphytica* **31**: 301-304.

Tjallingii, W.F. (1985). Stylet penetration activities by aphids. Ph.D. Dissertation, Agricultural University, Wageningen.

Van Helden, M. & W.F. Tjallingii (1990). Electrical penetration graphs of the aphid *Nasonovia ribisnigri* on resistant and susceptible lettuce (*Lactuca sativa*). *Proc. Int. Symp. Aphid-Plant Interactions: Population to Molecules*, p. 308, Stillwater: Oklahoma.

Proc. 8th Int. Symp. Insect-Plant Relationships, Dordrecht: Kluwer Acad. Publ.
S.B.J. Menken, J.H. Visser & P. Harrewijn (eds), 1992

Aphid pectinases, cell wall fragments, and biotype evolution

John C. Reese[1], Runlin Zhang Ma[2], and William C. Black IV[3]
[1] Kansas State University, Dept of Entomology, Manhattan, Kansas, USA
[2] University of Illinois, Dept of Entomology, Urbana, Illinois, USA
[3] Colorado State University, AIDL, Ft. Collins, Colorado, USA

Key words: Aphididae, host plant resistance, tolerance, toxins

A number of economically important aphid species exhibit the genetic capability to respond to selection pressure, thus forming new, more virulent biotypes, that are capable of overcoming previously resistant host germplasm and/or making species-level shifts in host range. Because the ability to enzymatically dissolve its way down to the sieve elements of the phloem is apparently critical to the feeding success of most aphid species, it is not surprising to find a relationship between pectinase chemistry of the insect and pectin chemistry of the plant (Campbell & Dreyer, 1990). We have elucidated some aspects of these very complex relationships with the greenbug (*Schizaphis graminum* (Rondani)) pectinase - sorghum pectin system (Reese *et al.*, 1990).

Experiments

When greenbugs were allowed to feed on an agarose gel laced with pectin, they injected polygalacturonase (PG) and pectinesterase (PE) (Ma *et al.*, 1990; Ma, 1991; Ma *et al.*, 1991). These enzymes were detectable by staining the gel with ruthenium red. Repeating the assay for greenbug biotypes C and E and for other species of aphids demonstrated the presence of these enzymes in salivary secretions of both greenbug biotypes, but not in secretions of species that elicit no apparent response by the plant.

Preliminary attempts to isolate, purify, and partially characterize PG suggest that it has a molecular weight of about 45,000 Daltons, an isoelectric point higher than 9.3, a pH optimum between 5.6 and 5.9, and maximum activity at 45°C. This protein can be renatured following SDS-PAGE by washing off with Triton X-100.

Plant response to pectinases and cell wall fragments was measured by a SPAD-METER^{TM} chlorophyll meter. Greenbug pectinases elicited a reduction in chlorophyll content when injected into the leaves of sorghum plants. Commercially available PG from the pathogen *Aspergillus niger* (van Tieghem), a recombinant PG from *Erwinia carotovora* (Jones) Bergey *et al.* purified from an *Escherichia coli* (Migula) Castellani & Chalmers clone, and greenbug PG were injected into greenbug-susceptible sorghum plants (Ma, 1991). All treatments caused significant reductions in chlorophyll. Similar responses were elicited by cell wall fragments produced by incubating sorghum pectin with greenbug pectinases, suggesting that a two-step process is involved in which greenbug enzymes release biologically active oligosaccharides, which, in turn, elicit the plant response. Thus, differences in specific plant pectins and insect pectinases may explain some of the observable differences among biotypes of aphid species.

Acknowledgements. This project was supported by the Kansas Agricultural Experiment Station and by a grant from Pioneer Hi-Bred International. Contribution 92-467-A from the Kansas Agricultural Experiment Station.

Voucher specimens, voucher No. 016, are located in the Kansas State University Research Collection of Insects.

References

Campbell, B.C. & D.L. Dreyer (1990). The role of plant matrix polysaccharides in aphid-plant interactions. In: R.K. Campbell & R.D. Eikenbary (eds), *Aphid-Plant Genotype Interactions*, pp. 149-170. New York: Elsevier.

Ma, R.Z. (1991). Pectic enzymes in the greenbug (*Schizaphis graminum*) saliva: identification, isolation, partial characterization, and their toxicity on sorghum plants. Ph.D. Dissertation. Kansas State University, Manhattan, KS., 133 pp.

Ma, R., J.C. Reese, W.C. Black IV & P. Bramel-Cox (1990). Detection of pectinesterase and polygalacturonase from salivary secretions of living greenbugs, *Schizaphis graminum* (Homoptera: Aphididae). *J. Insect Physiol.* **36**: 507-512.

Ma, R., J.C. Reese, W.C. Black IV & P. Bramel-Cox (1991). Detection of plant cell wall-degrading enzymes in greenbug saliva. In: D.C. Peters, J.A. Webster & C.S. Chlouber (eds), Proceedings, Aphid-Plant Interactions: *Populations to Molecules*, p. 291. MP-132, Oklahoma State University, Stillwater, OK.

Reese, J.C., P. Bramel-Cox, R. Ma, A.G.O. Dixon, T.W. Mize & D.G. Schmidt (1990). Greenbug and other pest resistance in sorghum. In: *44th Annual Corn Sorgh. Indust. Res. Conf.* pp. 1-12.

Proc. 8th Int. Symp. Insect-Plant Relationships, Dordrecht: Kluwer Acad. Publ.
S.B.J. Menken, J.H. Visser & P. Harrewijn (eds), 1992

Resistance of lettuce to the leaf aphid *Macrosiphum euphorbiae*

Kees Reinink[1] and Frans L. Dieleman[2]
[1] *DLO-Centre for Plant Breeding and Reproduction Research, Wageningen, The Netherlands*
[2] *Wageningen Agricultural University, Dept of Entomology, Wageningen, The Netherlands*

Key words: Cultivar × clone interaction, genetics, potato aphid

The potato aphid (*Macrosiphum euphorbiae* Thomas) is commonly found on lettuce (*Lactuca sativa* L.) and causes economic damage. The aphids, although vectors of virus diseases do not cause the main damage nor do they directly affect crop production. The damage is done by the mere presence of insects in the head, rendering the crop unmarketable. To control the infestation, lettuce growers spray several times a week during periods favourable for aphid attack. Introduction of lettuce cultivars with resistance to aphids is an environmentally safe alternative to reduce both aphid damage and the use of insecticides.

Research on host-plant resistance to *M. euphorbiae* was started by screening a collection of 90 *Lactuca* accessions. No absolute resistance was detected, but several cultivars showed interesting levels of partial resistance (Reinink & Dieleman, 1989). Two criteria were used to specify the level of resistance of a cultivar: the weight of eight day-old nymphs grown for seven days on the test plants and the reproduction of apterae of synchronized age. The weight test was very efficient in identifying resistant genotypes: all genotypes with low nymphal weights also showed low fecundity. The expression of partial resistance was dependent on environmental conditions and is in inverse proportion to the level of light intensity.

Genetics of resistance

The segregation of resistance to *M. euphorbiae* was studied in F_2 populations from crosses between susceptible and partially resistant cultivars. The weight of eight day-old nymphs was taken as a criterion for resistance. Aphid weights on susceptible parents were three to four times higher than on resistant parents. Most F_1 and all F_2 progenies were intermediate between their parents. Thus, effects of dominance seem to be of minor importance. Estimates of heritability in the F_2 population were low (h^2_{wide} = 0.00 - 0.21) in crosses with parents of the same resistance type (both susceptible or both partially resistant) and high (h^2_{wide} = 0.67 - 0.76) in crosses with parents of a different resistance type. This indicates that the parents represented the extremes of resistance to *M. euphorbiae*. The frequency distributions of the F_2 populations indicated a quantitative inheritance, possibly determined by only a few genes.

Selection for resistance in crosses between partially resistant and susceptible genotypes should not be too difficult because estimates of heritability are high in these crosses. However, because no segregation was detected in a cross between two resistant parents, any increase in resistance above the level of the most resistant cultivar would seem unlikely.

Lettuce cultivar × aphid clone interactions

Interactions between cultivars and biotypes were observed in experiments using five clones of *M. euphorbiae* and a set of ten cultivars covering the whole range of resistance. This interaction differed significantly from cultivar × clone interactions observed earlier for *Myzus persicae* (Sulz.) on lettuce (Reinink *et al.*, 1989). Marked differences were found in the general level of aggressiveness on lettuce between clones of *M. persicae*. The reproduction of less aggressive clones was relatively poor both on susceptible and on partially resistant cultivars. In contrast, the clones of *M. euphorbiae* had about the same overall aggressiveness on the set of lettuce cultivars. The cultivar × clone interaction could largely be attributed to the behaviour of one of the clones, WMe3, which performed better than average on resistant cultivars, and less than average on susceptible cultivars. This interaction was found both in experiments for which nymphal weight and reproduction were used as criterion. There could be a trade-off mechanism involved in these results: clone WMp3 could have a physiological characteristic making it less sensitive to the resistance in lettuce, but negatively affecting its weight increase and reproduction on susceptible cultivars.

Apart from the general type of interaction shown by clone WMe3, specific interactions were also found: a specific combination of a clone and a cultivar performing better or worse than expected in an additive model.

The interactions observed between clones of *M. euphorbiae* and cultivars of lettuce could present problems in breeding programmes for resistance to this aphid. The most resistant cultivars still reduce the weight increase and reproduction of clone WMe3 considerably. However, almost nothing is known about the occurrence and frequency of clones with different aggressiveness in the field. There may be other clones of *M. euphorbiae* that are less sensitive to the resistance in lettuce.

Conclusions

We found a large variation in resistance to *M. euphorbiae* in lettuce. Because of high estimates of heritability in F_2 populations it is expected that the resistance can be handled efficiently in breeding programmes. Interactions between clones of *M. euphorbiae* and cultivars of lettuce could be problematic in resistance breeding programmes. More information regarding the occurrence and frequency of clones with different aggressiveness in the field is needed.

References

Reinink, K. & F.L. Dieleman (1989). Resistance in lettuce to the leaf aphids *Macrosiphum euphorbiae* and *Uroleucon sonchi*. *Ann. appl. Biol.* **115**: 489-498.

Reinink, K., F.L. Dieleman, J. Jansen & A.M. Montenarie (1989). Interactions between plant and aphid genotypes in resistance of lettuce to *Myzus persicae* and *Macrosiphum euphorbiae*. *Euphytica* **43**: 215-222.

Proc. 8th Int. Symp. Insect-Plant Relationships, Dordrecht: Kluwer Acad. Publ.
S.B.J. Menken, J.H. Visser & P. Harrewijn (eds), 1992

Expression of resistance in lettuce (*Lactuca sativa*) to *Macrosiphum euphorbiae* based on a monogenic factor

B. Gabrys[1] and P. Harrewijn[2]
[1] *Dept of Agricultural Entomology, Agricultural University, Wroclaw, Poland*
[2] *DLO-Research Institute for Plant Protection, Physiology Section, Wageningen, The Netherlands*

Key words: Aphids, electrical penetration graph, plant resistance, probing behaviour

The selection of resistant lines of lettuce (*Lactuca sativa* L.) infested by several aphid species has resulted in the transfer of a factor, from wild species of *L. virosa*, that confers resistance to *Nasonovia ribisnigri* (Mosley). Isogenic plant material with complete resistance to this aphid could be obtained based upon a dominant gene (Nr-gene). As *Macrosiphum euphorbiae* (Thomas) can thrive on lettuce selections with a considerable level of resistance to both *N. ribisnigri* and *Myzus persicae* (Sulzer), the objective of this study was to determine the effect of the Nr-gene on the probing behaviour of *M. euphorbiae*.

Materials and methods

The conditions for cultured plants, aphids and experiments were 18 ± 2°C, 60% RH and a photoperiod of L16:D8. *M. euphorbiae* (biotype 1) was maintained on lettuce plants of *cv.* "Snijsla". *Nasonovia*-resistant (NR), *Nasonovia*-susceptible (NS) and the *cv.* "Batavia Chou de Napoli" (BN) were grown in soilless culture as described by Harrewijn & Dieleman (1984). The last mentioned plants were used as a control because of their high susceptibility to *M. euphorbiae*. For all experiments, 3-4 week old plants with 3-4 true leaves were used.
Experiment 1. Development of *M. euphorbiae*. Mean relative growth rate (MRGR) was calculated from aphids separated in clip cages using a Cahn 4700 microbalance. Reproduction and mortality were recorded for seven days after the onset of reproduction.
Experiment 2. Probing behaviour of *M. euphorbiae*. Apterous aphids reproducing for 2-4 days were starved for 2 h after which a gold wire of 20 μm was glued with silver paint to the dorsum. After 15 min they were given access to the underside of a leaf. EPG (electrical penetration graph) recordings were made on tape according to the DC system of Tjallingii (1990), to be analysed using an "ASYST" computer programme.

Results and discussion

Experiment 1. Growth and fecundity were low on both NR and NS lines, MRGR being 0.20 and daily offspring not exceeding 0.93, with a larval mortality of 73 and 63%, respectively. On BN the aphids performed much better with a MRGR of 0.29, daily offspring of 2.27 and larval mortality of 27%. These results suggest a partial resistance of NR and NS lines to *M. euphorbiae*.
Experiment 2. During the first 4 h after the aphids had been placed on test plants, the duration of penetration activity (pattern C) was similar on both NR and BN, but almost twice as long as on NS. Phloem ingestion was twice as short on NR and BN as it was on

NS. This changed during the next 4 h. Aphids produced 10 times more pattern C on NR and NS than on BN, on which almost 100% of the recorded patterns was Epd (Tjallingii, 1990), *i.e.*, phloem ingestion (Fig. 1). Number of probes and time to first Epd from start of the experiments were similar on all lines. This is in contrast to the performance of *N. ribisnigri* on NR and NS lines (Van Helden & Tjallingii, 1990). Once *M. euphorbiae* started

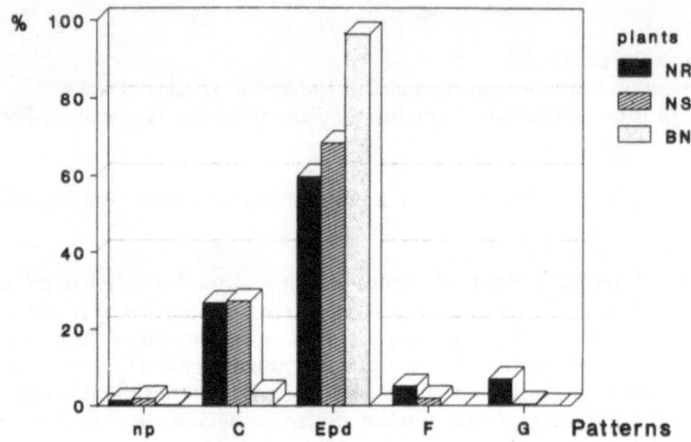

Figure 1. EPG recording of *M. euphorbiae* during 4-8 h after access to lettuce plants. NR - Nasonovia-resistant; NS - *Nasonovia*-susceptible; BN - Batavia Chou de Napoli.

to ingest phloem sap, its stylets remained in one cell on BN, whereas phloem sap ingestion was frequently interrupted on NS and NR. The similar performance of the aphids on NR and NS lines demonstrates that there is no effect of the Nr-gene on probing behaviour and development of *M. euphorbiae*, suggesting an independant resistance factor in both lines. This factor expressing itself during prolonged access of the aphids to the plants, is located in the mesophyll and possibly in the phloem. High frequency microcautery of the stylets during Epd registration did not result in phloem sap exudation on NR and NS plants, although this can easily be obtained on BN.

The present study may serve to stress the importance of extended EPG recordings when partial plant resistance is involved.

References

Harrewijn, P. & F.L. Dieleman (1984). The importance of mineral nutrition of the host plant in resistance breeding to aphids. *Proc. VIth Int. Congress on Soilless Culture*, Lelystad, pp. 235-244.

Helden, M. van & W.F. Tjallingii (1991). Electrical penetration graphs of the aphid *Nasonovia ribisnigri* on resistant and susceptible lettuce (*Lactuca sativa*). In: D.C. Peters, J.A. Webster & C.S. Chlouber (eds), *Proc. Aphid-Plant Interactions; Populations to Molecules*, p. 308.

Tjallingii, W.F. (1990). Stylet penetration parameters from aphids in relation to host-plant resistance. *Symp. Biol. Hung.* **39**: 411-419.

Proc. 8th Int. Symp. Insect-Plant Relationships, Dordrecht: Kluwer Acad. Publ.
S.B.J. Menken, J.H. Visser & P. Harrewijn (eds), 1992

Feeding behaviour of the aphid *Sitobion avenae* on resistant and susceptible wheat

C.M. Caillaud[1], J.P. Dipietro[1], B. Chaubet[1] and C.A. Dedryver
[1]   *École Nationale Supérieure d'Agronomie, Chaire de Zoologie, Rennes, France*
[2]   *INRA, Laboratoire de Zoologie, Le Rheu, France*

Key words: Antibiosis, discriminant analysis, electronic monitoring, stylet penetration

The English grain aphid *Sitobion avenae* F. has sporadical outbreaks causing yield losses of up to 10% in Northern Europe, and is one of the important vectors of Barley Yellow Dwarf virus (BYDV). Numerous tests have been done on wheat genotypes to assess their potential resistance to aphids. These studies revealed an interesting source of antibiosis in *Triticum monococcum* L. lines, characterized by its effect on nymphal survival, adult longevity and fecundity, associated with or without the expression of necrotic lesions surrounded by chlorotic halos (Dipietro *et al.*, in prep.). Efforts are now focused on understanding the mechanisms by which this resistance operates, through the study of 5 wheat lines: *T. monococcum* no. 44, 45 and 46 (Resistant R, intrinsic rate of increase r_m ranges from 0.09 to 0.12), *T. monococcum* no. 47 (Intermediate resistance R/S, r_m equals 0.2) and *T. aestivum* cultivar Arminda (Susceptible S, r_m equals 0.27). Since probing behaviourof aphids plays a key role in host selection, our objectives are to find the possible location of the resistance mechanism in the plant, by quantifying the effects of wheat genotype on aphid feeding behaviour.

Methods

We used the Direct Current Electrical Penetration Graph technique as described by Tjallingii (1988). The electrical signals or EPG patterns caused by aphid stylet penetration into the plant have been correlated with stylet tip position and aphid activities such as salivation and ingestion. The typical waveforms obtained were: NP for non penetration; A for first contact stylet-plant epidermis; B for salivary sheath formation; C for tissue penetration associated with stylet penetration and cell membrane puncture (A, B and C often overlap in time); E1 for start of sieve ingestion; E2 for sustained sieve ingestion; F for stylet movement of unknown origin; G for xylem ingestion.

Apterous adults were used 4-7 days after moulting, and monitored on 2-week-old plants for eight successive hours under standard conditions (20°C, 1500 lux). Eleven recordings were made on either the susceptible *T. aestivum* cultivar Arminda or each of the four *T. monococcum* lines (TM44, TM45, TM46, and TM47). For each recording, we analyzed a total of 21 parameters linked with probing behaviour, time from start of penetration to first access to phloem, and duration of the different EPG patterns. All data were submitted to a one-way analysis of variance and a factorial discriminant analysis.

Results and conclusions

Although aphids probed within the first 10 min defying the resistance of the plant, after numerous probes no access to either phloem or xylem was observed on TM 44 and TM 45 (R). Nearly all aphids were able to reach the sieve elements, but a longer time was needed to reach a sieve element and for a sustained ingestion to occur on resistant plants. The number of phloem access attempts was equal on resistant and susceptible plants.

On Arminda (S) and TM47 (R/S) aphids spent nearly 50% of their time ingesting plant sap whereas the E2 pattern represented only 15% of the total recording time in the case of TM44 (R). On resistant plants, aphid behaviour was mainly characterized by non-penetration activities (NP, linked with walking behaviour and search for other feeding sites along the leaf) and plant penetration in non-nutritional tissues (ABC). E1, F, and G durations showed no significant difference although the F pattern was made more often and for a longer total duration on *T. monococcum* lines.

In the discriminant analysis graph, the first two axes accounted for 63% of the total variance. Axis I was mainly formed by sieve ingestion pattern duration (E2) and the time preceding the sieve element first access (time to E2). The number of phloem ingestion attempts and the total number of penetrations were correlated with the second axis. A clear opposition appeared between E2 duration and all the other EPG events. F duration was correlated with such parameters as the number of probes and the time taken to reach a sieve element (time to E1). Neither the time to first probe, nor the E1 and F duration seemed to play a role in the formation of the two discriminant axis: they did not account for data variability.

The plane defined by axes I and II separated the three plant groups according to their effects on *S. avenae* main feeding features although 19% of the individuals could not be correctly assessed: group I made by the two resistant *T. monococcum* lines 44 and 45; group II formed by the resistant *T. monococcum* line 46; group III constituted by TM 47 (R/S) and the susceptible genotype Arminda.

Although TM 47 was found to be less suitable for aphids than Arminda, aphid feeding behaviour appeared to be rather similar, suggesting a possible difference in plant sap quality. TM46 differs from TM 44 and TM45 mainly in the parameters, such as, total number of probes, registration time preceding the first sieve element access and non-penetration duration: *T. monococcum* lines may contain 2 kinds of resistance mechanisms. This work suggests that EPG technique could help in understanding resistance mechanisms and could be used as a plant screening method if followed by a multivariate analysis. Resistance in *T. monococcum* lines mainly affects the occurrence of a prolonged sieve ingestion and histological studies are now needed to compare aphid stylet pathway and physical differences between the varieties.

Reference

Tjallingii, W.F. (1988). Electrical recording of stylet penetration activities. In: A.K. Minks & P. Harrewijn (eds), *Aphids, their Biology, Natural Enemies and Control*, Vol 2B, pp. 95-108. Amsterdam: Elsevier.

Proc. 8th Int. Symp. Insect-Plant Relationships, Dordrecht: Kluwer Acad. Publ.
S.B.J. Menken, J.H. Visser & P. Harrewijn (eds), 1992

Russian wheat aphid and drought stresses in wheat: tritrophic interactions with *Diaeretiella rapae* and plant resistance

R.K. Campbell[1], D.K. Reed[2], J.D. Burd[2], and R.D. Eikenbary[1]
[1] *Dept of Entomology, Oklahoma State University, Stillwater, Oklahoma, USA*
[2] *Plant Science Research Laboratory, USDA-ARS, Stillwater, Oklahoma, USA*

Key words: Biological control, Hymenoptera, Braconidae, parasitoids, water stress

The Russian Wheat Aphid (RWA), *Diuraphis noxia* (Mordvilko) (Homoptera: Aphididae), has become a serious pest of grain production in North America. Unexpanded (rolled) leaves caused by RWA feeding serve as refugia for the aphid colonies and may limit effective search and attack by aphid parasitoids. Initial investigations on tritrophic interactions with resistant wheat ('TAM W-107'), RWA, greenbug, *Schizaphis graminum* (Rondani) and the aphid parasitoid *Lysiphlebus testaceipes* (Cresson) were completed (Campbell *et al.*, 1990). More recently, Reed *et al.* (1991) found that a tolerant RWA resistant wheat (PI 372129) was beneficial for parasitoid action because the extent of leaf rolling was substantially reduced. Virtually nothing is known of the interactions of drought, RWA, plant resistance and biological control agents. This study was undertaken to investigate such interactions using seedlings of susceptible ('TAM W-101') or resistant (PI 372129) wheat, RWA and the parasitoid *Diaeretiella rapae* McIntosh.

Methods

The experiment was conducted in a greenhouse using individual plants grown in a fritted clay medium. Plants were infested with 25 mature apterous RWA at the 3-leaf stage and caged. Treatments were the six combinations of water stressed or not, RWA present or not, and parasitoid introduced or not. A completely randomized design was used with 15 replications/treatment. Parasitoid treatments were infested with one mating pair of *D. rapae* nine days after aphid infestation and removed after 24 h. The experiment continued a sufficient time to allow the formation of all F1 generation parasitoid mummies. Leaf water status was measured at the times of aphid and parasitoid infestation and at harvest using leaf-cutter psychrometers. The following parameters were measured: plant chlorosis, stunting, leaf rolling, numbers of leaves and tillers, total leaf lengths and areas, root and shoot dry weights; RWA population development and age structure; parasitoid fecundity, % parasitization, development time, sex ratio, and adult size measurements.

Results

A substantial level of drought stress was imposed by the drought treatments. Reduction in leaf turgor in the droughted plants was exacerbated by RWA feeding. RWA tolerant wheat was more capable of maintaining leaf turgor when compared to 'TAM W-101'. Aphid feeding pressure alone was incapable of significantly reducing turgor in the resistant plants, but did affect the susceptible ones. Introduction of the parasitoid appeared to spare the 'TAM W-101'. A sparing effect also occurred in the parasitoid treated resistant wheat; there was a significant reduction in the amount of leaf rolling and chlorosis when compared to RWA only treatments. No such effect was seen for the 'TAM W-101' entry. Plant stunting was the most sensitive parameter. Counts of tillers and leaves and measurements of total leaf length were severely affected in all droughted plants. Additional stress from RWA had no greater detrimental effect. Again, inclusion of the parasitoid in the well watered treatments spared those plants some damaging effects due to RWA. Similar findings were exhibited by total leaf area and root and shoot biomass measurements. In general, higher aphid populations were found on drought stressed plants.

The parasitoid usually was capable of significantly lowering the total numbers of aphids. The drought stressed resistant wheat treatment significantly lengthened the developmental period of both male and female parasitoids. In contrast, drought had no effect on parasitoid development time with the RWA susceptible wheat. It is unknown whether this is due to some nutritional deficiency or some drought stress induced toxic principle. Under well watered conditions the developmental period was quite similar in both PI 372129 and 'TAM W-101'. No differences were found between treatments in any of the size measurements of the parasitoids. Parasitoids placed on drought stressed resistant wheat had a significantly lower parasitization level than those on well watered plants. Aphids on these plants were less widely dispersed (less available) and more concentrated within the new rolled leaf areas where seclusion limited parasitization even though overall these plants exhibited less tightly rolled leaves than 'TAM W-101'. There was also a significant male-biased sex ratio on the droughted PI 372129 when compared to well watered plants. Again, this effect was not seen on susceptible 'TAM W-101'. None of the other parameters exhibited significant differences between treatments. The parasitoid's developmental period, parasitization rate and sex ratio were found to be detrimentally altered on RWA resistant plants when subjected to severe drought stress.

This is Professional Paper No. 3665 of the Agricultural Experiment Station, Oklahoma State University, Stillwater, Oklahoma.

References

Campbell, R.K., C.E. Salto, L.C. Sumner & R.D. Eikenbary (1990). Tritrophic interactions between grains, the greenbug (*Schizaphis graminum*) and entomophaga. In: A. Szentesi & T. Jermy (eds), *Insects-Plants '89*, pp. 393-401. Symp. Biol. Hungary 39. Budapest: Akadémiai Kiadó.
Reed, D.K., J.A. Webster, B.G. Jones & J.D. Burd (1991). Tritrophic relationships of Russian wheat aphid (Homoptera: Aphididae), a hymenopterous parasitoid (*Diaeretiella rapae* McIntosh), and resistant and susceptible small grains. *Biol. Control* 1: 35-41.

Proc. 8th Int. Symp. Insect-Plant Relationships, Dordrecht: Kluwer Acad. Publ.
S.B.J. Menken, J.H. Visser & P. Harrewijn (eds), 1992

Performance of pea aphid clones in relation to amino acid composition of phloem sap and artificial diets

Jonas Sandström
Dept of Plant and Forest protection, Swedish University of Agricultural Sciences, Uppsala, Sweden

Key words: *Acyrthosiphon pisum*, aphids, intraspecific variation, resistance

Aphids feed exclusively on sap from vascular tissues of plants, obtaining most of their nitrogen from the phloem in the form of free amino acids. Auclair *et al.* (1957) and Weibull (1988) suggested that the concentrations and composition of amino acids are factors that mediate aphid resistance in plant genotypes and restrict the host plant range of aphids.

The aim of this study was to determine whether different clones of the pea aphid, *Acyrthosiphon pisum* (Harris), are physiologically adapted to the amino acid composition of their food plants. This knowledge is needed before addressing the question of whether the amino acid composition of phloem sap acts as a factor mediating resistance in peas *Pisum sativum* (L.) genotypes.

Aphid performance

The performance of three parthenogenetic pea aphid clones, collected from different host plants, was evaluated on five different genotypes of pea, *P. sativum*, and one genotype each of broad bean, *Vicia faba* L., lucerne, *Medicago sativa* L., and red clover, *Trifolium pratense* L. Large differences in performance were observed among the pea aphid clones. All clones performed well on the plant species from which they were collected. None of the clones performed well on all plants. These results correspond with other findings indicating that the pea aphid shows a large degree of intraspecific variation concerning host utilization (Via, 1991).

Concentrations and composition of free amino acids in the phloem

Pure phloem sap samples were collected from the above-mentioned pea genotypes and plant species. The samples were obtained from cut aphid stylets using the radio frequency microcautery method. Concentrations of individual amino acids were determined by means of high performance liquid chromatography with the o-phtalaldehyde method. There was no significant difference in the mean total amino acid concentration between pea genotypes or plant species. The total amino acid concentration in phloem sap can therefore be excluded as a factor that mediates pea aphid resistance in the investigated pea genotypes.

A statistical analysis of the free amino acid composition of the phloem sap did not reveal any differences in composition among the pea genotypes. The lack of distinct amino acid profiles in the pea genotypes suggests that the amino acid composition is not an important factor mediating pea aphid resistance.

299

In contrast to the pea genotypes, I found among the plant species examined more or less distinct differences in the amino acid profiles. These differences in amino acid profiles could explain the observed differences in host range between pea aphid clones. However, when analyzed in a multivariate model, there was no significant correlation between any amino acid variable and the performance of individual pea aphid clones.

Aphid performance on artificial diets

Four different artificial diets were prepared. The nutrient compositions of the diets were identical except for the amino acids, which were added in proportions to mimic the phloem sap composition of each of the four plant species. The performances of the above-mentioned pea aphid clones on each of the diets were evaluated.

One of the clones performed better than the others, regardless of the amino acid composition of the diets. Nevertheless, in general, performance of the pea aphid clones on the artificial diets was not related to performance on the corresponding plant species.

These results suggest that it is unlikely that pea aphid clones are physiologically adapted to the specific amino acid composition of their host plants' phloem sap. Furthermore, differences in amino acid composition of the phloem sap probably have little or no influence in determining the host ranges of the tested pea aphid clones.

References

Auclair, J.L., J.B. Maltais & J.J. Cartier (1957). Factors in resistance of peas to the pea aphid, *Acyrthosiphon pisum* (Harr.) (Homoptera: Aphididae). II. Amino acids. *Can. Entomol.* **89**: 457-464.

Via, S. (1991). The genetic structure of host plant adaption in a spatial patchwork: Demographic variability among reciprocally transplanted pea aphid clones. *Evolution* **45**: 827-852.

Weibull, J. (1988). Free amino acid in the phloem sap from oats and barley resistant to *Rhopalosiphum padi*. *Phytochemistry* **27**: 2069-2072.

Proc. 8th Int. Symp. Insect-Plant Relationships, Dordrecht: Kluwer Acad. Publ.
S.B.J. Menken, J.H. Visser & P. Harrewijn (eds), 1992

Effect of certain proteins on *Acyrthosiphon pisum* growth and development. Potential influence on aphid-plant interactions

Yvan Rahbé and Gérard Febvay
INRA, INSA Villeurbanne, France

Key words: Lectin, phloem protein, protease inhibitor, toxin, transgenic plant

Aphids are phloem feeders, and phloem sap is generally poor in proteins (individual levels < 100 µg/ml). However, aphids encounter substantial protein concentrations on certain host plants, such as members of the Cucurbitaceae (Sabnis & Hart, 1979), or in other tissular compartments such as plant cell walls. In this paper, we describe the effect that certain proteins might have on aphid physiology in a normal plant-aphid interaction. The *in vitro* assay enabled us to test the effect of soluble proteins, at "physiological" concentrations, on aphid growth and mortality.

Material and methods

Neonate *Acyrthosiphon pisum* (Harris) were reared on artificial diets containing different concentrations of purified proteins (10-50-250 µg/ml). The test lasted one week and mortality checks were noted on days 1, 3 and 7. Aphids were weighed individually at day 7. To determine LC50s, mortality data were fitted by a standard Probit (% corrected mortality) = f(Log(Protein dose)). Growth inhibition was modelled by a Michaelis equation of weights *vs* protein doses:

$$\text{Weight} = \text{Ctrl. Weight} - \frac{P_1 \times \text{dose}}{P_2 + \text{dose}}$$

This model allows the determination of the concentrations giving 50% or 20% growth inhibitions when compared to the control, *i.e.*, IC50 & IC20, or the P_1 & P_2 parameters (maximal potential inhibition -P_1-, and dose of half maximal inhibition -P_2-; see Rahbé & Febvay, submitted).

Results and discussion

Among the proteins tested, proteases were the most toxic (lethal), with "side-effect" growth-inhibition properties (often not dose-responsive). Compared to a chemical aphicide, toxicity was ≈ 50 times lower on a molar basis (for example NTN-types *vs* Trypsin).

Some proteins, such as the legume lectins Concanavalin, Pisum and Lens lectins (CON, PIS and LEN) showed both lethal and inhibitory properties. These lectins (man and glc binding) are thus good candidates for a potential antibiosis function in aphid-plant interactions (natural situation or transgenic plants).

Table 1 shows the features of some characteristic proteins scattered in the global toxicity graph displayed in Fig. 1. A number of different proteins (various protease

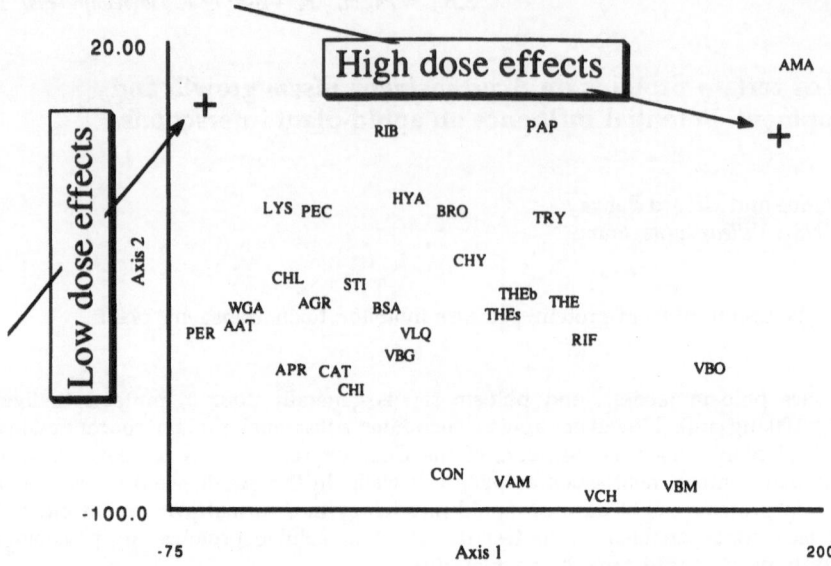

Figure 1. Principal component analysis of mortality and growth inhibition data of various proteins on *A. pisum*. Proteins highly toxic at high doses are at the right side of the graph. Proteins (slightly) toxic at low doses are at the upper left side. Proteins at the bottom of the graph are toxic at high doses and inactive at low doses.

inhibitors, STI, CHI, and BSAlbumin) exhibited low lethality and medium growth inhibition activity. The other proteins (PER, WGA, and Pectinase, Hyaluronidase, Ribonuclease and Lysozyme) formed a group exhibiting low biological activity under the conditions used.

Table 1. Mortality (LC50) and Growth inhibition (IC50 and IC20) for the 10 polypeptides tested in Fig. 1

Label	Protein	Mortality	Growth inhibition	
		LC50 (µg/ml)	IC50 (µg/ml)	IC20 (µg/ml)
AMA	α-amanitine	4.5	nd	nd
TRY	Trypsin	22	103	10
VBM	Toad venom	35	nd	27
RIF	Rifampicine	84	21	3
VAM	Honey Bee venom	87	222	61
CON	Concanavaline A	134	182	53
WGA	Wheat Germ Agglutinin	> 500	not dose-resp.	not dose-resp.
STI	Trypsin Inhibitor	> 500	93	15
CHI	Chitinase	not lethal	367	83
PER	Peroxydase	not lethal	> 500	> 500

302

A. pisum was shown to be affected by a number of proteins at concentrations within the range occurring in phloem sap of a common plant. This finding arouses interest in the functions of phloem proteins in an aphid-plant interaction, as yet largely ignored (Sabnis & Hart, 1979).

None of the proteins tested exhibited any antixenosis activity in the concentration range used (as measured by the fixation rate one hour after starting the test). This is in concordance with the commonly observed absence of phagostimulatory activity of polypeptides (with some very specific exceptions).

A class of proteins known widely to occur in phloem sap of very different plant species (anti glc-NAc lectins) proved to be very well tolerated by *A. pisum*, possibly as a result of an adaptation. Contrastingly, mannose-binding lectins (D-man, D-glc) showed interesting toxic properties, that should confer antibiosis to plants able to express such polypeptides at sufficient levels in their phloem sap. A hypothesis for toxicity is their interaction with some man(or glc)-glycoprotein from the digestive tract of aphids.

Proteases are quite toxic to aphids. This is compatible with the absence of endo-proteasic activities in aphid midguts (namely, peptidases), resulting in a lack of protective mechanisms of the intestinal mucosa towards these enzymes.

This work concentrated on "short-term" antibiosis (one larval period). Long-term aphid-protein interactions should also be investigated. There may be effects on reproduction and on aphid feeding mechanisms: the interactions of enzymes and structural proteins of the cell-wall or the phloem with salivary secretions are still largely unknown (Miles, 1990). Unlike the described "antibiosis approach", the "behavioural approach" should first be investigated *in situ*, with the occasional support of *in vitro* experiments.

References

Miles, P.W. (1990). Aphid salivary secretions and their involvement in plant toxicoses. In: R.K. Campbell & R.D. Eikenbary (eds), *Aphid-Plant Genotype Interactions*, pp. 131-147. Amsterdam: Elsevier.

Rahbé, Y. & G. Febvay. Testing protein toxicity towards aphids: an *in vitro* test on *Acyrthosiphon pisum*. Submitted to *Entomol. exp. appl.*

Sabnis, D.D. & J. Hart (1979). Heterogeneity of phloem protein complement from different species. Consequences to hypotheses concerned with P-protein function. *Planta* **145**: 459-466.

Proc. 8th Int. Symp. Insect-Plant Relationships, Dordrecht: Kluwer Acad. Publ.
S.B.J. Menken, J.H. Visser & P. Harrewijn (eds), 1992

The influence of some non-protein amino acids on winter wheat resistance to the grain aphid

A.P. Ciepiela, S. Niraz and C. Sempruch
Agricultural and Pedagogic University, Institute of Biology, Department of Biochemistry, Poland

Key words: Antibiosis, L-3,4-dihydroxy-phenylalanine, ornithine, γ-aminobutyric acid, *Sitobion avenae*

The grain aphid *Sitobion avenae* (F.) (Homoptera: Aphididae) is a common pest of winter wheat in Poland. One of the most important cereal resistance mechanisms in winter wheat is antibiosis. The level of antibiosis is usually associated with the quantitative and qualitative composition of different nutritional and allelochemical components in the host plant tissues. Numerous reports have indicated that some non-protein amino acids play an important role in plant resistance (Romero, 1984; Weibull, 1988). The aim of this study was to compare some correlations between the concentrations of L-3,4-dihydroxyphenyl-alanine (L-DOPA), ornithine and γ-aminobutyric acid (GABA) in moderately resistant and susceptible winter wheat varieties with their antibiotic resistance to *S. avenae*.

Materials and methods

We tested the antibiotic properties to *S. avena* of four winter wheat cultivars: Grana and Saga (moderately resistant) and Emika and Liwilla (susceptible). The tests were done in the field using the method described by Leszczynski *et al.* (1989). Based on available data on each variety, life tables for *S. avenae* were constructed and intrinsic rates of natural increase (r_m) calculated. Non-protein amino acids were extracted from flag leaves with 80% ethanol and purified on ion exchange resins, using the method described by Ciepiela (1989). Amino acids were separated and quantitatively determined using a completely automated AAA - 339 amino acid analyzer. All analyses were replicated three times with each winter wheat variety. The linear correlations between concentrations of amino acids and r_m's of *S. avenae* on tested cultivars of wheat were calculated.

Results and discussion

The r_m values of the aphids varied from 0.1799 for the moderately resistant variety Saga to 0.2144 for the susceptible one Liwilla. The results showed that high antibiotic properties of moderately resistant cultivars were positively correlated with the greater concentrations of L-DOPA (r = -0.937; Fig. 1) and ornithine (r = -0.907; Fig. 2), but not with concentrations of GABA (r = -0.363; Fig. 3). Moreover, there is a definite link between the contents of L-DOPA and ornithine in flag leaves of the resistant wheat and the growth and development rate of the grain aphid. It seems that presence of these two non-protein amino acids in moderately resistant cultivars of winter wheat is one of the most important factors influencing their level of antibiosis to *S. avenae*. The results will be further evaluated by *in vitro* tests using synthetic diets.

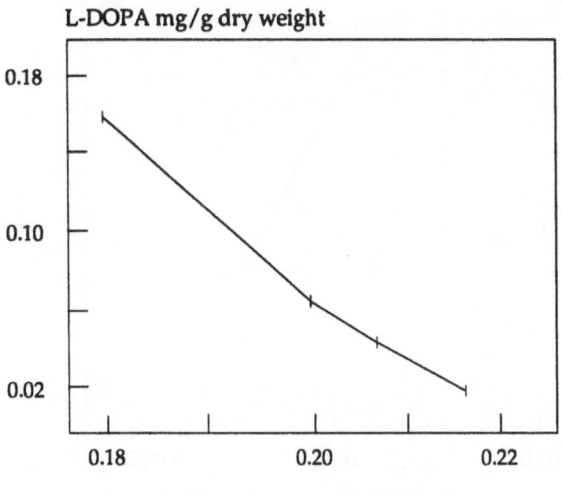

Figure 1. The relationships between concentration of L - DOPA in flag leaves of winter wheat studied and the intrinsic rate of natural increase r_m of *S. avenae*. Vertical bars are SE.

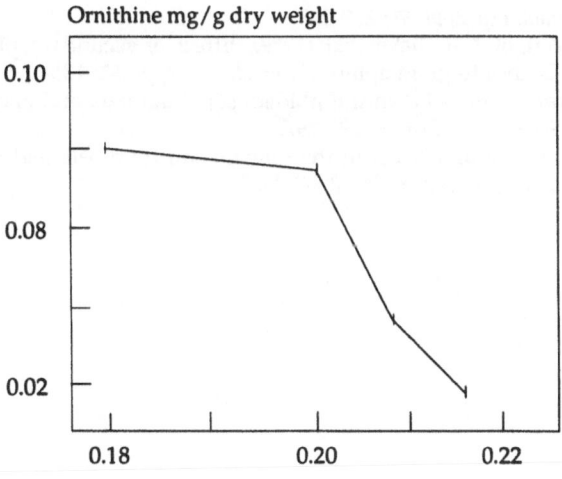

Figure 2. The relationships between concentration of ornithine in flag leaves of winter wheat studied and the intrinsic rate of natural increase r_m of *S. avenae*.

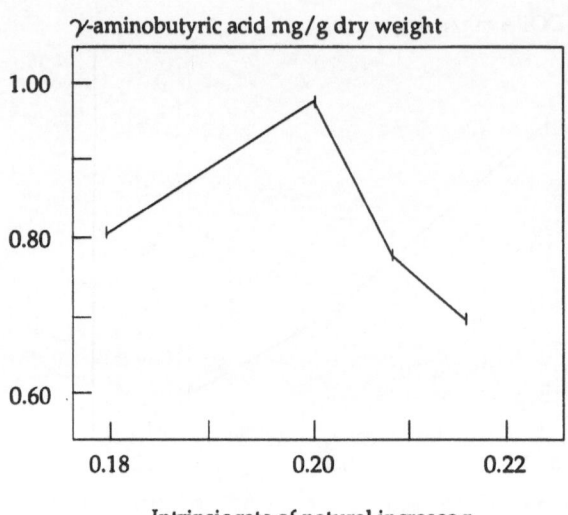

γ-aminobutyric acid mg/g dry weight

Intrinsic rate of natural increase r_m

Figure 3. The relationships between the level of GABA in flag leaves of winter wheat studied and the intrinsic rate of natural increase r_m of *S. avenae*.

References

Ciepiela, A.P. (1989). Biochemical basis of winter wheat resistance to the grain aphid, *Sitobion avanae. Entomol. exp. appl.* **51**: 269-275.

Leszczynski, B., L.C. Wright & T. Bakowski (1989). Effect of secundary plant substances on winter wheat resistance to grain aphid. *Entomol. exp. appl.* **52**: 135-139.

Romero, J.T. (1984). Free amino acids in the phloem sap from oats and barley resistant to *Rhopalosiphum padi. Bioch. Syst. Ecol.* **3**: 293-297.

Weibull, J.H.W. (1988). Free amino acids in the phloem sap from oats and barley resistant to *Rhopalosiphum padi. Phytochemistry* **27**: 2069-2072.

Proc. 8th Int. Symp. Insect-Plant Relationships, Dordrecht: Kluwer Acad. Publ.
S.B.J. Menken, J.H. Visser & P. Harrewijn (eds), 1992

Presence of hydroxamic acids in the honeydew of aphids feeding on wheat seedlings, and its significance for plant resistance and virus transmission

Arturo Givovich and Hermann M. Niemeyer
Departamento de Quimica, Facultad de Ciencias, Universidad de Chile, Santiago, Chile

Key words: Aphididae, DIMBOA, *Rhopalosiphum padi, Triticum aestivum*

Cereal aphids damage wheat plants through direct feeding, promotion of fungal and transmission of viral diseases such as barley yellow dwarf virus (BYDV). Internal resistance factors in the plant may prevent virus transmission. Wheat contains hydroxamic acids (Hx), a family of compounds playing a key role in the defence of wheat seedlings against cereal aphids (Niemeyer, 1988). The main Hx in wheat is 2,4-dihydroxy-7-methoxy-1,4-benzoxazine-3-one (DIMBOA). It occurs in the intact wheat plant as a 2-β-O-D-glucopyranoside (DIMBOA-Glc), which is hydrolysed by endo-glucosidases when the plant is injured. Both DIMBOA and DIMBOA-Glc are aphid feeding deterrents and antibiotics, but the glucoside is less effective than the aglucone. Wheat seedlings with higher Hx concentrations induce a decrease in aphid performance and a decrease in BYDV transmission (Givovich & Niemeyer, 1991). To assess the relative importance of DIMBOA and DIMBOA-Glc in modifying aphid behaviour and fitness, we have complemented previous analyses of Hx in aphid whole bodies with Hx analysis of honeydew.

Methods

Seedlings of wheat (*Triticum aestivum* (L.) *cvs* Platifen, Millaleu, Mexifen, Nobo, Anza and Maiten) at the one-leaf stage (growth stage 10, decimal code), were infested with twenty aphid nymphs (*Rhopalosiphum padi* (L.)) confined to a clip cage. A previously weighed piece of aluminium foil was placed in the bottom of each cage. After 36 h, the disks were weighed and washed successively with methanol and distilled water. The washings were evaporated to dryness, redissolved in 200 µl of methanol and analyzed by HPLC. This was repeated seven times with each wheat line.

Results and discussion

DIMBOA-Glc was present in all the honeydew samples, but neither the aglucone nor its main breakdown product were found in any of them. Both honeydew production and concentration of DIMBOA-Glc in honeydew followed a biphasic curve when plotted against Hx concentration in the wheat seedlings used: as the seedling Hx concentration increased, both dependant variables first increased and then decreased.

These results suggest that Hx are present in the phloem of wheat seedlings. Furthermore, the biphasic curves suggest passive ingestion of Hx from the phloem at low seedling Hx concentration and limited ingestion of Hx at high seedling Hx concentration due to feeding deterrency of Hx in mesophyll cells. Ingestion of Hx from the phloem would lead to reduced aphid performance. Detection by the aphid of Hx in the mesophyll would lead to feeding deterrency and decreased virus transmission.

Acknowledgements. We are indebted to International Program in the Chemical Sciences (IPICS), Fondo Nacional de Desarrollo Científico y Tecnológico (FONDECYT), Dirección Técnica de Investigación de la Universidad de Chile (DTI-UCh), Agency for International Development (AID) and International Foundation for Science (IFS), for their financial support.

References

Givovich, A. & H.M. Niemeyer (1991). Hydroxamic acids affecting barley yellow dwarf virus transmission by the aphid *Rhopalosiphum padi. Entomol. exp. appl.* **59**: 79-85.
Niemeyer, H.M. (1988). Hydroxamic acids (4-hydroxy-1,4-benzoxazin-3-ones), defence chemicals in the Gramineae. *Phytochemistry* **27**: 3349-3358.

Proc. 8th Int. Symp. Insect-Plant Relationships, Dordrecht: Kluwer Acad. Publ.
S.B.J. Menken, J.H. Visser & P. Harrewijn (eds), 1992

Changes in phenolic compounds in cucumber leaves infested by the two-spotted spider mite (*Tetranychus urticae*)

A. Tomczyk
Dept of Applied Entomology, Warsaw Agricultural University, Warsaw, Poland

Key words: Mite infestation, phenols, resistance

Spider mites can induce defence reactions in their host plants. These reactions can occur after a few days of infestation even in susceptible plants. Phenols were found to be important chemical compounds for resistance of many host plants to insects and mites. However, very little information has been available up to now about changes in phenolic compounds induced by spider mite feeding.

Material and methods

The experiment was conducted with two cultivars of cucumber, Atos and Wilanowski, cultivated under greenhouse conditions. The plants were infested with the two-spotted spider mite (*Tetranychus urticae* Koch). Mite females were transferred from bean leaves to 6 week old cucumber plants (development stage of the plant - 7 leaves). The initial spider mite population of 0.5 females per cm^2 did not change during the first 12 days of the experiment. After 42 days the mite populations increased to 1.6 and 1.8 mites per cm^2 on *cvs* Wilanowski and Atos, respectively.

Leaf samples were taken from both infested and noninfested plants for biochemical analyses at 3, 6, 12, 35, and 42 days after infestation with mites. The concentration of total phenols, monophenols, polyphenols and phenol glucosides - derivatives of dihydroquinone, were estimated. Fresh leaf material was used for most of the analyses. The leaves were dried at 60°C for the estimation of phenol glucosides. These compounds were extracted from the dry leaf material using boiling water and estimated after Emerson Reaction. The amounts of total phenols and monophenols were estimated in 80% methanol extracts, after reaction with Folin-Ciocalateau Reagent. Polyphenols were removed from the extracts by absorption on Polyclar AT prior to estimating the monophenols. The amount of polyphenols was calculated as a difference between total phenols and monophenols.

Results and conclusions

We found that changes in concentration of phenolic compounds in infested cucumber leaves depended on the length of mite feeding. After the first three days of infestation a small increase in total phenol content in the leaves of cv. Wilanowski was observed but not in the leaves of the cv. Atos, which is more suitable for spider mites. This small increase was found in both monophenols and polyphenols.

After six days of spider mite feeding the total phenol concentration increased in the infested plants of both cucumber cultivars studied by 10-12%. An increase in concentration of total phenols in the leaves of *cv*. Atos six days after infestation with spider mites, was found to be related to the increase in polyphenols. The concentration of monophenols was not affected by the infestation. In the infested leaves of the *cv*. Wilanowski, less suitable for spider mites, an increase in both monophenols and polyphenols was observed. The content of phenol glucosides was the same in infested and noninfested leaves of *cv*. Wilanowski. Contrastingly, the damaged leaves of *cv*. Atos had 23% more phenol glucosides than the undamaged leaves.

After 12 days of infestation, the concentration of total phenols began to decrease in both cultivars of cucumber in relation to the control plants. The ratio of polyphenols to monophenols was also lower. After 42 days of mite feeding, compared to the noninfested ones, the content of phenol glucosides was 13% and 26% higher in the infested leaves of of *cv*. Wilanowski and *cv*. Atos, respectively.

It can be concluded that an increase in the concentration of the total phenols at the beginning of spider mite feeding is probably not connected with the release of phenols from complexes with sugars, because the concentration of phenol glucosides was higher in the infested leaves than in those not infested. The decrease in the concentration of the total phenols after a longer period of spider mite feeding could be connected with a decrease in phenol synthesis or with increased oxidation of phenols to quinones.

References

Tomczyk, A. (1989). Physiological and biochemical responses of different host plants to infestation by spider mites (Acarina: Tetranychidae). Treatises and Monographs, Warsaw Agricultural University, 112 p.
Tomczyk, A. & D. Kropczynska (1984). Feeding effects of *Tetranychus urticae* Koch on the physiology of some plants. *Acarology* 2: 740-746.

Proc. 8th Int. Symp. Insect-Plant Relationships, Dordrecht: Kluwer Acad. Publ.
S.B.J. Menken, J.H. Visser & P. Harrewijn (eds), 1992

Effects of spider mite infestation on biochemical characteristics of different gerbera cultivars

M. Kiełkiewicz [1] and M. Dicke [2]
[1] Dept of Applied Entomology, Warsaw Agricultural University, Warsaw, Poland
[2] Dept of Entomology, Wageningen Agricultural University, Wageningen, The Netherlands

Key words: *Gerbera jamesoni*, leaf biochemistry, leaf morphology

Until very recently, little data dealing with interactions between gerbera and spider mites were available. Results of our preliminary experiments indicated that the two-spotted spider mite (*Tetranychus urticae* Koch) performed better on Estelle and Donna Tella gerbera *cvs* than on Fame *cv*. In this study gerbera plants were morphologically and biochemically compared before and after spider mite infestation. Leaves from two-month-old micropropagated gerbera (*Gerbera jamesoni* Bolus, *cvs* Fame, Donna Tella and Estelle) plants were used in all experiments. Type of leaf hairs as well as their density were examined using a stereomicroscope. Some biochemical analyses were performed after three days of spider mite feeding on a third gerbera leaf. The initial number of females was 3.2 per cm^2 of leaf. The infested area of leaves and the adjacent area were cut and soluble protein content, total phenolic concentration and phenylalanine ammonia lyase (PAL) activity were determined.

We observed that there was a higher density of trichomes on the more suitable gerbera leaves (Estelle and Donna Tella *cvs*). The less suitable gerbera *cv*. Fame was less hairy. Three morphological types of trichomes were present on gerbera leaves: 1. long, simple (*ca*. 2.2 - 4.8 mm in length), 2. short, simple (*ca*. 0.3 - 0.6 mm in length) and 3. glandular (*ca*. 0.15 - 0.25 mm in length, Fig. 1). The glandular type of trichomes was found only on the leaves of Fame *cv*. However, in a later study similar glandular trichomes were found on the upper leaf surface of Pascal and Maria *cv*. A pinkish liquid substance was observed in the glandular hairs and in some short, simple ones. The nature of this substance remains unknown.

We found that a short period of mite feeding increased the content of soluble protein in both directly damaged and adjacent leaf areas. The increase of soluble protein in the leaves of Estelle and Donna Tella *cvs* was significantly higher in the surrounding tissues than in directly injured tissue. Enhanced soluble protein content may be the result of protein synthesis which can be seen as one of the early damage - recognition processes. The synthesized proteins may consist of enzymes which activate the synthesis of protective chemicals.

Phenolic compounds have been pointed out as an important factor mediating plant resistance to spider mites (Kiełkiewicz, 1990; Kiełkiewicz & Van de Vrie, 1990). In this study, spider mite infestation of gerbera plants was associated with a marked initial increase in total phenolic content of infested leaves of Estelle and Donna Tella *cvs*. This increase indicates that phenolics are either being synthesized or they are being transported from other leaves. If *de novo* synthesis had occurred, PAL activity should have increased. This enzyme is involved in the biosynthetic pathway of secondary products

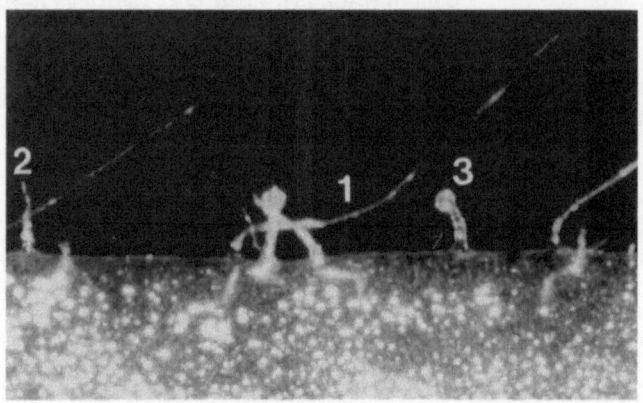

Figure 1. Three types of trichomes on the foliage of gerbera *cv*. Fame.

derived from phenylpropanoids. Our investigation revealed that 3 days of spider mite infestation lead to a greater increase of PAL activity in infested Fame *cv*. leaves than in the leaves of Estelle or Donna Tella *cvs*. Thus, there was no correlation between accumulation of total phenolics and the increase of PAL activity in infested gerbera plants. The results suggest that an increase of PAL activity in spider mite infested Fame *cv*. leaves may be connected with synthesis of certain other secondary metabolities like flavonoids (or isoflavonoids), some alkaloids, esters or polyacetylene compounds, but this has not yet been studied. Future studies of gerbera - mite relationships should focus on the role of these compounds.

Acknowledgements. This research was carried out at the Department of Entomology, Wageningen Agricultural University, The Netherlands and financed by NWO. M. Kiełkiewicz's participation in SIP was supported by the European Science Foundation (Strassbourg, France) and Batory Foundation (Warsaw, Poland).

References

Kiełkiewicz, M. (1990). Metabolic consequences of stress induced by the feeding of *Tetranychus cinnabarinus* on tomato plants. *Symp. Biol. Hung.* **39**: 485-486.
Kiełkiewicz, M. & M. van de Vrie (1990). Within-leaf differences in nutritive value and defence mechanism in chrysanthemum to the two-spotted spider mite (*Tetranychus urticae*). *Exp. Appl. Acarol.* **10**: 33-43.

Proc. 8th Int. Symp. Insect-Plant Relationships, Dordrecht: Kluwer Acad. Publ.
S.B.J. Menken, J.H. Visser & P. Harrewijn (eds), 1992

Pleiotropic effects of genes in glossy *Brassica oleracea* resistant to *Brevicoryne brassicae*

Rosemary A. Cole and Wendy Riggall
Horticulture Research International, Wellesbourne, Warwick, UK

Key words: Aphids, EPG, epicuticular wax, glossy

A series of mutations have been identified in *Brassica oleracea* L. that change the normal whitish bloom of leaves so that they appear 'glossy' (Macey & Barber, 1970). The glossy character is caused by a reduction in the paraffin wax fraction of the leaf cuticle. Glossy brassicas are less preferred by *Brevicoryne brassicae* (L.) (Stoner, 1990) in the field but not under glasshouse conditions (K. Stoner, pers. comm.). As resistance to aphids is diminished under glasshouse conditions, it is unclear to what extent resistance is determined by the physical characteristic of glossiness, or by more subtle pleiotropic effects of genes for glossiness. Chambers and Possingham (1963) proposed that the architecture of epicuticular wax is important in reducing cuticular transpiration. During the summer crops are usually subject to some degree of water stress whereas under glasshouse conditions plant growth is luxuriant. This paper sets out to compare aphid performance on glossy and glaucous lines grown under field, glasshouse and under reduced watering regimes.

Methods

Feeding behaviour of *B. brassicae* was recorded using the DC modification of electronic monitoring (EPG) described by Tjallingii (1987), using computerised data-acquisition (Cole *et al.*, unpubl. results). Differences in probing behaviour are related to cuticular waxes analysed from plants grown under different conditions.

Isogenic lines differing only in the genes for glossiness were used in these experiments. Brassica plants were grown in a field experiment at Wellesbourne during July/August 1991. Glasshouse plants were raised in 12 cm pots containing Levington compost watered by capillary matting. Finally, plants were also raised in a Saxile controlled environment cabinet where they were only watered after the compost had dried but before wilting occurred.

Two *B. brassicae* apterae were confined in clip cages on the underside of leaves for three days, the two initial apterae were removed and the colonies left to develop. After 14 days, aphids were counted.

Leaves were examined under a Cambridge 'Stereoscan 200'. The wax was extracted and analysed on a 25 m x 0.32 mm id, CP-SIL 5 CB (Chromopack) capillary column programmed from 150-250°C at 10°C/min.

The amplified EPG signal was stored on the hard disk of an IBM PC/AT using an analog to digital conversion board PC26AT (Amplicon) and Microscope data-acquisition software program (G.A. Wilson, Amplicon).

Results and discussion

The glossy line grown in the glasshouse was as susceptible to the build-up of aphid populations as the susceptible glaucous line. There were no significant (P = 0.05) differences between the size of colonies with a mean of 25 (± 5) aphids produced on glossy plants in the glasshouse, and on glaucous plants in all environments. In the field and under reduced watering aphid numbers (9 ± 1) were greatly reduced on the glossy plants.

Electron micrographs confirmed the reduced coverage of wax on all glossy leaves (1-2 mg/kg of leaf) and the increased production of wax (100-143 mg/kg of leaf) on glaucous leaves when watering was reduced both in the field and in a controlled environment. Leaf wax was also reduced (22 mg/kg of leaf) on glaucous leaves grown under glasshouse conditions. Glossy and glaucous lines grown with reduced watering were more comparable to field than glasshouse grown lines, both in cover of epicuticular wax and resistance to *B. brassicae*. Aphids thrived on glasshouse grown glossy and glaucous lines even though there was reduced coverage of wax compared to field-grown glaucous lines. This confirmed that the genes for glossiness were not directly related to aphid resistance.

Table 1. Means (10 plants) of electronically recorded (5 h) probing events for *Brevicoryne brassicae* on glaucous and glossy brassica plants

	Time in minutes of			Time to 1st probe	Number of probes	Number of cell pene-trations
	Non probing	pathway Stylet	Phloem ingestion			
Glasshouse conditions						
Glaucous plants	11	197	93	3	4	108
Glossy plants	13	190	110	5	4	124
Field conditions						
Glaucous plants	15	196	98	4	5	141
Glossy plants	38	214	44	3	14	116
Controlled environment						
Glaucous plants	10	184	125	8	6	182
Glossy plants	34	208	58	5	12	120
L.S.D. (P < 0.05)	13.6	36.6	37.4	3.1	3.4	41.0

On all leaves probing began almost immediately followed by one or two cell penetrations, this behaviour was similar on both glossy and glaucous leaves under all growing conditions. EPG monitoring indicated no significant (P = 0.05) differences in behaviour on glasshouse grown glossy and glaucous leaves (Table 1). However, although aphids were not deterred from making initial probes on field and controlled environment grown glossy lines many of these were soon terminated. Aphids made significantly more probes on these leaves before eventually successfully penetrating to the phloem. No significant differences were encountered in time of stylet pathway to phloem on these leaves but the time before phloem uptake was delayed and the duration of phloem uptake was reduced. These differences may account for the reduced numbers of aphids indicating the presence of feeding deterrents induced by the stress of increased cuticular transpiration.

References

Chambers, T.C. & J.V. Possingham (1963). Studies of the fine structure of the wax layer of sultana grapes. *Aust. J. Biol. Sci.* **16**: 818-825.

Macey, M.J.K. & H.N. Barber (1970). Chemical genetics of wax formation on leaves of *Brassica oleracea. Phytochemistry* **9**: 5-23.

Stoner, K. (1992). Density of imported cabbageworms (Lepidoptera: Pieridae), cabbage aphids (Homoptera: Aphididae) and flea beetles (Coleoptera: Chrysomelidae) on glossy and trichome bearing lines of *Brassica oleracea. Ann. Entomol. Soc. Am.* **85**: 1023-1030.

Tjallingii, W.F. (1987). Stylet penetration activities by aphids: new correlations with electrical penetration graphs. In: *Proc. 6th Int. Symp. on Insect-Plant Relationships*, pp. 301-306. Pau, France.

Proc. 8th Int. Symp. Insect-Plant Relationships, Dordrecht: Kluwer Acad. Publ.
S.B.J. Menken, J.H. Visser & P. Harrewijn (eds), 1992

The role of brassica leaf surface chemicals in antixenotic resistance to *Delia floralis*

R.J. Hopkins[1,2], A.N.E. Birch[2], D.W. Griffiths[2] and R.G. McKinlay[3]
[1] *Institute of Ecology and Resource Management, University of Edinburgh, Edinburgh, UK*
[2] *Scottish Crop Research Institute, Invergowrie, Dundee, UK*
[3] *Scottish Agricultural College, Edinburgh, UK*

Key words: Allelochemicals, behaviour, oviposition

Turnip root fly, *Delia floralis* (Fall.), is a major pest of brassicas in Northern Europe. Understanding the mechanisms of host plant resistance is important to the development of integrated pest management. Antixenosis is a key component of resistance (Birch, 1988) because damage is reduced by deterring oviposition. The oviposition behaviourial sequence of *D. floralis* was characterised by Havukkala & Virtanen (1985). Later work by Städler & Schöni (1990) demonstrated the importance of leaf surface chemicals in host plant selection by cabbage root fly, *Delia radicum* (L.). The objective of this work is to isolate the chemical stimuli which influence the host selection of *D. floralis*.

Materials and methods

Two swede genotypes (*cv.* Doon Major and SCRI breeding line GRL aga) and two kales (*cvs* Fribor and Dwarf Green Curled) were selected to represent a range of antixenotic resistance to turnip root fly. Choice tests were carried out on 8-10 true leaf stage plants randomly arranged on a turntable in an illuminated test chamber at 20°C for 24 h. In each replicate four plants, one of each genotype, were exposed to 30-40 gravid females aged 7-17 days. The host plant selection sequence of gravid females was observed in a perspex cage and behavioural events were recorded using Noldus "Observer 2" software. Behaviour was categorised into six classes: landing; exploration of leaf surface; stem run; walking on soil surface; pre-oviposition behaviour (probing with ovipositor or digging) and oviposition. Leaf surface extracts were made by dipping leaves of the four brassicas into dichloromethane and then methanol for five seconds each, with a ten second interval between solvents. Extracts of 1.25 g leaf equivalents (gle) were sprayed on to dummy plants and were tested in a choice experiment, as outlined above.

Results

Behaviourial analysis demonstrated that discrimination between resistant and susceptible plants took place immediately after landing and during leaf exploration (Fig. 1). Bioassays of whole plants in the acceptance tests showed a range of antixenosis as measured by egg counts (Table 1). Bioassays of leaf surface extracts showed that the oviposition stimulant activity was predominantly present in the methanol fractions. The activities in the fractions appeared to reflect the level of antixenosis in the whole plants (Table 1).

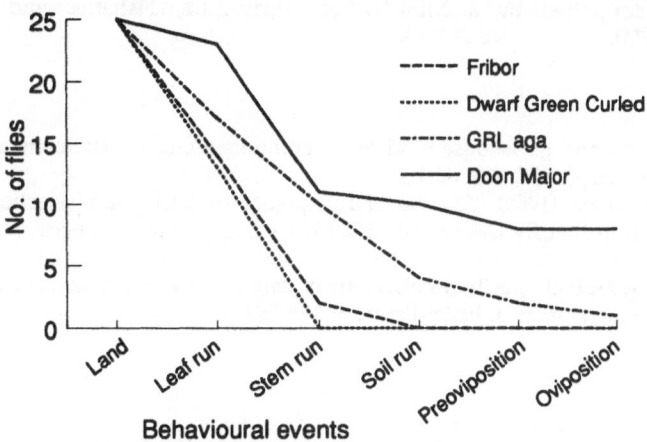

Figure. 1. Post-landing behaviour of turnip root fly on four *Brassica* genotypes.

Table 1. Percentage of total oviposition on whole and dummy plants, in choice tests

Crop	Genotype	Whole Plants (n = 10)	Sprayed Dummy Plants (n = 13)	
			Methanol	Dichloro-methane
Swede	Doon Major	77.0	40.6	1.7
Swede	GRL aga	18.9	32.5	0.8
Kale	Dwarf Green Curled	3.3	19.0	0.1
Kale	Fribor	0.8	4.4	0.2
	Control		0.5	0.2
Total		100%	100%	

Conclusions

High levels of antixenotic resistance to turnip root fly were confirmed in the two kales tested. A lower level of antixenosis was found in swede breeding line GRL aga. Behaviourial analysis of gravid female *D. floralis* on whole plants indicated the importance of the leaf surface in the acceptance of a plant for oviposition. Four percent and 40% of flies progressed beyond the leaf exploration stage on kales and swedes, respectively. In contrast to GRL aga, most flies reaching the stem on *cv.* Doon Major went on to oviposit. This may indicate that the stem surface chemicals are also important in the selection of some brassicas, like GRL aga. The dummy plant bioassays clearly demonstrated the importance of leaf surface chemicals in the oviposition behaviour of *D. floralis*. Current research is targeted at identifying the active components in the methanol fractions.

Acknowledgements. Work was carried out at the Scottish Crop Research Institute, Dundee. Richard Hopkins is supported by a Ministry of Agriculture, Fisheries and Food Studentship (No. AE8373).

References

Birch, A.N.E. (1988). Field and glasshouse studies on components of resistance to rootfly attack in swedes. *Ann. Appl. Biol.* **113**: 89-100.

Havukkala, I. & M. Virtanen (1985). Behaviourial sequence of host plant selection and oviposition in the turnip root fly *Delia floralis* (Fall.) (Anthomyiidae). *Z. angew. Entomol.* **100**: 39-47.

Städler, E. & R. Schöni (1990). Oviposition behaviour of cabbage root fly *Delia radicum* (L.) influenced by host plant extracts. *J. Insect Behav.* **3**: 195-209.

Proc. 8th Int. Symp. Insect-Plant Relationships, Dordrecht: Kluwer Acad. Publ.
S.B.J. Menken, J.H. Visser & P. Harrewijn (eds), 1992

Resistance to insects in glossy genetic lines of cole crops (*Brassica oleracea*)

Kimberly A. Stoner
Connecticut Agricultural Experiment Station, New Haven, CT, USA

Key words: *Brevicoryne brassicae*, epicuticular wax, *Pieris rapae*, plant resistance, *Plutella xylostella*

The glossy cauliflower PI 234599 has been used for many years in breeding programs as a source of resistance to diamondback moth *Plutella xylostella* (L.) and other lepidopterous pests. Only recently, however, have other genes for glossiness in cole crops (*Brassica oleracea* L.) been tested for their effects on insect resistance (Stoner, 1990).

Glossy lines that have been tested, in addition to PI 234599, are the broccoli Broc3 and the collards Green Glaze, White's Green Glaze and South Carolina Glaze, each of which has a single dominant gene for glossiness (Stoner, 1990; Stoner, unpubl. results), and the broccoli Broc5, the cauliflower Glossy Andes, and the kales KCR4 and Glazed Vates, each of which has a single recessive gene for glossiness (Stoner, 1990; Stoner, unpubl. results). The genes for glossiness in Broc5 and Glossy Andes appear to be non-allelic to each other and to the gene for glossiness in PI 234599. The genes for glossiness in PI 234599, KCR4 and Glazed Vates are allelic (Stoner, unpubl.).

These lines were compared with normal-wax standard varieties of broccoli, cauliflower, collards and kale under natural insect infestations (Stoner, 1990) of imported cabbageworm (*Pieris rapae* (L.)), diamondback moth, cabbage aphid (*Brevicoryne brassicae* (L.)) and the flea beetles *Phyllotreta cruciferae* (Goeze) and *Phyllotreta striolata* (Fabricius). Artificial infestations of the imported cabbageworm were also created by attaching sheets with known numbers of eggs to the leaves of the plant (Stoner, 1992). Data on artificial infestations of diamondback moth are from Eigenbrode *et al.* (1991). I calculated the percentage reduction in insect numbers on glossy lines compared to the normal-wax standard varieties in the same crop for all the tests over a two-year period.

Results and discussion

Summaries of the data show that glossy lines other than PI 234599 could be used as sources of resistance to these three pests. All glossy lines except KCR4 reduced numbers of imported cabbageworm by more than 40% in every planting, under both artificial and natural infestations. All glossy lines except the two collards had a mean reduction in cabbage aphid of over 90%. (The collards did not have higher numbers of cabbage aphids than other glossy lines, but the standard variety with which they are compared, Vates collard, is more resistant to cabbage aphids than other normal-wax varieties.) Although resistance to diamondback moth is more variable, both among lines and within lines among plantings, lines such as Broc3 and Green Glaze appear to have levels of resistance comparable to those of PI 234599.

Broc3 appears to be particularly promising as an additional source of insect resistance. Broc3 had 90% fewer imported cabbageworms than normal-wax varieties of broccoli under natural infestations in seven plantings (range: 81-96%), and 84% fewer under artificial infestation. Broc3 also had 78% fewer diamondback moth larvae (range: 68-83% over three plantings) and 54% fewer diamondbacks under artificial infestation, and 99.6% (range: 99-100% over three plantings) fewer cabbage aphids. This line was also quite vigorous, unlike some of the other glossy lines, and has a dominant gene for glossiness. If insect resistance is also dominant, making insect-resistant hybrid varieties would be much easier than using recessive genes for resistance. One drawback of all glossy lines studied to date, however, is that they tend to be more susceptible to flea beetles than lines with normal wax.

To obtain these genetic lines: Broc3, KCR4, White's Green Glaze, South Carolina Glaze, Glossy Andes and Broc5 have been deposited in the Plant Germplasm Resource Unit, at the New York Agricultural Experiment Station, Geneva, NY, 14456, with the Geneva identification numbers of G-30009 to G-30014, respectively. PI 234599 can be obtained from the same source. Green Glaze and Glazed Vates were obtained from Alf Christianson Seed Co., P.O. Box 98, Mount Vernon, WA 98273.

References

Eigenbrode, S.D., K.A. Stoner, A.M. Shelton & W.C. Kain (1991). Characteristics of glossy leaf waxes associated with resistance to diamondback moth in *Brassica oleracea*. *J. Econ. Entomol.* **84**: 1609-1618.

Stoner, K.A. (1990). Glossy leaf wax and plant resistance to insects in *Brassica oleracea* under natural infestation. *Environ. Entomol.* **19**: 730-739.

Stoner, K.A. (1992). Density of imported cabbageworms, cabbage aphids, and flea beetles on glossy and trichome-bearing lines of *Brassica oleracea* L. *J. Econ. Entomol.* **85** (in press).

Proc. 8th Int. Symp. Insect-Plant Relationships, Dordrecht: Kluwer Acad. Publ.
S.B.J. Menken, J.H. Visser & P. Harrewijn (eds), 1992

Breeding potato for resistance to insect pests

A.M. Golmirzaie and J. Tenorio
CIP, Lima, Peru

Key words: Insecticide, Liriomyza huidobrensis, Phthroimaea operculella, Solanum

Insects become a big problem when they develop resistance to insecticides. Plant breeders are developing a series of strategies (conventional and nonconventional breeding methods) that increase plant resistance to insects when used in combination with other control methods (e.g., biological). Development of plant resistance to insects must start by identifying sources of resistance in cultivated and wild materials. Consequently, one of CIP's (Centro Internacional de la Papa (potato)) priority areas of research is the utilization of cultivated and wild genetic resources to develop potato populations with good agronomic characteristics and resistance to important pests, such as potato tuber moth (*Phthroimaea operculella* (Zeller), leaf-miner fly (*Liriomyza huidobrensis* (Blanchard)), aphids (*Myzus persicae* (Sulzer)) and mites.

Source of resistance

Around 3747 accessions of cultivated native varieties, compromising mostly *Solanum tuberosum* ssp. *andigena* Juz. et Buk., *S. chauncha* Juz. et Buk., *S. curtilobum* Juz. et Buk., *S. juzepczukii* Buk. and *S. stenotomum* Juz. et Buk., have been evaluated to identify sources of resistance to insects.

Other sources of resistance are the elite populations (LM86B and LM87B) from the program for development of true potato seed (TPS) parental lines. These last were evaluated for field resistance to leaf-miner fly (LMF). The 450 clones of populations LM86 and LM87 evaluated for pest resistance had, in their background, mainly *S. tuberosum* ssp. *tuberosum*, *S. phureja* Juz. et Buk. and *Neotuberosum*.

Results

Potato tuber moth. Because most of the damage from potato tuber moth (PTM) is reported in stored potatoes, a laboratory test to identify resistance in tubers was developed with the objective of identifying those genotypes that show less damage and reduced reproduction of PTM (Paman & Palacios, 1982). Twenty-two highly resistant clones were identified in the cultivated material, and twenty-five clones from wild species including three resistant clones from an interspecific hybrid population involving *S. chacoense* Bitt. (chc), *S. sparsipilum* (Bitt.) Juz. et Buk. (spl) and *S. phureja* (phu).

The resistant clones were crossed with three susceptible clones from the same population, and the progenies of the crosses were tested using the closed-container method. Based on the frequency of resistance in the progeny, it was concluded that resistance was mostly dominant and was controlled by few genes. Resistance was transferred to the tetraploid level through clones with 2n eggs. Twenty clones of the tetraploid progeny showed resistance. This population was one of the base populations for the tetraploid breeding population (Raman *et al.*, 1981).

Work with diploid populations is continuing. The objectives are to further elucidate the inheritance of resistance, to increase the frequency of resistant genes, and to obtain diploids with 2n gametes in order to incorporate resistance in the tetraploid breeding population.

Leaf-miner fly. For LMF, five clones showed a good level of resistance and good agronomical traits after several cycles of selection. These clones are: C662LM86, C136LM86, C282LM87, C641LM86 and C40LM86. Interestingly, all clones identified include *Neotuberosum* in their background. *S. tuberosum* ssp. *andigena* is a source of resistance for LMF.

The mechanisms of resistance to this pest were studied in greenhouse tests of clones C136LM86B and C662LM86. Oviposition, development time (egg to adult), and mortality of larva and pupa were affected, indicating a high level of antibiosis in these clones.

During 1990 and 1991, 4320 seedlings of 36 progenies of resistant x resistant and resistant x susceptible clones were evaluated in diverse locations in Peru where the leaf-miner fly is endemic. The selected clones and populations are being used, first, to improve the agronomical character of other populations and, second, to combine resistance for different insects.

References

Raman, K.V. & M. Palacios (1982). Screening potato for resistance to potato tuberworm. *J. Econ. Entomol.* **75**: 47-49.
Raman, K.V., M. Iwanga, M. Palacios & R. Egusquiza (1981). Breeding for resistance to potato tuberworm *Phthroimaea operculella* (Zeller). *Am. Potato J.* **58**: 516 (abstract).

Proc. 8th Int. Symp. Insect-Plant Relationships, Dordrecht: Kluwer Acad. Publ.
S.B.J. Menken, J.H. Visser & P. Harrewijn (eds), 1992

Modelling the impact of cotton fruiting phenology on pink bollworm population dynamics in Egypt

D.A.Russell[1] and S.M.Radwan[2]
[1] Natural Resources Institute, Chatham Maritime, Kent, UK
[2] Plant Protection Research Institute, Dokki, Cairo, Egypt

Key words: Gossypium barbadense, Pectinophora gossypiella, simulation

Pink bollworm (Pectinophora gossypiella (Saunders)) is a major pest of Egyptian cotton fruiting forms. As the pressure to reduce toxic chemical applications increases, novel ways of limiting its impact are being sought. Crop/pest models are one way of exploring management options.

A time varying distributed delay simulation model for Upland cotton (Gossypium hirsutum L.) and Pink bollworm was developed in California (Stone & Gutierrez, 1986; Gutierrez & Curry, 1989). It has been initialised for use in a number of countries with a range of cottons including extra long staple (G. barbadense L.) varieties (Gutierrez et al., 1991). This metabolic pool model covers the dynamics of photosynthesate allocation to respiration, reproduction, vegetative growth and reserves as well as the stochastic processes of ageing and mortality of plant parts. Growth is scaled by plant density and by water and nitrogen stress. The insect model uses temperature and the availability of cotton fruiting forms as determined by the plant model, to simulate the dynamics of both individual and population growth in pink bollworm. Diapause entry and exit are controlled by routines modelling the complex interaction of temperature and daylength.

Results

Work carried out in three governorates of the Nile delta, between 1989 and 1991 has provided a numerical description of G. barbadense var. Giza-75, the major Egyptian variety. Using data on soil type, soil nitrogen status, weather and agronomic practices (such as the frequency and quantity of irrigations and fertilisations), the model now closely simulates plant growth throughout the season in the major Egyptian cotton growing regions.

The temporal distribution of the spring emergence of pink bollworm in Egypt (peaking in early May) is such that less than a third of the overwintering insects have still to emerge at the time of the production of the first cotton buds, c. 60 days post planting (later in the North, earlier in the South). Fewer still are present when the bolls become available c. 25 days later (bolls are more suitable as oviposition and feeding sites but are susceptible to larval penetration only in their first c. 21 days). However, early planting is currently recommended by the Egyptian Ministry of Agriculture, which also organises the destruction of cotton trash after harvest. Both of these strategies are aimed at minimising the diapausing population and hence the colonising population in the following spring.

Model simulations for a range of weather and plant management scenarios indicate that even a 95% reduction in the overwintering population of pink bollworm, although slightly delaying the initial build up of numbers by reducing the first generation of fruit feeding larvae (around day 100 to 120 post planting), has little effect on the magnitude of subsequent generations. In the absence of control measures, larval numbers are controlled largely by the availability of susceptible green bolls, commencing around day 80 and peaking at day 130 post planting.

Late planting has a rather different effect. Delaying planting by one month results in a smaller first, post-diapause, generation resulting from a lower founding population, but higher second generation larval numbers as susceptible green boll numbers are at a maximum during this period. However, the dangerous third and fourth generations are suppressed by the rapid decline in the availability of young green bolls and the reduction in adult numbers caused by increasing larval entry into diapause. Infestations at harvest, although still unacceptable in the absence of control practices, are only half those following a normal planting date. Unfortunately, the harvest period would be drawn out by the cooler weather of late summer and late planting is not likely to be adopted.

The fruiting season may also be affected by changes in the flowering phenology of the variety. For example, if the plants were to flower from sympodia growing from earlier (lower) nodes, they would reach the control point more rapidly. When this occurs aggregate demand for photosynthesate by vegetative and fruiting parts is greater than that which the current leaf area can supply. As photosynthesate allocation is controlled by the priority order: Respiration > Fruit growth > Vegetative growth > Reserves; vegetative growth would slow earlier and a smaller plant, producing bolls earlier, would result.

If Giza-'75 plants are modelled on the assumption that the mean first fruiting node will be number 4.0 and not 7.5 as is currently the case, simulations show that the resulting early fruiting and prospective early harvest allow a shorter season. Advantages in reduced pink bollworm pressure may then be gained by the reduction in the period over which susceptible fruit are available for attack. Simulations for a number of sites and years suggest that damaging attacks in the six weeks prior to harvest would be reduced by 40-60%.

Conclusion

Preliminary suggestions from the model are that a short season variety developed in this way would not suffer from yield reductions, but the interaction of factors giving rise to the premium extra long staple fibres are complex and much further work would be required before breeding recommendations could be made.

References

Gutierrez, A.P. & G.L. Curry (1989). Conceptual framework for studying crop-pest systems. In: R.E. Frisbie, K.M. El-Zik & L.T. Wilson (eds), *Integrated Pest Management Systems and Cotton Production*. New York: John Wiley and Sons.

Gutierrez, A.P., W.J. Dos Santos, A. Villacorta, M.A. Pizzamiglio, C.K. Ellis, L.H. Carvalho & N.D. Stone (1991). Modelling the interaction of cotton and the cotton boll weevil. I. A comparison of growth and development of cotton varieties. *J. Appl. Entomol.* 28: 371-97.

Stone, N.D. & A.P. Gutierrez (1986). Pink bollworm control in southwestern desert cotton. I. A field oriented simulation model II. A strategic management model. *Hilgardia* 54: 1-41.

Proc. 8th Int. Symp. Insect-Plant Relationships, Dordrecht: Kluwer Acad. Publ.
S.B.J. Menken, J.H. Visser & P. Harrewijn (eds), 1992

The susceptibility of maize to *Prostephanus truncatus* infestation: development of a bioassay

V. Pike, I. Gudrups, J. Padgham, N.A. Bosque-Pérez, N. Denning and J. Orchard
The Natural Resources Institute, Chatham Maritime, Chatham, UK

Key words: Bostrichidae, resistance

The Larger Grain Borer, *Prostephanus truncatus* (Horn), a bostrichid beetle, was originally a minor neo-tropical pest. In the late 1970's it was accidently introduced into Africa and has since spread into several countries.

Maize in storage is highly susceptible to damage by *P. truncatus* due to its considerable boring capabilities. In Africa, on-farm losses of 10% and more have been recorded in less than six months. Improved maize varieties exhibit a higher degree of susceptibility to insect attack than traditional, low-yielding African varieties.

Screening for resistance

NRI has implemented a programme to screen for host-plant resistance against primary storage pests. The objectives are to isolate and characterise the mechanisms responsible for resistance, so that the identified components can be incorporated into the preferred varieties.

Usually, bioassays for testing varietal resistance/susceptibility involve placing adults on whole grains for a specific duration and monitoring the F_1 emergents (Dobie and Urrelo methods).

P. truncatus bores deeply into grains, laying its eggs in blind-ended tunnels. The F_1 adults normally remain inside the grains and cannot be removed without disrupting the development of the juveniles. This prevents accurate assessment of maize kernel susceptibility.

A different bioassay

At NRI, an alternative approach has been developed employing gelatine capsules evenly packed with a known weight of finely ground whole maize flour. A single 0-2 day-old larva is placed in each capsule. The mortality and development rates on a range of accessions can easily be monitored. Removal of the developing adult can be readily performed without damage to the insect. This technique ensures the sole assessment of biochemical attributes; variables such as grain hardness and topography are eliminated.

A further benefit is the ease whereby putative antifeedant compounds can be incorporated into the capsules. Initial studies using this method will incorporate known biochemical resistance factors; phenolics, lectins and proteinase inhibitors.

Multitrophic Interactions

Proc. 8th Int. Symp. Insect-Plant Relationships, Dordrecht: Kluwer Acad. Publ.
S.B.J. Menken, J.H. Visser & P. Harrewijn (eds), 1992

Microbial brokers of insect-plant interactions

A.E. Douglas
Dept of Zoology, University of Oxford, Oxford, UK

Key words: Allelochemicals, cellulases, essential amino acids, phytophagous insects, symbiosis

Abstract

Many phytophagous insects form symbioses with microorganisms that possess metabolic capabilities absent from the insect. These microorganisms can be considered as 'brokers' which enable insects to overcome the biochemical barriers to herbivory. Some microbial capabilities are of significance to a relatively limited range of insects. For example, microbial cellulose degradation is utilised by a minority of termite species, but not by the phytophagous Lepidoptera and Orthoptera; microbial sterols are important only to insects with eukaryotic symbionts (*e.g.*, fungal ectosymbionts of ambrosia beetles, yeasts in some planthoppers) because bacteria do not synthesise sterols at substantial rates; and very few insects use microorganisms to detoxify plant allelochemicals.

A microbial function of significance to many insect herbivores may be related to the 'nitrogen barrier' to herbivory, *i.e.*, plants have a low N content with a different amino acid composition from insects. Symbiotic microorganisms have been implicated in the N nutrition of several groups, including the cockroaches, termites, homopterans and some beetles, and they have been demonstrated to synthesise essential amino acids in cockroaches and aphids. Furthermore, comparisons of larval pea aphids (*Acyrthosiphon pisum* (Harris)), containing and lacking their symbiotic bacteria, suggest that the bacteria promote weight gain on diets of low essential amino acid content (low N-quality) but do not buffer aphids against low total amino acid content (low N-quantity). Taken with the generality that the performance of insects, both containing and lacking symbionts, is limited by plant nitrogen content, these data suggest that microbial brokers may contribute to insect conquest of the N-quality barrier to herbivory, but not the N-quantity barrier.

Symbiotic microorganisms as microbial brokers

Plants are not 'easy meat' for animals, for two reasons. First, the chemical composition of plants and animals differ. Cell walls, containing cellulose and other refractory polysaccharides, account for a high proportion of plant biomass; and plants have lower protein and lipid contents than animals, and may contain significant concentrations of compounds toxic to animals. Second, animals have a very limited repertoire of metabolic capabilities. For example, insects cannot synthesize sterols, vertebrates cannot degrade cellulose, and no animals can synthesise certain unsaturated fatty acids and 10 of the 20 protein-amino acids.

Several authors (*e.g.*, Jones, 1984; Southwood, 1985) have pointed out that many plant-animal trophic links include third party, namely symbiotic microorganisms associated with the animal. By forming a symbiosis, the animal gains access to microbial metabolic capabilities, such as cellulose degradation or essential amino acid synthesis (Douglas, 1992). In the terminology of Southwood (1985), the microorganisms are 'brokers', enabling animals to overcome the biochemical barriers to herbivory.

The microbial brokers in phytophagous insects can be classified according to their location relative to the insect body. **Ectosymbionts** are external to the insect, frequently in the nest or brood chamber. They are invariably fungi, and many are known only in association with insects (*e.g.*, *Attamyces* cultivated by the leaf-cutting Attinine ants, and *Termitomyces* maintained by termites of the subfamily Macrotermitinae). **Endosymbionts** are borne within the insect body. They include two major groups: first, the gut symbionts, tens-to-hundreds of microbial species at densities up to 10^9-10^{10} cells/ml in the lumen of the alimentary tract (Bignell, 1984); and, second, the mycetocyte symbionts, usually of one or two morphological forms located in specialised insect cells, called mycetocytes (Douglas, 1989).

The functions of microbial brokers

The capabilities of symbiotic microorganisms utilised by insects are divisible into three broad categories: detoxification of plant allelochemicals; degradation of cellulose; and biosynthesis of nutrients in short supply in the diet. The significance of these functions to insects is considered in this section.

Detoxification of plant allelochemicals. Plant allelochemicals, including flavonoids, tannins and alkaloids, can play a major role in limiting insect herbivory. Many insects which utilise plants rich in these compounds possess intrinsic detoxification or sequestration systems, but detoxifying microorganisms have been identified in a few insects. For example, the symbiotic yeasts enable larvae of the cigarette beetle *Lasioderma serricorne* F. to grow well on tannin- and flavonoid-supplemented diets (Dowd & Shen, 1990); and the fungal ectosymbionts of leaf-cutting ants (the Attinini) have a variety of detoxifying enzymes which are believed to increase the variety of plants that the ants can utilise (Cherrett *et al.*, 1989).

Cellulose degradation. The cell wall fraction, which includes cellulose, accounts for 40-70% of the biomass of food ingested by phytophagous insects. Despite this, some phytophagous lepidopteran larvae are 'cellulase-independent herbivores', *i.e.*, they ingest large amounts of plant material, utilise the easily-assimilated nutrients and void the plant cell wall fraction in their frass. Cellulose-degrading bacteria have been isolated from the guts of some Orthoptera, *e.g.*, *Schistocerca gregaria* (Forskål) and *Acheta domesticus* L., but they do not appear to be significant to the nutrition of these insects (Charnley *et al.*, 1985; Kaufman *et al.*, 1989).

Virtually all carbon in wood, especially heartwood, is in the form of lignocellulose and the concentration of soluble sugars is very low. Linked to this, xylophagous insects can degrade cellulose.

From the microbiological stand-point, the best-studied group of xylophagous insects are the termites, and three strategies by which these insects utilise cellulose-rich diets are recognised:
1. Cellulolytic gut microbiota. The lower termites (5 families, accounting for *ca.* 550 species) possess cellulolytic hypermastigote protists in the hindgut. These microorganisms utilise cellulose as a carbon source, releasing short chain fatty acids (SCFAs), especially acetic acid, as waste products of their fermentative metabolism. The SCFAs are absorbed

across the gut wall and used as a source of energy by the aerobic tissues of the insect. The microbial cellulolysis is supplemented by cellulases of insect origin (Hogan et al., 1988b).

2. Intrinsic cellulases. All termites of the family Termitidae ('higher termites' comprising 75% of all termite species) degrade dietary cellulose by intrinsic enzymes (i.e., of insect origin) secreted into the lumen of the midgut (Hogan et al., 1988a). Higher termites lack cellulose-degrading microorganisms.

3. Ingested microbial cellulases. Higher termites of the subfamily Macrotermitinae (the fungus-growing termites) feed on the cellulase-rich nodules of their ectosysmbiotic fungus Termitomyces. The fungal cellulases are retained in the insect midgut where they can potentially degrade ingested cellulose to glucose (Martin & Martin, 1978). However, the quantitative significance of these ingested fungal enzymes is uncertain; in Macrotermes michaelsoni (Sjöstedt), fungal-derived cellulases contribute less than 0.03% of the total cellulase activity (Veivers et al., 1991).

Many wood-feeding beetles, including cerambycids, anobiids and bupestrids, and the siricid wasps have cellulase activity in the midgut, but the relative contribution of intrinsic enzymes and ingested fungal enzymes to this activity is uncertain (Martin, 1991).

The midguts of some detritivores (e.g., larvae of the stonefly Pteronarcys proteus Newman and caddisfly Pycnopsyche luculenta (Krafka)) have detectable cellulase activity, derived from fungi which contaminate ingested food; but these enzymes do not significantly increase the digestibility of the food and are not required by the insect (Sinsabaugh et al., 1985; Martin, 1991).

Synthesis of nutrients. Many plant tissues have low concentrations of various nutrients that insects cannot synthesise; these nutrients include sterols, B vitamins and essential amino acids. Only eukaryotes can synthesise sterols at substantial rates and consequently, symbiotic bacteria do not contribute sterols to insects. Various fungi provide sterols, e.g., ectosymbiotic Fusarium associated with the ambrosia beetles Xyleborus (Chu et al., 1970), and yeast endosymbionts in some planthoppers (Noda et al., 1979).

Two beetles, the anobiid L. serricorne and the weevil Sitophilus oryzae (L.), are independent of a dietary supply of B vitamins, but this independence is lost when their mycetocyte symbionts are eliminated (Jurzitza, 1969; Baker, 1975), suggesting that their symbiotic microorganisms provide vitamins to these species. However, these associations may be exceptional, for the performance of several other beetles (e.g., Oryzaephilus) and also cockroaches and various homopteran groups with intact symbiotic microorganisms is depressed by the omission of dietary B vitamins (reviewed in Douglas, 1989).

The provision of essential amino acids may be the most widespread function of microbial symbionts. Chewing insects require a different balance of amino acids from those in plant tissues, largely because the major proteins of insects and plants are different. Plant sap-sucking insects may experience particular difficulties because the essential amino acid content of phloem and xylem sap rarely exceeds 25 mol%, while a 'balanced diet' for animals contains ca. 50 mol% essentials (Brodbeck & Strong, 1987).

Additionally, insects need relatively large amounts of the aromatic essential amino acids, tryptophan and phenylalanine, used in cuticle synthesis (Bernays & Woodhead, 1984). The discrepancy between the essential amino acid requirements of insects and the supply of these nutrients in plants is compounded by the considerably higher total nitrogen content of animals (7-14%) than plants (< 5%, and in wood and plant sap < 0.5%) (Mattson, 1980).

These fundamental differences between animals and plants represent the **qualitative** and **quantitative** aspects, respectively, of the nitrogen barrier to herbivory.

In general, microbial brokers do not appear to contribute to insects' conquest of the 'nitrogen-quantity' barrier. The performance of phytophagous insects is closely linked to the nitrogen content of plants, whether or not the species contains symbiotic

331

microorganisms (Scriber, 1984), and insects which utilise very low-nitrogen diets (*e.g.*, folivores on woody plants, xylophages, xylem-suckers) have low basal metabolic rates and low growth rates, whether or not they are associated with microorganisms (Mattson & Scriber, 1987).

The role of microbial brokers in phytophagous insects' response to diets of low nitrogen quality has been examined in aphids (Prosser & Douglas, 1992). Aphids are amenable to study because the nitrogen in their diet of phloem sap is almost exclusively in the form of free amino acids, and several aphids can be maintained on chemically-defined diets of varying amino acid content and composition.

The data in Table 1a indicate that the symbiotic bacteria in the pea aphid *A. pisum* can 'buffer' aphids against low dietary essential amino acid content (*i.e.*, low nitrogen quality). The weight gain by aphids containing bacteria differed by 12% or less between balanced diets (with 50 mol% essential amino acids) and unbalanced diets similar to phloem sap (20 mol % essentials); but the weight gain of bacteria-free aphids was depressed by 30-50% on diets of low essential amino acid content. This difference between the response of the two groups of aphids to variation in dietary nitrogen quality is statistically significant (see interaction 'antibiotic' x 'essential amino acids' in Table 1b). The data in Table 1 are also consistent with the generalisation that microorganisms do not assist in insect utilisation of low-nitrogen diets. Weight gain of pea aphids, both containing and lacking bacterial symbionts, was reduced on low-nitrogen diets, and the interaction between antibiotic treatment and amino acid concentration is not significant.

The role of microbial symbionts in the response of other phytophagous insects to variation in dietary nitrogen quality has not been addressed. Experimentally, it is difficult to manipulate dietary nitrogen quality for 'chewing' phytophagous insects because most of the utilisable nitrogen is protein, of amino acid composition, genetically-fixed by the plant. One method is to vary the proportions of dietary proteins which differ in nutritional value (*e.g.*, see experiments of Karowe & Martin (1989) on *Spodoptera* larvae, which lack microbial symbionts).

How insects with microbial brokers overcome the 'nitrogen-quality' barrier to herbivory

The principal mechanism by which microorganisms enable insects to utilise diets of low nitrogen quality is the provision of essential amino acids to the insect.

The microbial symbionts in several insects are known to synthesis essential amino acids. For example, ^{35}S-sulphate applied to the cockroach *Blattella germanica* L. and aphid *Myzus persicae* (Sulzer) is incorporated into sulphur amino acids, including the essential amino acid, methionine (Henry & Block, 1960; Douglas, 1988); and ^{15}N-glutamine injected into the pea aphid *A. pisum* is recovered from several essential amino acids (Sasaki *et al.*, 1991). However, these experiments merely demonstrate that bacteria can synthesise essential amino acids; this capacity is of no significance to the insect unless the essential amino acids are translocated to the insect tissues.

Direct evidence for the transport of two essential amino acids, methionine and tryptophan, is available for aphids. The bacteria in the pea aphid *A. pisum* have high activities of the enzyme tryptophan synthetase, and the honeydew of aphids reared on tryptophan-free diets contain appreciable concentrations of tryptophan (Douglas & Prosser, 1992). This suggests that the bacteria can both synthesise tryptophan and release the amino acid to the insect at rates in excess of aphid requirements. Twenty per cent of methionine, synthesised from ^{35}S-sulphate by the bacteria in the green peach aphid *M. persicae*, is transported to portions of the insect body (head and thorax) that lack bacteria (Douglas, 1988).

Table 1. Effect of amino acid concentration and composition on the performance of larval pea aphids *Acyrthosiphon pisum*

a) Final weight of aphids (mean ± SE (number of replicates))

Dietary essential amino acid content (mol%)	Concentration of dietary amino acids					
	220 mM		110 mM		55 mM	
	Untreated aphids	Antibiotic-treated aphids	Untreated aphids	Antibiotic-treated aphids	Untreated aphids	Antibiotic-treated aphids
50	301±9 (27)	211± 8 (22)	271±10 (22)	168±9 (7)	219± 7 (24)	135±11 (13)
35	292±9 (24)	170±11 (15)	255±14 (22)	145±6 (22)	197±12 (21)	100± 4 (13)
20	294±8 (26)	106± 9 (23)	238± 9 (23)	99±1 (20)	212±10 (20)	92± 4 (17)

b) F values for ANOVA of data

Source of variation	df	F value	P
Main effects	5	197.90	< 0.001
antibiotic treatment	1	730.22	< 0.001
amino acid concentration	2	91.55	< 0.001
% essential amino acids	2	29.07	< 0.001
2-way interactions	8	7.01	< 0.001
antibiotic x amino acid concentration	2	2.05	> 0.05
antibiotic x % essential amino acids	2	17.27	< 0.001
amino acid concentration x % essential amino acids	4	2.41	> 0.05
3-way interactions	4	1.96	> 0.05
Explained	17	61.96	< 0.001
Total	360		

The aphids were raised for five days from birth on chemically-defined diets of different nitrogen quantity (55-220 mM amino acids) and quality (20-50 mol% essentials). Half the aphids were treated with the antibiotic chlortetracycline (50 μg/ml diet) to disrupt the symbiotic bacteria. (Data from Prosser & Douglas, 1992)

Discussion in the literature of the source of nitrogen from which symbiotic microorganisms may synthesise essential amino acids has concentrated on two microbial metabolic capabilities: (1) nitrogen fixation - microbial reduction of dinitrogen to ammonia, and (2) nitrogen recycling - microbial utilisation of animal nitrogenous waste compounds, *e.g.*, uric acid, ammonia.

Nitrogen fixation consumes considerable amounts of energy (25 moles ATP per mole N_2 fixed), and the product, ammonia, is, in any case, generated at considerable rates by animals. From the perspective of animals, nitrogen fixation is an energetically-costly way to synthesise a nitrogen waste product. It is therefore not surprising that nitrogen-fixing symbioses are rare in insects, and certainly absent from phytophagous groups. The sole properly-documented example is in termites, many of which contain nitrogen-fixing bacteria, *e.g.*, *Citrobacter freundii* (Braak) Werkman & Gillen and *Enterobacter agglomerans* (Beijerinck) Ewing & Fife, in their hindgut. However, the significance of these bacteria to the nitrogen nutrition of the insect varies widely between species and with environmental conditions, including diet and season. For example, microbial nitrogen fixation provides for a doubling within one year of the nitrogen content of wood-feeding *Nasutitermes* and *Coptotermes* colonies, but nitrogen fixation rates are low in soil-feeding *Rhynchotermes* and barely detectable in the fungus-growing Macrotermitinae (Breznak, 1982).

Nitrogen recycling has been demonstrated in the mycetocyte symbionts of the cockroach *Periplaneta americana* L. and pea aphid *A. pisum* and in the bacterial gut symbionts of the termite *Reticulotermes flavipes* (Kollar) (Wren & Cochran, 1987; Potrikus & Breznak, 1981; Wilkinson *et al.*, submitted). In all of these insects, the concentration of nitrogenous waste compounds (ammonia or uric acid) is elevated when the microbial symbionts are eliminated. Furthermore, uric acid is degraded by isolated symbionts of *P. americana* and *R. flavipes*, and the bacteria in *A. pisum* can utilise exogenous ammonia.

Concluding remarks

Insect utilisation of microbial metabolic capabilities is just one of several strategies by which insects have overcome the biochemical barrier to herbivory. There is now persuasive evidence that microorganisms contribute to the capability of some (but not all) insects to degrade cellulose, and of some Homoptera to utilise phloem sap of low nitrogen quality. In addition, symbiotic fungi may contribute to the sterol nutrition of a few insects, and to detoxification of certain allelochemicals, but many insects utilise cellulose-rich, allelochemical-rich or micronutrient-poor plant tissues without the 'assistance' of microorganisms.

Are there drawbacks of possessing microbial brokers? Some tentative indications come from the taxonomic distribution of certain symbioses. In particular, microbial symbionts have been lost within the radiation of several homopteran groups, the ants and the termites (Buchner, 1965). Among the termites, the most primitive species, *Mastotermes darwiniensis* Froggatt, has mycetocyte symbionts and both bacterial and protist gut symbionts. All other termites lack mycetocyte symbionts, and the most specious and advanced termites, the Termitidae, have abandoned symbiotic cellulolysis by protists in favour of intrinsic process (see section 2b). The disadvantages of possessing microbial symbionts may include the costs of the nutrients consumed and space occupied by the microorganisms. Insect herbivores which require microbial brokers also face the dual (and contrary) hazards of, first, becoming separated from their partners and, second, being colonised by deleterious microorganisms masquerading as microbial symbionts.

Acknowledgements

I thank Dr J.B. Searle and Dr S.J. Simpson for helpful comments on the manuscript. This article was written while in receipt of a University Research Fellowship of the Royal Society of London.

References

Baker, J.E. (1975). Vitamin requirements of larvae of *Sitophilus oryzae. J. Insect Physiol.* **21**: 1337-1342.

Bernays, E.A. & S. Woodhead (1984). The need for high levels of phenylalanine in the diet of *Schistocerca gregaria* nymphs. *J. Insect Physiol.* **30**: 489-493.

Bignell, D.E. (1984). The arthropod gut as an environment for microorganisms. In: J.M. Anderson, A.D.M. Rayner & D.W.H. Walton (eds), *Invertebrate-Microbial Interactions*, pp. 206-227. Cambridge: Cambridge University Press.

Breznak, J.A. (1982). Intestinal microbiota of termites and other xylophagous insects. *Annu. Rev. Microbiol.* **36**: 323-343.

Brodbeck, B. & D. Strong (1987). Amino acid nutrition of herbivorous insects and stress to host plants. In: P. Barbosa & J.C. Schultz (eds), *Insect Outbreaks*, pp. 347-364. Academic Press.

Buchner, P. (1965). *Endosymbioses of Animals with Plant Microorganisms.* Chichester, England: Wiley & Sons.

Charnley, A.K., J. Hunt & R.J. Dillon (1985). The germ-free culture of desert locusts, *Schistocerca gregaria. J. Insect Physiol.* **31**: 477-485.

Cherrett, J.M., R.J. Powell & D.J. Stradling (1989). The mutualism between leaf-cutting ants and their fungus. In: N. Wilding, N.M. Collins, P.M. Hammond & J.F. Weber (eds), *Insect-Fungus Interactions*, pp. 93-120. London: Academic Press.

Chu, H.M., D.M. Norris & L.T. Kok (1970). Pupation requirements of the beetle *Xyleborus ferrugineus*: sterols other than cholesterol. *J. Insect Physiol.* **16**: 1379-1387.

Douglas, A.E. (1988). Sulphate utilisation in an aphid symbiosis. *Insect Biochem.* **18**: 599-605.

Douglas, A.E. (1989). Mycetocyte symbiosis in insects. *Biol. Revs* **64**: 409-434.

Douglas, A.E. (1992). Symbiosis in evolution. In: *Oxford Surveys in Evolutionary Biology* **8**: (in press).

Douglas, A.E. & W.A. Prosser (1992). Synthesis of the essential amino acid tryptophan in the pea aphid (*Acyrthosiphon pisum*) symbiosis. *J. Insect Physiol.* (in press).

Dowd, P.F. & S.K. Shen (1990). The contribution of symbiotic yeast to toxin resistance of the cigarette beetle (*Lasioderma serricorne*). *Entomol. exp. appl.* **56**: 241-248.

Henry S.M. & R.J. Block (1960). The sulphur metabolism of insects. IV. The conversion of inorganic sulfate to organic sulfur compounds in cockroaches. The role of intracellular symbionts. *Contrib. Boyce Thompson Inst.* **20**: 317-329.

Hogan, M.E., P.C. Veivers, M. Slaytor & R.T. Czolij (1988a). The site of cellulose breakdown in higher termites (*Nasutitermes walkeri* and *Nasutitermes exitosus*). *J. Insect Physiol.* **34**: 891-899.

Hogan, M.E., M.W. Schulz, M. Slaytor, R.T. Czolij & R.W. O'Brien (1988b). Components of the termite and protozoal cellulases of the lower termite *Coptotermes lacteus* Froggatt. *Insect Biochem.* **18**: 45-51.

Jones C.G. (1984). Microorganisms as mediators of plant resource exploitation by insect herbivores. In: P.W. Price, C.N. Slobodchikoff & W.S. Gaud (eds), *A New Ecology*, pp. 53-99. New York: Wiley.

Jurzitza, G. (1969). Der Vitaminbedarf normaler und aposymbiontischer *Lasioderma serricorne* F. (Coleoptera, Anobiidae) und die Bedeutung der symbiontischen Pilze als Vitaminquelle fur ihre Wirte. *Oecologia* **3**: 70-83.

Karowe, D.N. & M.M. Martin (1989). The effects of quantity and quality of diet nitrogen on the growth, efficiency of food utilisation, nitrogen budget, and metabolic rate of fifth-instar *Spodoptera eridania* larvae (Lepidoptera: Noctuidae). *J. Insect Physiol.* **35**: 699-708.

Kaufman, M.G., M.J. Klug & W. Merritt (1989). Growth and food utilization parameters of germ-free house crickets, *Acheta domesticus. J. Insect Physiol.* **35**: 957-967.

Martin, M.M. (1991). The evolution of cellulose digestion in insects. *Phil. Trans.* **333**.

Martin, M.M. & J.S. Martin (1978). Cellulose digestion in the midgut of the fungus-growing termite *Macrotermes natalensis*: the role of acquired digestive enzymes. *Science* **199**: 1453-1455.

Mattson, W.J. (1980). Herbivory in relation to plant nitrogen content. *Annu. Rev. Ecol. Syst.* **11**: 119-161.

Mattson, W.J. & J.M. Scriber (1987). Nutritional ecology of insect folivores of woody plants: nitrogen, water, fiber, and mineral considerations. In: F. Slansky & J.G. Rodriguez (eds), *Nutritional Ecology of Insects, Mites, Spiders, and Related Invertebrates*, pp. 105-146. New York: Wiley.

Noda, H., K. Wada & T. Saito (1979). Sterols in *Laodelphax striatellus* with special reference to the intracellular yeast-like symbiotes as a sterol source. *J. Insect Physiol.* **25**: 443-447.

Potrikus, C.J. & J.A. Breznak (1981). Gut bacteria recycle uric acid nitrogen in termites: a strategy for nutrient conservation. *Proc. Natl Acad. Sci. USA* **78**: 4601-4605.

Prosser, W.A. & A.E. Douglas (1992). A test of the hypotheses that nitrogen is upgraded and recycled in an aphid symbiosis. *J. Insect Physiol.* (in press).

Sasaki, T., H. Hayashi & H. Ishikawa (1991). Growth and reproduction of the symbiotic and aposymbiotic pea aphids, *Acyrthosiphon pisum* maintained on artificial diets. *J. Insect Physiol.* **37**: 749-756.

Scriber, J.M. (1984). Host plant suitability. In: W.J. Bell & R.T. Cardé (eds), *Chemical Ecology of Insects*, pp. 159-202. London: Chapman & Hall.

Sinsabaugh, R.L., A.E. Linkens & E.F. Benfield (1985). Cellulose digestion and assimilation by three leaf-shredding aquatic insects. *Ecology* **66**: 1464-1471.

Southwood, T.R.E. (1985). Interactions of plants and animals: patterns and processes. *Oikos* **44**: 5-11.

Veivers, P.C., M. Muhlemann, Slaytor, R.H. Leuthold & D.E. Bignell (1991). Digestion, diet and polyethism in two fungus-growing termites: *Macrotermes subhyalinus* Rambur and *M. michaelseni* Sjostedt. *J. Insect Physiol.* (in press).

Wren, H.N. & D.G. Cochran (1987). Xanthine dehydrogenase activity in the cockroach endosymbiont *Blattabacterium cuenoti* (Mercier, 1906) Holland and Favre 1931 and in the cockroach fat body. *Comp. Biochem. Physiol.* **88B**: 1023-1026.

Proc. 8th Int. Symp. Insect-Plant Relationships, Dordrecht: Kluwer Acad. Publ.
S.B.J. Menken, J.H. Visser & P. Harrewijn (eds), 1992

Microorganisms and kairomone production from *Allium* chemicals in a host-parasitoid relationship

Eric Thibout[1], Jean-François Guillot[2] and Jacques Auger[1]
[1] Institut de Biocénotique Expérimentale des Agrosystèmes, Tours, France
[2] Microbiologie, IUT, Tours, France

Key words: *Acrolepiopsis assectella, Diadromus pulchellus,* disulfides, Lepidoptera, tritrophic interaction

The specialist ichneumonid *Diadromus pulchellus* Wesmael parasitizes the pupae of its host, the leek moth *Acrolepiopsis assectella* Zeller, a phytophagous specialist of *Allium* plants. The caterpillar frass of the moth emits kairomones previously identified as disulfides (S_2), which stimulates host-seeking in the parasitoid (Auger *et al.*, 1989). Auger *et al.* (1990) has shown that after the addition of either propyl or methyl disulfide or of their precursor sulfur amino acids to the caterpillar's artificial diet, dimethyl, dipropyl and methyl-propyl S_2 are always found in the frass. This implies the transformation from an S-methyl to an S-propyl moiety and conversely, through a pathway in which microorganisms could mediate. We investigated the potential activity of microorganisms.

Methods

The host orientation of *D. pulchellus* was observed in the presence of frass from the host caterpillar and from a non-host generalist caterpillar, the carnation tortrix *Cacoecimorpha pronubana* Hübner. Both species were reared on artificial septic or aseptic diet with or without the sulfur amino acids usually present in *Allium*. The presence of microorganisms was studied in the gut and frass of the two caterpillar species, by scanning electron microscopy and by numeration after dilution and plating on trypticase soy-agar followed by incubation. The S_2 emission was studied by head-space GC analysis 1) from frass produced by caterpillars fed on various diets, and 2) by culturing microorganisms present in the frass on agar, or agar supplemented with sulfur amino acids.

Results

The kairomones emitted by frass were not specific to the leek-moth. *C. pronubana* frass also stimulated *D. pulchellus*. Microorganisms observed in the gut and frass of the two caterpillars were numerous in frass from leek plant origin, less abundant in frass from an artificial diet and absent in frass from a diet containing antibiotics like amoxycilline. The three S_2 were always present together in the frass of the two caterpillars species fed on *Allium* plants and on artificial diets containing sulfur chemicals but devoid of antibiotics. When several microorganisms of the frass were cultured on agar the S_2 were identified in the head-space solely if methyl-propyl sulfur precursors had been present in the caterpillar food. Each time, the presence of both dimethyl and dipropyl S_2 was observed, but methyl-propyl S_2 was not found. This difference with frass emissions could be the result of a difference in developmental conditions of frass bacteria when cultured on agar (pH, texture, humidity).

Conclusions

Microorganisms present in the frass of the two caterpillars appeared to be capable of metabolizing sulfur amino acids like propyl-cysteine sulfoxyde which naturally occur in *Allium*. Furthermore, thiosulfinates, naturally produced when *Allium* is broken, are absent in the frass odour and the kairomone production is non-specific. Consequently the S_2 emission responsible for the host-seeking in the female wasp evolves from the enzymatic activity of an extraneous organism and not from the plant alliinase or from a specific leek moth larval enzyme. To our knowledge, this is one of the few examples where microorganisms interfere in tritrophic relationships between the second and the third level. Furthermore, it is probably the first example where microorganisms have a role in interactions in strictly phytophagous insects, and not microbivores as defined by Dicke (1988).

References

Auger, J., C. Lecomte, J. Paris & E. Thibout (1989). Identification of leek moth and diamondback moth frass volatiles that stimulate parasitoid *Diadromus pulchellus*. *J. Chem. Ecol.* **15**: 1391-1398.
Auger, J., C. Lecomte & E. Thibout (1990). Origin of kairomones in the leek moth (*Acrolepiopsis assectella*, Lep.) frass. Possible pathway from methylthio to propylthio compounds. *J. Chem. Ecol.* **16**: 1743-1750.
Dicke, M. (1988). Microbial allelochemicals affecting the behavior of insects, mites, nematodes, and Protozoa in different trophic levels. In: P. Barbosa & D. Letourneau (eds), *Novel Aspects of Insect-Plant Interactions*, pp. 125-163. New York: John Wiley & Sons.

Proc. 8th Int. Symp. Insect-Plant Relationships, Dordrecht: Kluwer Acad. Publ.
S.B.J. Menken, J.H. Visser & P. Harrewijn (eds), 1992

Plant resistance versus parasitoid attack in the evolution of the gall-forming fly *Lipara lucens*

L. de Bruyn
Dept of Biology, University of Antwerp (RUCA), Antwerpen, Belgium

Key words: Chloropidae, Diptera, nutrition, predator, reed

The high variation in gall morphology observed in nature has repeatedly raised the question as to the adaptive nature of insect galls (Price *et al.*, 1987). In many cases, the gall contents can serve as an important food resource for gall-making insects. Sometimes however, certain characteristics of gall morphology have no apparent relationship to the nutritional status. This is usually explained by the fact that galls surrounding the herbivore should deter parasitoid and predator attack.

Lipara lucens Meigen, induces terminal stem galls on the common reed, *Phragmites australis* (Cav.) Trin. ex Steud. (Poaceae) (Chvála *et al.*, 1974). The female flies oviposit on the surface of the reed shoots. The young larvae enter the shoot and crawl to the growing point where they feed on the young, newly formed leaves. As a result, the newly formed internodia do not elongate any more. The gall chamber that is formed is filled with a dense mass of parenchymatous tissue with a tough, hard wall. After a while the larva gnaws through the growing point and enters the gall chamber where it feeds on the parenchymatous tissue. Another species, *Lipara pullitarsis* Doskočil & Cvála, does not form a gall chamber. Throughout their life cycle, the larvae feed on the young enwrapped leaves above the growing point.

During the developmental cycle, *L. lucens* is attacked by a number of parasitoids (De Bruyn, 1987). The parasitoid *Stenomalina liparae* (Giraud), in particular, may act as a selective force on the gall-maker. To reach its host, the parasitoid has to penetrate the plant wall with its ovipositor. Field experiments revealed that infested hosts are only found in galls formed on the thinnest shoots while no mortality occurs on the thicker shoots. The shoot diameter is an important feature of a reed shoot. Earlier studies have revealed that there is a strong correlation between the basal shoot diameter and growth, final shoot length and the possibility to produce an ear. Consequently, it is a measure of the strength of the shoot.

In the present study we investigated the adaptive significance of the galling habit of *L. lucens*. Two possible alternative hypothesis were tested, *viz.*, the enemy hypothesis and the nutritional hypothesis.

Results and discussion

In a first experiment the survival up to gall formation was tested in an experimental reed bed. The highest proportion of infested shoots was found on the thinnest shoots. As the diameter increased, the proportion of galls decreased. No galls were formed where the diameter was more than 8 mm. So, thicker shoots were found to be more resistent to herbivore attack.

The relationship between shoot and gall diameter was evident: thicker shoots produce thicker galls. However, the diameter ratio decreases significantly when shoots become thicker. Therefore, the quantitative increase of nutritive tissues is much higher on thinner shoots.

In addition, qualitative analyses of the gall tissues revealed that the nutritional value (water, protein and sugar contents) of the parenchymatous tissues in the gall chamber was significantly higher than that of the young enwrapped leaves above the growing point.

A combined field and laboratory experiment was carried out to assess the impact of parasitoids on the survival of *L. lucens*. Almost 17% of the *L. lucens* larvae were killed by the parasitoid *S. liparae*. Moreover, about 7% was killed by *Polemon liparae* (Giraud), an egg-larva parasitoid. The female wasps attacked the eggs of *Lipara* while they were attached on the outer surface of the reed shoot. Another 2.63% of the larvae were killed by other, less successful parasitoid species. The same laboratory experiment was carried out with *L. pullitarsis*. Here the total amount killed by parasitoids did not exceed 3% and more than 91% of the galls gave rise to adult flies.

Rearing experiments under laboratory conditions showed that *L. pullitarsis* can escape parasitoid attack by early adult emergence. At a constant room temperature of 22°C, *L. pullitarsis* emerged 3 to 4 weeks earlier than *L. lucens*. Emergence of the parasitoid species is clearly synchronous with the emergence of *L. lucens*.

Oviposition observations in the field and oviposition experiments in the laboratory revealed that parasitoid attack occurs well before gall formation takes place. *P. liparae* attacks the eggs of *L. lucens*, while *S. liparae* oviposits in a *Lipara* larva when the latter is feeding on the young enwrapped leaves above the growing point.

When the galls were collected at the end of the winter, some of them showed a hole in the gall wall and the larva had disappeared. These galls were attacked by bird predators, probably the blue tit, *Parus caeruleus* L., which can enter reed beds during winter to search for food. The impact of bird predation on *Lipara* survival is not very clear at present. Most larvae killed were seemingly those of *L. pullitarsis*. Practically no gall of *L. lucens* was found to be opened. However, the predation level is very variable, both in time and space. Bird attack is very local and no opened galls were found in many reed beds. High attack rates can be attained during harsh winters.

Conclusion

To conclude, we can state that gall formation in *L. lucens* supports the nutrition hypothesis. The high proliferation of nutritive rich parenchymatous tissues in the galls increases the survival rate on the thinner shoots. Parasitoid attack does not support the enemy hypothesis. The mortality rate due to parasitization is even much higher in the gall forming *L. lucens* than in *L. pullitarsis*. Only predator attack by birds may play a possible role as a selective pressure on gall formation. More experimental work is necessary to confirm this.

References

Chvála, M., J. Doskočil, J.H. Mook & V. Pokorný (1974). The genus *Lipara* Meigen (Diptera, Chloropidae), systematics, morphology, behaviour and ecology. *Tijdschr. Ent.* **117**: 1-25.

De Bruyn, L. (1987). The parasite-predator community attacking *Lipara* species in Belgium. *Bull. Annls Soc. r. ent. Belg.* **123**: 346-350.

Price, P.W., G.W. Fernandes & G.L. Waring (1987). Adaptive nature of insect galls. *Environ. Entomol.* **16**: 15-24.

Proc. 8th Int. Symp. Insect-Plant Relationships, Dordrecht: Kluwer Acad. Publ.
S.B.J. Menken, J.H. Visser & P. Harrewijn (eds), 1992

Genetic relationships between *Oreina* species with different defensive strategies

M. Rowell-Rahier[1] and J.M. Pasteels[2]
[1] *Zoologisches Institut, Universität Basel, Basel, Switzerland*
[2] *Lab. Biologie animale, Fac. Sciences, ULB, Brussel, Belgium*

Key words: Allozymes, Chrysomelinae, *Coleoptera*, plant toxins

Oreina is a central European genus of leaf beetles often occurring sympatrically in moist, high elevation habitats, feeding on plants belonging to either the Asteraceae or the Apiaceae. To get a better understanding of how the utilization of plant toxins for defence has evolved, we studied the food plant preferences and genetic relationships of several species, as well as their modes of chemical defence.

Genetic relationships between species and populations

We estimated genetic distances between the different beetle populations by starch gel electrophoresis of six allozyme loci. The genetic relationships indicated by this method corresponds with their grouping by host plant family.

Chemical nature of the defensive secretion

The questions we needed answering were: (1) How diverse is the defensive chemistry of the adults at the species level, and at the population level? (2) Do the host plants and their secondary compounds influence the beetles' defensive chemistry?
   a. Cardenolides. The adults of the Apiaceae-feeding species synthesize cardenolides *de novo*, which they store only in defensive glands (and not in the beetle body), and subsequently liberate. The mixture is very complex. We used principal component (PC) analysis on both genotype frequency data and the cardenolide composition of the secretion for each population studied. The first PC based on the proportion of cardenolides (38 variables) in the defensive mixture explained 46% of the variation among populations. The first PC based on allozyme genotype frequencies (42 variables) explained 27% of the variation between populations; with the addition of the second component, 50% of the variation was explained. The first PC of the chemical data and the first PC of the allozyme data were highly significantly correlated (Pearson coefficient 0.96). The mixtures of components secreted are species specific and chemical similarity between species parallels genetic similarities.
   b. Pyrrolizidine N-oxides. *Oreina cacaliae* (Schrank) secretes only pyrrolizidine (= PA) N-oxides sequestered from the food plants. *O. speciosissima* (Scopoli) sequesters PAs and biosynthesizes cardenolides. Sequestering of these compounds by the beetles from their host plants was demonstrated unambiguously by feeding the beetles radioactively labelled PA N-oxides and testing the secretion collected thereafter for radioactivity. When *O. cacaliae* or *O. speciosissima* were fed with labelled senecionine N-oxides, or when it was

injected directly into the hemolymph, significant radioactivity was detected in the secretion. In the laboratory, *O. speciosissima* was as efficient as *O. cacaliae* at sequestering PA N-oxides. Only the N-oxide form, which is the form in which the PAs are found in the food plants, was encountered in the secretion and no tertiary alkaloids were detected. In both feeding and injection experiments more radioactivity was recovered from the body of the beetles than in the secretion.

PAs are abundant in *Adenostyles alliariae* (Gouan) Kerner leaves (22 mg/g dry wt), less so in those of *Senecio fuchsii* C.C. Gmelin (circa 10 mg/g dry wt), and absent from those of *Petasites paradoxus* (Retz.) Baumg. In the wild, *O. cacaliae* adults are found in early spring on *Petasites*. Later, they switch to either *Senecio* or *Adenostyles*. On *Senecio*, larval growth is better but adult defence is weaker than on *Adenostyles*. This suggests that sequestration is not always the best solution and that trade-offs exist between phenology, growth and defence.

When *O. speciosissima* sequestrates PAs, it biosynthesizes less cardenolides. Whether this reflects a possible cost of sequestration in this species, or whether the two types of compounds compete for transport or space in the defensive glands, remains unknown. The larval growth rate is significantly higher on *P. paradoxus* than on *A. alliariae*. Both observations suggest that for this species the costs of sequestering PAs are high.

In the wild, *O. speciosissima* does not feed on plants rich in PA N-oxides even when they are available. Thus, although this species is physiologically able to sequester toxins it does not utilize this possibility in the field, perhaps because of the high metabolic costs involved.

Genetic differentiation between populations of *O. cacaliae* and of *O. speciosissima*

The between-population component of the overall inbreeding coefficient, F_{st}, was significantly greater than zero for both *O. cacaliae* (0.234 ± 0.035) and *O. speciosissima* (0.051 ± 0.011), indicating substantial genetic differentiation. Interestingly, the F_{st} of *O. cacaliae* was greater than that of *O. speciosissima*. This suggests a greater diversity within the *O. cacaliae* populations which may have evolved more specific adaptations to their local host plants. The within-population component of the overall inbreeding coefficient, F_{is}, was significantly greater than zero for both *O. cacaliae* (0.326 ± 0.021) and *O. speciosissima* (0.332 ± 0.039), suggesting inbreeding.

Conclusions

Sequestration is a secondary event following a host-plant shift and occurred only once in the lineage studied in *Oreina*. Defence due to sequestration is unreliable, since it is dependent on host-plant allelochemicals. In both sequestrating species, there are trade-offs between defence, larval growth and phenology. Based on the defensive chemical data, *O. cacaliae* is more different from the other beetle species than *O. speciosissima*.

Acknowledgement. Part of the chemical work was done in collaboration with A. Ehmke and T. Hartmann (Braunschweig).

Proc. 8th Int. Symp. Insect-Plant Relationships, Dordrecht: Kluwer Acad. Publ.
S.B.J. Menken, J.H. Visser & P. Harrewijn (eds), 1992

Effect of host plants on growth and defence in two *Phratora* (Coleoptera: Chrysomelidae) species

Päivi Palokangas, Seppo Neuvonen and Susanna Haapala
University of Turku, Dept of Biology and Kevo Subarctic Research Institute, Turku, Finland

Key words: Leaf beetles, predation, salicin, spider, tritrophic level

We studied the differences in host-plant preference, and the effect of host-plant species on growth and defence in two closely related leaf beetles that live in subarctic areas. *Phratora polaris* (L.) is a common mountain birch (*Betula pubescens* ssp. *tortuosa* Ehrh.) herbivore. In our study area (69°45′ N, 27°01′ E) *Phratora vitellinae* L. feeds on willow (*Salix borealis* Fries) or aspen (*Populus tremula* L.). *P. vitellinae* bases its defence on chemicals which it derives from precursors in the food plant (*e.g.*, Pasteels *et al.*, 1988), while *P. polaris* obviously synthesizes its defensive chemicals *de novo*. We asked the following questions: (1) Which food plant species do *P. polaris* and *P. vitellinae* larvae prefer in a choice experiment? (2) Does the growth and survivorship of larvae on different food plants correlate with the results of the choice experiment? (3) Is the defence of *P. vitallinae* more efficient than that of *P. polaris* when exposed simultaneously to predators?

Results

Food preference. We offered 60 *P. polaris* larvae, originating from mountain birches, five different food plants: aspen, *S. borealis*, *S. phylicifolia* L., mountain birch and dwarf birch (*Betula nana* L.). We also used 40 *P. vitellinae* larvae originating from aspen and *S. borealis*. We offered these four food plants: *S. borealis*, *S. phylicifolia*, aspen and mountain birch. The number of independent replicates in each trial was 10. After the experiment which lasted 24 h, we calculated the leaf area that each plant species had eaten. *P. polaris* larvae preferred both mountain and dwarf birch leaves. Plants with high salicin contents were avoided (P < 0.001). Regardless of their original host species *P. vitellinae* larvae preferred both aspen and *S. borealis* to *S. phylicifolia* and especially mountain birch (P < 0.001).

Growth trials. We randomized larvae of *P. polaris* and *P. vitellinae* to grow in the laboratory on different food plants. *P. polaris* originating from mountain birch were reared on mountain birch, dwarf birch, *Salix phylicifolia*, *S. borealis* or aspen. *P. vitellinae* originating from *S. borealis* were reared on *S. borealis*, *S. phylicifolia*, aspen or mountain birch. *P. polaris* reached significantly higher body masses, and a higher proportion of them developed into adults when reared on *Betula* or *S. phylicifolia* compared to plants with high salicin contents (P < 0.01). *P. vitellinae* had a better performance on *S. borealis* and aspen than on plants with low salicin (P < 0.05).

Predation. The predation experiment was executed in transparent plastic tubes in the laboratory (ø 18 cm, height 50 cm). Each tube contained mountain birch, *S. borealis* and aspen twigs, the positions of which were randomized. Three second instar larvae, which had been reared on the same food plants in their earlier stages were placed on the leaves of each plant species (*i.e.*, 3 *P. polaris* and 3+3 *P. vitellinae* larvae). We used spiders as predators (10 wolf spiders, Lycosidae and 6 crab spiders, Thomisidae). They were released into the tubes an hour after the larvae had settled. For 24 h, we recorded the plant species on which, or in the vicinity of which, the spider stayed. We then counted the number of live larvae on each host plant. Both wolf and crab spiders ate significantly more *P. polaris* living on birch than *P. vitellinae* living on aspen or willow ($P < 0.001$). Wolf spiders were observed more often on or close to birch rather than aspen or willow, while the opposite was true of crab spiders.

Conclusions

P. polaris and *P. vitellinae* larvae showed distinct differences in host preference. The latter preferred plant species with high salicin content in their foliage (aspen and *S. borealis*), while the former preferred birches and *S. phylicifolia*, *i.e.*, hosts with low salicin. The performance of both species of leaf beetle on different host plants correlated with the ranking of host species in the preference experiment. Furthermore, when exposed to predation by spiders, the survival of larvae living on different host plants varied. More *P. polaris* living on plants with low salicin were eaten than *P. vitellinae* feeding on salicin-rich food plants.

Reference

Pasteels, J.M., M. Rowell-Rahier, J.C. Braekman & D. Daloze (1984). Chemical defences in leaf beetles and their larvae: The ecological, evolutionary and taxonomic significance. *Biochem. Syst. Ecol.* **12**: 395-406.

Proc. 8th Int. Symp. Insect-Plant Relationships, Dordrecht: Kluwer Acad. Publ.
S.B.J. Menken, J.H. Visser & P. Harrewijn (eds), 1992

Effects of early season foliar damage on *Phratora polaris* (Chrysomelidae) and its defensive ability

Seppo Neuvonen, Päivi Palokangas and Susanna Haapala
University of Turku, Dept of Biology and Kevo Subarctic Research Station, Turku, Finland

Key words: Ant predation, *Epirrata autumnata*, mountain birch, tritrophic level

The leaf beetle *Phratora polaris* (L.) feeds mainly on mountain birch foliage in northern Fennoscandia. Vast areas of mountain birch forests are damaged at 9-10 year intervals by early season geometrid larvae (*e.g.*, *Epirrita autumnata* (Bkh.)). Early season foliar damage has negative effects on the growth and/or survival of lepidopteran and sawfly larvae feeding on birches (Neuvonen & Haukioja, 1991). Just how birch-feeding leaf beetles respond to foliage damage, however, is not well known.

Leaf beetles can defend themselves chemically against natural enemies (Pasteels *et al.*, 1988). However, little is known about how induced changes in plant quality affect the defensive ability of leaf beetle larvae. We tested how partial damage to growing birch leaves affected the growth and defensive ability of *P. polaris* larvae.

Material and methods

Two experiments were done in Finnish Lapland: 1) Growth assay (12 days) on control (undamaged) *vs* induced foliage, *i.e.*, undamaged leaves adjacent to a branch on which larvae of *Epirrita autumnata* had previously been reared. Tree-specific mean weights of *P. polaris* larvae were used in the analyses (10 replicate trees/group). 2) Predation experiment, in which larvae reared on control *vs* induced foliage were exposed to predation by ants. The same trees as above were used. Small twigs with leaves were taken from the birches and larvae were placed on them. Pairs of twigs (control & induced) were placed close to each other near an ant mound, and we counted the number of larvae killed by ants during one day of exposure.

Results and discussion

Growth on induced vs *control foliage.* Larvae reared on control foliage were 18% heavier than those reared on induced foliage, but the difference was not significant. However, the result corresponded with earlier experiences with mountain birch herbivores (Neuvonen & Haukioja, 1991). In our experiment, larvae on induced foliage were not fed on the actual damaged leaves, but on leaves adjacent (some tens of centimeters) to the branch with previous insect damage.

Predation by ants on larvae reared on induced vs *control foliage.* Twenty-one per cent more larvae reared on control foliage were killed by ants compared to those reared on induced foliage, but the difference was not significant.

Conclusions

Although the differences were not significant the results from these two experiments were opposite. The growth of larvae was better when reared on control foliage, but larvae fed on induced foliage suffered less from predation. Thus, when estimating the net effects of induced foliage changes on herbivores, the third trophic level should also be considered.

References

Neuvonen, S. & E. Haukioja (1991). The effects of inducible resistance in host foliage on birch-feeding herbivores. In: D.W. Tallamy & M.J. Raupp (eds), *Phytochemical Induction by Herbivores*, pp. 277-291. New York: John Wiley & Sons.
Pasteels, J.M., J.C. Braekman & D. Daloze (1988). Chemical defence in the Crysomelidae. In: P. Jolivet, T.H. Hsiao & E. Petitpierre (eds), *The Biology of Chrysomelidae*, pp. 231-250. Dordrecht: Kluwer Academic Publishers.

Proc. 8th Int. Symp. Insect-Plant Relationships, Dordrecht: Kluwer Acad. Publ.
S.B.J. Menken, J.H. Visser & P. Harrewijn (eds), 1992

Simulated acid rain and the susceptibility of the European pine sawfly (*Neodiprion sertifer*) larvae to nuclear polyhedrosis virus

Kari Saikkonen and Seppo Neuvonen
Dept of Biology, University of Turku, Turku, Finland

Key words: Fungi, hymenoptera, Scots pine, virus efficacy

Little attention has been paid to the effects of pollutants via natural enemies of herbivores, and the effects on interactions among tritrophic levels (host plants - herbivores - natural enemies) have been almost totally neglected. However, the experiment by Neuvonen *et al.* (1990) suggested that simulated acid rain (SAR) with pH 3 applied to host trees (Scots pine, *Pinus sylvestris* L.) might reduce the susceptibility of young (first and second instar) European pine sawfly (*Neodiprion sertifer* (Fourcroy)) larvae to a nuclear polyhedrosis virus (NsNPV). On the other hand, there were no significant effects of SAR with pH 4 on the efficacy of NsNPV, and the effect of SAR was only on the timing of mortality of the larvae reared on pH 3 treated trees: all the larvae died toward the end of the larval period. These results were obtained after the host trees had been treated with SAR for only two growing seasons, which may be insufficient for the expression of many processes.

The aims of this study are: 1) to test the effects of simulated acid rain on the efficacy of NsNPV after the treatment of host foliage for four summers and to compare these results with those obtained after two growing seasons (Neuvonen *et al.*, 1990); 2) to find out whether older larvae exhibit the same responses as very young larvae.

Materials and methods

The study was conducted at the Kevo Subarctic Research Station (69° 45′ N, 27° E) in Finnish Lapland. The host plant treatments (irrigation control, pH 4 and pH 3 irrigation) were conducted three times a week (5 mm per occasion) during five growing seasons. For a more detailed description of the experimental set up, treatments, vegetation and background deposition levels, see Neuvonen *et al.* (1990) and Neuvonen & Suomela (1990).

The viral infection was caused as follows: pine needles given to the larvae were sprayed for two days (groups A1 and A2) or eight days (groups B1 and B2) with diluted virus suspension (± 200 polyhedrosis inclusion bodies (PIBs) per ml) after the larvae began to feed on the experimental foliage. After that, surviving larvae and cocoons were counted 11 times and fed with foliage of experimental trees when necessary, about once a week.

Results and discussion

The susceptibility to the virus was strongly reduced by ageing of the larvae. Most larvae treated with virus two days after they started feeding on experimental foliage (group A), died rapidly 10 days after the virus treatment. Larvae treated with virus one week later (group B) responded remarkable slower to it. Survival at the end of the larval period was only 2 - 22% in group A, while in group B it was 27 - 61%. In group A1 the larval survival on pH 3 treatment was significantly higher than in the other treatments; at the end of the larval period the difference was four-fold. Moreover, the survival of A1 on pH 4 treated foliage was initially significantly higher than on control foliage, but the difference vanished later. With later virus infections (= older larvae; group B) the effects of SAR on the susceptibility to NsNPV were not clear. Our results show that SAR can reduce the susceptibility of very young *N. sertifer* larvae to the virus.

The mechanisms which reduce the efficacy of NsNPV on acid rain treated pine foliage are unknown, but there are several hypotheses worth studying:

(1) The simplest hypothesis is that the dose of NsNPV on SAR treated needles ingested by *N. sertifer* larvae was smaller due to the erosion of surface waxes and changes in the wettability of the needles.

(2) The virions are released from PIBs under alkaline conditions and the efficacy of NsNPV (as well as other baculoviruses) is reduced by higher acidity (lower pH) (Keating *et al.*, 1989, 1990). Whether acidity on needle surfaces, internal needle pH's or pH of the insect's midgut is critical, depends on details of the NsNPV's infection process, the degree of interdependence of pH's at these sites and the buffering capacity of the "targets" (see *e.g.*, Keating *et al.*, 1990).

(3) Changes in the nutritive value of SAR exposed trees (treatments included nitric acid) may have decreased the window of larval susceptibility to the polyhedrosis virus.

(4) Possible changes in the allelochemical profiles produced by SAR treated pine foliage may interfere with the effectiveness of the virus. For example, it is known that many phenolic substances (tannins) decrease the susceptibility of lepidopteran larvae to (baculo)viruses (Keating *et al.*, 1989).

(5) Allelochemicals produced by host plant associated microfungi may also interfere with the virus; it is well known that SAR affects epiphytic fungi rapidly (Ranta, 1990), while the case with endophytic fungi is not so clear (but see Barklund, 1989).

References

Barklund, P. (1989). Occurrence of and interaction between *Gremmeniella abietina* and endophytic fungi in two conifers. Ph. D. Dissertation, University of Uppsala, Sweden.

Keating, S.T., W.J. McCarthy & W.G. Yendol (1989). Gypsy moth (*Lymantria dispar*) larval susceptibility to a baculovirus affected by selected nutrients, hydrogen ions (pH), and plant allelochemicals in artificial diets. *J. Invertebr. Pathol.* **54**: 165-174.

Keating, S.T., J.C. Schultz & W.G. Yendol (1990). The effect of diet on gypsy moth (*Lymantria dispar*) larval midgut pH, and its relationship with larval susceptibility to a baculovirus. *J. Invertebr. Pathol.* **56**: 317-326.

Neuvonen, S. & J. Suomela (1990). The effect of simulated acid rain on pine needle and birch leaf litter decomposition. *J. Appl. Ecol.* **27**: 857-872.

Neuvonen, S., K. Saikkonen & E. Haukioja (1990) Simulated acid rain reduces the susceptibility of the European pine sawfly (*Neodiprion sertifer*) to its nuclear polyhedrosis virus. *Oecologia* **83**: 209-212.

Ranta, H. (1990). Effects of simulated acid rain on epiphytic and endophytic microfungi of Scots pine (*Pinus sylvestris* L.). *Environ. Pollut.* **67**: 349-359.

Proc. 8th Int. Symp. Insect-Plant Relationships, Dordrecht: Kluwer Acad. Publ.
S.B.J. Menken, J.H. Visser & P. Harrewijn (eds), 1992

Identification of cereal aphid parasitoids by epicuticular lipid analysis

L.C. Irvin, R.D. Eikenbary, R.K. Campbell, W.S. Fargo and J.W. Dillwith
Dept of Entomology, Oklahoma State University, Stillwater, USA

Key words: Biochemical markers, hydrocarbons, Hymenoptera, tritrophic interactions,

Hymenoptera that parasitize aphids are being widely employed as biological control agents for a variety of species including important cereal pests like the greenbug, *Schizaphis graminum* (Rondani), and the Russian wheat aphid, *Diuraphis noxia* (Mordvilko). Identification of these parasitoids using morphological markers is often difficult and requires a trained expert. This presents a serious obstacle to the successful selection, importation, rearing, mass release and tracking of these insects in biological control programs. The characterization of convenient biochemical markers for rapid identification of parasitoids is an important goal. This study was undertaken to determine if parasitoid epicuticular hydrocarbons might serve as useful markers to distinguish parasitoid species.

Parasitoids and methods. Two aphid parasitoids, *Lysiphlebus testaceipes* (Cresson) and *Diaeretiella rapae* McIntosh, were used in most of the experiments. Three species of *Aphidius*, *A. picipes* (Nees), *A. matricariae* Hal. and *A. uzbekistanicus* Luzhetzki were also examined. Parasitoids were reared on either greenbugs or Russian wheat aphids growing on wintermalt barley. Rearing took place in growth chambers at 26.6°C light:dark 15:9. Adult parasitoids were collected shortly after emergence, and separated by sex. Individual samples contained ten adult insects. Cuticular lipids were extracted with two 1.5ml aliquoits of hexane for 30 sec each. Combined extracts were dried under nitrogen and the hydrocarbon fraction isolated using a column (6.0 cm x 0.5 cm i.d.) of Biosil A (Bio-Rad Laboratories, Richmond, CA) (Dillwith *et al.*, 1981). Isolated hydrocarbons were dissolved in 50µl of hexane and 1µl was analyzed by gas-liquid chromatography using a 100% methylsilicone column (DB-1, 15 m x 0.25 mm i.d., 0.1 µm film thickness, J and W Scientific, Folsom, CA). Samples were introduced using cool on-column injection at 55°C and an oven temperature program of 55°C (1.75 min), 25°C/min to 180°C, 8°C/min to 320°C (Hold 3 min). Hydrocarbon chain lengths were determined by comparison to authentic hydrocarbon standards of known chain length. Hydrocarbons were quantified by use of n-tetracosane as an internal standard. Data on percentage hydrocarbon composition were subjected to principle component analysis (Carlson & Brenner, 1988).

Results

All five species of parasitoids examined contained measurable amounts of cuticular hydrocarbon. *L. testaceipes* females contained 317 ± 83 ng/insect and males 163 ± 72 ng/insect. Similar quantities were present on other species. The hydrocarbon fraction contained multiple components with equivalent chain lengths from 23 to 33 carbons. Hydrocarbon profiles were unique for each of the five species and for males and females of a species. By inspection of chromatograms it was possible to easily distinguish all species examined. Principle component analysis applied to the data for *L. testaceipes* and

D. rapae produced distinct clusters for males and females of each species. Analysis of hydrocarbons from the host aphids and plants showed no relationship between parasitoid and host hydrocarbon compositions.

Discussion

The results of this study show that cuticular lipids are valuable biochemical markers for the identification of parasitic Hymenoptera. Cuticular hydrocarbon analysis has many advantages for routine monitoring of parasitoid colonies and field collected specimens. Hydrocarbons are stable both on insects and in extracts for many years at room temperature. Dried specimens can be stored and shipped without refrigeration. The method is non-destructive allowing extracted insects to serve as voucher specimens for further study. It involves simple rapid sample preparation and is relatively inexpensive. Principle component analysis can be used to objectively compare hydrocarbon profiles. This allows data acquisition and analysis to be automated increasing sample through-put.

References

Carlson, D.A. & R.J. Brenner (1988). Hydrocarbon-based discrimination of three North American *Blattella* cockroach species (Orthoptera: Blattellidae) using gas chromatography. *Ann. Entomol. Soc. Am.* 81: 711-723.
Dillwith, J.W., G.J. Blomquist & D.R. Nelson (1981). Biosynthesis of the hydrocarbon components of the sex pheromone of the housefly, *Musca domestica* L. *Insect Biochem.* 11: 247-253.

Proc. 8th Int. Symp. Insect-Plant Relationships, Dordrecht: Kluwer Acad. Publ.
S.B.J. Menken, J.H. Visser & P. Harrewijn (eds), 1992

Do plants reduce herbivore attack by providing pollen?

Paul C.J. van Rijn
University of Amsterdam, Dept of Pure and Applied Ecology, Amsterdam, The Netherlands

Key words: Alternative food, biological control, indirect defence, mutualism, thrips, tritrophic interactions

Plants may provide food to attract and arrest the natural enemies of their herbivores as a means of indirect defence. These natural enemies can then be referred to as bodyguards of the plant. Well-known examples of food types provided by plants are: extrafloral nectar, Beltian bodies and phloem exudate. Albeit mainly produced for other reasons, pollen is also a food source for potential bodyguards, such as predatory bugs and predatory mites.

Besides the beneficial effects, these food sources can potentially be used by herbivores as well. The question is, under which conditions is the provision of food beneficial to the plants? We studied the effects of pollen on a thrips - predatory mite system on greenhouse crops, where both the thrips and its predator can utilize pollen.

Results

(1) Adding pollen to a diet of cucumber leaves diminishes the juvenile period of the thrips, *Frankliniella occidentalis* (Pergande), by about 10%, and almost doubles its oviposition rate (Van Rijn, unpubl.).

(2) The predatory mite, *Amblyseius cucumeris* (Oudemans), can also utilize pollen for its development and reproduction (Van Rijn & Van Houten, 1991). Oviposition rate is slightly lower when pollen is offered instead of thrips larvae. Consequently, at low prey densities the presence of pollen will increase the population growth of the predator.

(3) At high thrips densities, thrips consumption by this predator is barely reduced when pollen is present (Van Rijn, unpubl.).

(4) In the greenhouse the predatory mites are arrested in areas with only pollen, as well as in areas with thrips alone (Van Rijn & Sabelis, 1990).

(5) Assuming homogeneous distributions, predator-prey models show that the equilibrium prey density will always be lower when pollen is present, since its equilibrium density depends solely on predator parameters (Holt, 1977). This equilibrium is stable for all realistic parameter values, due to invulnerability of the older thrips stages.

(6) On the other hand: thrips damage caused during the pre-equilibrium phase can be higher when pollen is available, depending on the relative effects of pollen on the properties of predator and prey, and on the initial numbers of predator and prey.

Discussion and conclusions

Assuming large, homogeneous populations and equilibrium conditions the model predicts that even though pollen promotes population growth of both thrips and predatory mite, the overall effect on thrips control is positive. The prospects for testing this prediction in greenhouse crops, such as sweet pepper and cucumber are good, especially because host plant varieties are available with low and high production of pollen.

When populations are not at equilibrium, predictions on the impact of pollen on thrips control are less straightforward. Whether the impact is positive or negative now depends on the extent to which prey populations growth is promoted relative to that of the predator, and on the initial numbers of predator and prey. This is because the prey population may grow larger by pollen feeding, before the predators become sufficiently numerous to suppress the prey population.

These results are of fundamental importance for understanding the evolution of pollen as a food source. While it is obvious that reproduction is the primary function of pollen, its accessibility as a food source can be viewed as part of the plant strategy to attract pollinators and bodyguards against insect herbivores.

Clearly, this strategy is vulnerable to invasion by herbivores that also consume the pollen. The model shows that pollen accessibility as a food source will evolve under equilibrium conditions, but may not be favourable under non-equilibrium conditions. If not favourable, pollen accessibility may evolve solely because of its positive effect on the pollinators.

References

Holt, R.D. (1977). Predation, apparent competition, and the structure of prey communities. *Theoret. Pop. Biol.* 12: 197-229.

Van Rijn, P.C.J. & Y.M van Houten (1991). Life history of *Amblyseius cucumeris* and *A. barkeri* (Acarina: Phytoseiidae) on a diet of pollen. *Modern Acarology*, Vol. 2, pp. 647-654. Prague: Academia and The Hague: SPB Academic Publishing.

Van Rijn, P.C.J. & M.W. Sabelis (1990). Pollen availability and its effect on the maintenance of populations of *Amblyseius cucumeris*, a predator of thrips. *Med. Fac. Landbouww. Rijksuniv. Gent* 55: 335-341.

Proc. 8th Int. Symp. Insect-Plant Relationships, Dordrecht: Kluwer Acad. Publ.
S.B.J. Menken, J.H. Visser & P. Harrewijn (eds), 1992

How cassava plants enhance the efficacy of their phytoseiid bodyguards

Frank M. Bakker and Martha E. Klein
*Dept of Pure and Applied Ecology, University of Amsterdam, Section Population Biology,
Amsterdam, The Netherlands*

Key words: Extrafloral nectar, plant defense, predation, Tetranychidae, tritrophic
interactions

Cassava (*Manihot esculenta* Crantz) exudes droplets rich in sugar at the base of its petioles
and sometimes at the midrib. This extrafloral nectar usually referred to as cassava
exudate, is composed of reducing sugars, fructofuranosides and some amino acids
(Perreira & Splitstoesser, 1987). Production sites are mainly found on the younger (1-7)
leaves. Organisms foraging for this exudate include ants, parasitoids, lacebugs and
phytoseiid mites (Bakker & Klein, unpubl. results). Phytoseiidae are small arthropods that
prey on pest organisms such as spider mites, rust mites and thrips. Laboratory studies
have shown that natural sugar solutions such as honey, honeydew and also plant
exudates have a positive effect on phytoseiid survival and, when combined with prey,
sometimes also on reproduction.

Typhlodromalus limonicus (Garman & McGregor) is a very common phytoseiid on
cassava in the Neotropics. In the absence of other food sources cassava exudate enhances
adult and juvenile survival of this species to a significant extent. Although oviposition
ceases on a diet of exudate, it is resumed as soon as prey is available again. Moreover,
juvenile development does take place with exudate, albeit only fractionally, whereas on a
diet of water (on clean cassava leaves) all juveniles die in the larval stage. Finally, two
choice experiments revealed that in the absence of prey the presence of exudate results in
an arrestment response of starved females and larvae of this species. The observation that
T. limonicus is restricted in its distribution to cassava together with these experimental
results, indicate that there might be a mutualistic relationship between cassava and *T.
limonicus* mediated by exudate. In this regard the following hypotheses were tested: (1) *T.
limonicus* benefits from choosing cassava as its host plant, because the presence of exudate
enables the predator to survive preyless periods or periods with very low prey density.
(2) *M. esculenta* benefits from the presence of *T. limonicus* because the latter relies on the
consumption of organisms harmful to cassava for its oviposition. (3) Presence of exudate
together with herbivores results in increased efficacy of the predator population. These
hypotheses were tested in a series of semi-field trials at four different prey densities, *viz.*,
(1) without prey, (2) at low prey density, (3) at intermediate prey density and (4) at high
prey density.

Methods and materials

The experiments in absence of prey were done in mite-proof tents at CIAT, Colombia.
Clean plants were taken from the nursery and trimmed to six leaves. All meristems were
removed to prevent the development of new exudate production sites. Exudate was made

inaccessible to the predatory mites by wrapping all petioles in PVC tubelets. On half of the plants three droplets of exudate were applied on each petiole after which field collected predators were released at a density of two per leaf. Population dynamics was monitored for a two week period.

The effect of combining a sugar source with prey was studied in a screenhouse at EMBRAPA, Bahia Brazil. Three-month-old plants were collected from the field and trimmed to ten leaves by removing the youngest leaves. Meristems were also removed. Thus we remained with plants that had lignified petioles only. These plants did not produce exudate which made the PVC-tubelets unnecessary. As it was impossible to collect sufficient amounts of exudate for this experiment, honey was used as a substitute. Three droplets were applied on each petiole on half of the plants. Field collected predators were released at a density of 1 per leaf. During the experiments there was a continuous inflow of spider mites in the screenhouse. By manually removing them from the leaves at variable intervals and by varying the moment of predator introduction it was possible to obtain three groups of plants with different prey densities.

Experiments lasted 40 days. There were 10 plants in each group.

Results and discussion

Predators did not persist on the plants in the absence of prey. Extinction rates were somewhat lower in the presence of exudate. In one replicate (out of 5) predators did persist. Here population density was about twice as high in the presence of exudate.

In the experiments where prey was also present, the predator population attained a consistently higher density throughout the experiment in the presence of the sugar source. This invariably coincided with lower herbivore abundance. Results were most pronounced at low prey densities. Despite the lower per capita access to prey items in the treated group, oviposition rates were similar in both groups. This study demonstrates the potential plants have to enhance the efficacy of their phytoseiid bodyguards by providing them with a (low quality) food source. It is to be expected that similar examples will be reported from other systems. We are intrigued by Janzen's (1966) observation of "...tiny output of tiny nectaries found on many plants that lack an obligate relationship with ants".

Acknowledgements. This study was financed by the International Institute of Tropical Agriculture, The Netherlands Organization for Scientific Research, Fundatie Vrijvrouwe van Renswoude and the Johanna Westerdijk Fonds. Logistic support was provided by the Empresa Brasileira de Pesquisa Agropecuaria and the Centro Internacional de Agricultura Tropical. Our special thanks go to Dr Gilberto deMoraes, Aloyseia Noronha, Dr Ann Braun and Nora Mesa.

References

Bakker, F.M. & M.E. Klein (1992). Transtrophic interactions in cassava. *Exp. Appl. Acarol.* 14 (in press).
Janzen, D.H. (1966). Coevolution of mutualism between ants and acacias in Central America. *Evolution* 20: 249-275.
Pereira, J.F. & W.E. Splitstoesser (1987). Exudate from cassava leaves. *Agric. Ecosys. Evironm.* 18: 191-194.

Proc. 8th Int. Symp. Insect-Plant Relationships, Dordrecht: Kluwer Acad. Publ.
S.B.J. Menken, J.H. Visser & P. Harrewijn (eds), 1992

Infochemicals that mediate plant-carnivore communication systemically induced by herbivory

Marcel Dicke

Wageningen Agricultural University, Dept of Entomology, Wageningen, The Netherlands

Key words: Elicitor, predatory mites, spider mites, tritrophic interactions, volatile synomone

Reliability-detectability problem of foraging natural enemies. Natural enemies of herbivores may use cues of the herbivore and/or cues of the food plant of the herbivore during long-range herbivore location. These two types of cues differ in their usability (Vet & Dicke, 1992). Cues of the herbivore have a high reliability in indicating herbivore presence and identity, but they are expected to have a low detectability because herbivores are a small component in a complex environment and it is thought that selection has reduced the emission of cues that enable the herbivore's enemies to locate the herbivore. On the other hand, plant cues are more detectable, but they are much less reliable indicators of herbivore presence and identity. One of the solutions to this reliability-detectability problem with which natural enemies are faced, is the use of herbivore-induced synomones that are produced in relatively large amounts by plants in response to herbivore damage, and are specific to the herbivore species that damages the plant (Vet & Dicke, 1992).

Induction of synomone production by herbivory. Many species of parasitoids and predators discriminate between uninfested plants and plants infested by their host/prey on the basis of volatile infochemicals (Vet & Dicke, 1992). Subsequent investigations in two systems revealed that these infochemicals are not produced by the herbivore but by the plant: (1) a system consisting of several plant species, the spider mite species *Tetranychus urticae* Koch and the predatory mite species *Phytoseiulus persimilis* Athias-Henriot (Dicke & Sabelis, 1988), and (2) a system consisting of corn plants, armyworm larvae and the parasitoid *Cotesia glomerata* (L.) (Turlings *et al.*, 1990). The conclusion was that the plant is the actual producer of the synomone and that synomone production is induced by herbivory. This was based on chemical identification of the infochemicals and the finding that the synomone components are emitted by plants infested with the herbivore, but not from artificially damaged or undamaged plants. Moreover, the herbivores themselves were not or hardly attractive to their natural enemies. All these investigations involved the emission of the synomone from herbivore-infested leaves. Furthermore, recent data show that uninfested leaves of bean plants that were partially infested with spider mites also attract predatory mites: synomone production occurs systemically in an infested plant (Dicke *et al.*, 1990). Chemical data confirmed the induced production in uninfested leaves of several synomone components that are also induced in the spider-mite infested leaves (Takabayashi *et al.*, 1991).

Isolation of an elicitor that systemically induces production of a synomone that attracts predatory mites. Takabayashi *et al.* (1991) demonstrated that synomone production was induced in uninfested Lima bean leaves that had been placed on wet cotton wool on which spider mite infested bean leaves had previously lain. This suggests that an elicitor was present in the wet cotton wool. We recently clearly demonstrated the presence of an elicitor. Lima bean leaves were incubated with their petiole in a vial with distilled water. One group of leaves was infested with spider mites and the other group was left uninfested. After 7 days the water of each leaf was transferred to new vials and uninfested Lima bean leaves were placed in the water with their petiole. After 3 days the leaves were used in a Y-tube olfactometer to investigate the response of the predatory mite *P. persimilis*: 83% of the predators preferred the odour of uninfested leaves that had been incubated in water in which infested leaves had been incubated previously and 17% of the predators preferred the odour of leaves that had been incubated in water in which uninfested leaves had been incubated previously (n = 117; P < $2x10^{-12}$). This is the first proof of the isolation of an elicitor that is systemically transported in plants and induces the production of carnivore attractants in uninfested plant parts.

Acknowledgements. Rob Wessels and Peter van Baarlen participated in the experiment on isolation of the elicitor.

References

Dicke, M. & M.W. Sabelis (1988). How plants obtain predatory mites as bodyguards. *Neth. J. Zool.* **38**: 148-165.

Dicke, M., M.W. Sabelis, J. Takabayashi, J. Bruin & M.A. Posthumus (1990). Plant strategies of manipulating predator-prey interactions through allelochemicals: prospects for application in pest control. *J. Chem. Ecol.* **16**: 3091-3118.

Takabayashi, J.M. Dicke & M.A. Posthumus (1991). Induction of indirect defence against spider-mites in uninfested lima bean leaves. *Phytochemistry* **30**: 1459-1462.

Turlings, T.C.J., J.H. Tumlinson & W.J. Lewis (1990). Exploitation of herbivore-induced plant odors by host-seeking parasitic wasps. *Science* **250**: 1251-1253.

Vet, L.E.M. & M. Dicke (1992). Ecology of infochemical use by natural enemies in a tritrophic context. *Annu. Rev. Entomol.* **37**: 141-172.

Proc. 8th Int. Symp. Insect-Plant Relationships, Dordrecht: Kluwer Acad. Publ.
S.B.J. Menken, J.H. Visser & P. Harrewijn (eds), 1992

Mite herbivory causes better protection in downwind uninfested plants

Jan Bruin[1], Astrid T. Groot[1], Maurice W. Sabelis[1] and Marcel Dicke[2]
[1] University of Amsterdam, Dept of Pure and Applied Ecology, Amsterdam, The Netherlands
[2] Wageningen Agricultural University, Dept of Entomology, Wageningen, The Netherlands

Key words: Plant defence, plant-to-plant communication, plant-to-plant contamination, spider mites, tritrophic interactions

Spider mites are ravenous herbivores. They tend to overexploit their food plants in the absence of natural enemies. After depletion of their food source, the spider mites disperse actively, ambulatorily, or passively, floating on air currents. The plants, however, are not as helpless as one would expect, for when infested by spider mites they start producing volatile infochemicals that act as a lure for predatory mites, the natural enemies of these herbivores (*e.g.*, Dicke, 1988). As these infochemicals are volatile, they can also reach downwind vegetation. This raises the question whether nearby uninfested plants can benefit from the airborne information, and take defensive measures prior to actual infestation. To investigate this we studied 1) the ovipositional rate of the spider mite *Tetranychus urticae* Koch, as a measure for prey performance in the absence of natural enemies, and 2) the olfactory response of the predatory mite *Phytoseiulus persimilis* Athias-Henriot. We used cotton (*Gossypium hirsutum* L., var. Acala SJ-2) and Lima bean (*Phaseolus lunatus* L., *cv.* Carolina or Sieva) as the host plant. For a more detailed report on part of the study on information transfer between infested and uninfested plants, see Bruin *et al.* (1992).

Materials and methods

All experiments took place in climate rooms (appr. 25°C, 70% RH). Uninfested plants or leaves were exposed to volatiles from plants or leaves that were infested with spider mites. Control plants or leaves were exposed to clean air or to substances from uninfested plants/leaves.

The spider mites' ovipositional rates were assessed on plants or leaves that had been previously exposed for 4-8 days. Adult females of the same age were allowed to oviposit for 3-4 days, during which time the plants or leaves remained exposed. On cotton, 9-10 females were confined to one leaf of the seedlings. On lima bean, single females were kept on leaf discs taken from previously exposed detached leaves.

The olfactory response of individual predatory mites was studied in a Y-tube olfactometer. The predators were offered a choice between plants or leaves that had been exposed to substances from spider-mite-infested or uninfested plants or leaves. For details of set up and procedures see Bruin *et al.* (1992).

Results

Herbivore performance. The experiment with cotton as a host plant was replicated four times. Each trial rendered two mean rates of oviposition, one for the treated and one for the control plants (Table 1).

Table 1. Mean oviposition rate of spider mites on cotton (eggs/mite/day)

Replicate	On control plants	On plants exposed to volatiles from infested plants
1	8.21	7.44
2	9.35	8.95
3	7.60	6.65
4	6.46	5.63

The means were based on 7-12 plants per treatment and 9-10 mites per plant. The trials can be viewed as true replicates. A paired t-test (n = 4) revealed a highly significant difference (P = 0.008; 2-tailed) between treated and control plants: the ovipositional rate of spider mites is lower on plants that are exposed to volatiles from spider-mite-infested neighbouring plants, compared to controls. The results of the experiment with Lima bean leaves, only done once, were consistent with the cotton results.

Predator performance. Predatory mites showed a clear preference for volatiles related to uninfested plants or leaves that had been exposed to volatiles from spider-mite-infested plants or leaves, compared to controls. This appeard to be true for both cotton (77.5%, n = 40, P < 0.001; binomial test) and Lima bean (69.0%, n = 87, P < 0.001).

Discussion

Our results show that exposure to volatiles from infested plants affects both herbivore and predator performance on downwind neighbouring plants. We conclude that uninfested plants on the leeward side of infested conspecifics, can gain protection against spider mite herbivory. Whether the exposed plants are actively involved (plant-to-plant *communication*) or passively (plant-to-plant *contamination*), however, remains to be investigated. Further experiments are required to elucidate the role of the plant.

References

Bruin, J., M. Dicke & M.W. Sabelis (1992). Plants are better protected against spider mites after exposure to volatiles from infested conspecifics. *Experientia* **48**: 525-529.

Dicke, M. (1988). Infochemicals in tritrophic interactions: origin and function in a system consisting of predatory mites, phytophagous mites and their host plants. Ph.D. Dissertation, Agricultural University, Wageningen, The Netherlands, 235 pp.

Proc. 8th Int. Symp. Insect-Plant Relationships, Dordrecht: Kluwer Acad. Publ.
S.B.J. Menken, J.H. Visser & P. Harrewijn (eds), 1992

Parasitoids foraging for leaf damage: do they see beyond the end of their antennae?

Felix Wäckers
Dept of Entomology, Wageningen Agricultural University, Wageningen, The Netherlands

Key words: Host location, prey recognition, search behaviour

Visual orientation in foraging parasitoids has recently receiving increasing attention. Wäckers & Lewis (1992) demonstrated that parasitoids can use visual information during various stages of the host-location process. Feeding damage, besides being a main source of plant derived volatiles, can be a distinct visual indicator of herbivore presence. The visual image of the damaged leaf could convey specific information to the parasitoid, such as the species of the herbivore and its stage of development. The role of behavioural chemicals in parasitoid foraging has been studied in great detail (Turlings *et al.*, 1992). We investigated whether *Cotesia rubecula* (Marshall), a solitary parasitoid of *Pieris* spp., uses visual information from herbivore damage during host foraging.

Material and methods

Flight-chamber experiments. We used Brussels-sprouts plants (*cv.* Titurel), a natural host plant for *Pieris* spp., for our experiments. Four plants of uniform age and size were placed in a flight-chamber to create a plant patch. The distance between pots was 35 cm, to allow the flying parasitoid sufficient space to move freely among the plants. Artificial damage was produced by either punching or cutting leaves one week before the experiment. Four leaf treatments were randomly assigned to the eight youngest leaves:
- Two leaves containing 30 small punch holes (Ø 0.3 cm);
- Two leaves containing 6 big punch holes (Ø 1.5 cm);
- Two leaves damaged by cutting a thin strip along the edge, to check possible damage-induced odour cues;
- Two undamaged leaves.

To avoid any possible differences in leaf-odour concentrations, the number of punched holes in the first two treatments covered an equal area of leaf damage. To determine the effect of an additional odour source on visual preferences, we studied the natural feeding damage of second instar *Pieris rapae* L. that had been feeding on Brussels sprouts leaves for at least 24 h. Leaf discs (Ø 1.5 cm) containing feeding damage were punched out just before the start of the experiment, and applied to the leaves on wetted filter paper of equal diameter to avoid drying out. Second instars of *P. rapae* were used as a reward in the training experiment.

Experiments and results

The initial response of *C. rubecula* to the visual component of feeding damage was examined by observing the foraging behaviour of unexperienced females within the plant patch, in the absence of natural damage. Females were released into the flight-chamber and allowed to forage for a maximum of 30 min. We recorded both the number of landings on the different leaf categories as well as the subsequent searching time.

The parasitoids exhibited an initial preference for leaves containing small punch holes. This preference was concluded from the significantly higher number of landings made on these targets. Searching time, however, did not differ between leaf categories.

Initial visual preference was not found when leaf discs containing natural leaf damage were added to the different leaf categories. This indicates that under our experimental conditions visual orientation was overruled by herbivore induced volatiles.

Next, we investigated whether foraging parasitoids could be conditioned to distinguish between different leaf-damage patterns according to their profitability.

Parasitoids with repeated oviposition experiences on leaves with small punch holes demonstrated a conditioned preference for this leaf category. This preference expressed itself both in a higher number of landings and in a longer search time on this leaf type.

Discussion and conclusions

Our experiments show that *C. rubecula* has an innate preference for the visual image of leaf damage. This initial preference was overruled when larval feeding damage was added.

Parasitoids can be conditioned to the visual component of leaf damage. This associative learning of a rewarding leaf-damage image could enable parasitoids to concentrate their foraging on the most profitable host sites. Consequently, the visual component of leaf damage should not be overlooked when studying tritrophic interactions. We believe that cryptic feeding strategies found in various caterpillar species (Heinrich & Collins, 1983) may have developed not only to reduce predation by visually hunting birds, but may also be adaptive in reducing parasitization.

References

Heinrich, B. & S. Collins (1983). Catterpillar leaf damage, and the game of hide and seek with birds. *Ecology* **64**: 592-602.

Turlings, T.C.J., F.L. Wäckers, L.E.M. Vet, W.J. Lewis & J.H. Tumlinson (1992). Learning of host-location cues by hymenopterous parasitoids. In: A.C. Lewis & D.R. Papaj (eds), *Insect Learning: Ecological and Evolutionary Perspectives.* Chapman and Hall, New York. (in press).

Wäckers, F.L. & W.J. Lewis (1992). Olfactory and visual learning and their interaction in host site location by *Microplitis croceipes*. *Biocontrol* (in press).

Proc. 8th Int. Symp. Insect-Plant Relationships, Dordrecht: Kluwer Acad. Publ.
S.B.J. Menken, J.H. Visser & P. Harrewijn (eds), 1992

Do plants manipulate the third trophic level?

Stanley H. Faeth
Dept of Zoology, Arizona State University, Tempe, Arizona, USA

Key words: *Cameraria*, herbivory, induced responses, tritrophic level interactions, natural enemies, plant defenses

Recent evidence suggests that many plants respond chemically, morphologically, and structurally to herbivory. These plant responses are thought to deter subsequent herbivores by killing them or reducing growth and fecundity. Some researchers have proposed that plants have evolved these responses as induced defenses against herbivores. Induced plant responses may also alter attack by natural enemies of herbivores by providing search cues for parasites and predators (Faeth, 1988). Likewise, some have used this evidence to suggest that plants have evolved induced responses to manipulate the third trophic level as another line of defense against herbivores (Faeth, 1988). Support for the induced defense via natural enemy hypothesis (hereafter referred to as IDNE) requires that 1) post-herbivory chemical or structural changes occur, 2) natural enemies respond to these changes (Turlings *et al.*, 1990), and 3) the effects of 1) and 2) are translated into discernible effects on population dynamics of phytophagous insect species such that herbivory on the plant is reduced. While there is evidence for the first two requisites of IDNE, support for the third is generally lacking, particular in field studies involving natural populations.

I tested the IDNE hypothesis by experimentally manipulating levels of folivory (0%, 25%, 50%, and 75% of leaves damaged) on branches of eight trees of *Quercus emoryi* (L.) in a complete randomized block design. I monitored changes in colonization, survival and mortality (including attack by natural enemies) of a dominant leaf-mining species, *Cameraria* sp. nov. (Davis) in the first and second growing season, and endemic herbivory by leaf-chewing insects and plant growth in the second growing season.

In the first growing season, mortality from predators, parasitoids, and natural enemies (predation and parasitism) did not vary among treatments (Fig. 1). Death from premature leaf abscission increased significantly with increasing numbers of damaged leaves within branches, while mortality from other causes, including death from bacterial and fungal attack, decreased significantly with levels of folivory. Neither of these changes in mortality, however, resulted in changes to overall survival of larval and pupal leaf miners. By far, the most important factor affecting survival and causes of mortality was the presence of co-occurring larvae within leaves (*i.e.*, intraspecific competition).

Folivory in the first growing season did not affect colonization by leaf miners or leaf chewers in the next growing season. Survival of leaf miners and rate of attack by their natural enemies were not associated with levels of natural leaf-chewing herbivory in the second growing season. Simulated folivory in the first growing season did, however, reduce leaf production in the second season.

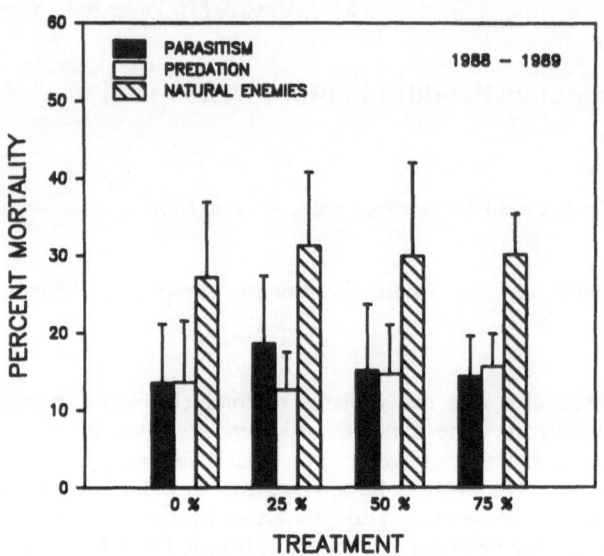

Figure 1. Mean (± SE) percent rate of parasitism, predation, and natural enemy attack (predation plus parasitism) for larval and pupal leaf miners by treatments. Differences among treatments are nonsignificant (parasitism, $F_{3,52}$ = 1.35, P = 0.27; predation, $F_{3,52}$ = 1.32, P = 0.28; natural enemy attack, $F_{3,52}$ = 0.29, P = 0.83).

These results do not support either the basic premise that induced responses protect trees from future herbivory or the extended IDNE hypothesis that plants use induced responses to reduce herbivory by enhancing attack by natural enemies of herbivores, despite that the first two requisites for the IDNE hypothesis are present. Folivory results in phytochemical (increased tannins), phenological (increased leaf abscission), and structural (jagged leaves) changes that have been shown in other studies to attract natural enemies. These changes, however, do not decrease survival of the leaf miner and do not decrease colonization or damage by leaf miners and leaf chewers. Folivory does not decrease future folivory but does reduce plant growth suggesting folivory reduces plant fitness. The intriguing notion that plants have evolved induced responses to manipulate the third trophic remains unsupported in natural settings where many factors affect dynamics of insect populations (Faeth, 1992).

References

Faeth, S.H. (1988). Plant-mediated interactions between seasonal herbivores: enough for evolution or co-evolution? In: K.C. Spencer (ed), *Chemical Mediation of Coevolution*, pp. 391-414. New York: Academic Press.

Faeth, S.H. (1992). Interspecific and intraspecific interactions via plant responses to folivory: an experimental field test. *Ecology* **73** (in press).

Turlings, T.C., J.H. Tumlinson & W.J. Lewis (1990). Exploitation of herbivore-induced plant odors by host-seeking parasitic wasps. *Science* **250**: 1251-1253.

Proc. 8th Int. Symp. Insect-Plant Relationships, Dordrecht: Kluwer Acad. Publ.
S.B.J. Menken, J.H. Visser & P. Harrewijn (eds), 1992

The role of plant cues in mediating host and food searching behavior of parasitoids: importance to biological control

W. Joe Lewis
USDA, ARS, Tifton, Georgia, USA

Key words: *Cotesia marginiventris, Microplitis croceipes*

Parasitic wasps of plant-feeding insects are important tools in the quest for safe and effective means of crop protection. The parasitic wasps rely on plants as sources of nectar and pollen as adult food as well as insect hosts. An understanding of the mechanisms that govern how the parasitoids locate their host and food resources is important to their reliable use. Wind tunnel and small field-plot studies with the larval parasitoids, *Microplitis croceipes* (Cresson) and *Cotesia marginiventris* (Cresson), have served as model systems to elucidate how chemical and visual cues from the plant aid the parasitoid in locating hosts and food (Tumlinson et al., 1992). The studies showed that plant chemicals induced by the host's feeding and by-products of the host derived from the plant are vital to the parasitoid's host-finding success (Lewis & Martin, 1990; Turlings et al., 1990). Visual cues associated with the host are also important (Wäckers et al., 1992). Moreover, learning by the parasitoid serves a crucial role in their ability to exploit these chemical and visual cues (Lewis & Tumlinson, 1988; Lewis et al., 1990; Vet et al., 1990).

Learned use of plant-related chemical and visual cues were found to be important also in the parasitoid's location of its adult food needs. Further, the parasitoid demonstrated a very sophisticated ability to associatively link arbitrary cues to successful location of host and food and subsequently use the cues based on current needs of hosts versus food (Lewis & Takasu, 1990).

These findings can be used to improve biological control by both improving the quality of the parasitoid and by enhancing the suitability of the target environment (Table I). Genetic, behavioral and physiological state qualities of laboratory-produced parasitoids

Table 1. Ways for enhancing biological control by parasitoids

I.	Parasitoid quality control	
	A.	Genetic attributes
	B.	Informational state
	C.	Physiological state
II.	Management of the environment	
	A.	Artificial addition of cues
		1. Host cues
		2. Plant cues
	B.	Optimize the inherent attributes of crop
		1. Genetic enhancement
		2. Agronomic management

can be improved through the use of this information as criteria for selection of founder colonies, design of rearing procedures, and by use of cues for pre-release conditioning (Lewis et al., 1990). Studies in small plots show that formulations of selected cues from the plant and/or hosts can be applied to the crop to lure the parasitoid and retain them in an effective behavioral state (Lewis & Martin, 1990). Traps, using the cues as baits, can be developed for monitoring the presence and density of parasitoids. Also, technology to breed and ergonomically manage plants in order to optimize their inherent attributes in support of the effectiveness of parasitoids are being explored.

References

Lewis, W.J. & W.R. Martin, Jr. (1990). Semiochemicals for use with parasitoids: status and future. *J. Chem. Ecol.* **16**: 3067-3089.

Lewis, W.J. & Keiji Takasu (1990). Use of learned odours by a parasitic wasp in accordance with host and food needs. *Nature* **348**: 635-636.

Lewis, W.J. & J.H. Tumlinson (1988). Host detection by chemically mediated associated learning in a parasitic wasp. *Nature* **331**: 257-259.

Lewis, W.J., L.E.M. Vet, J.H. Tumlinson, J.C. van Lenteren & D.R. Papaj (1990). Variations in parasitoid foraging behavior: essential element of a sound biological control theory. *Environ. Entomol.* **19**: 1183-1193.

Tumlinson, J.H., W.J. Lewis & L.E.M. Vet (1992). How beneficial wasps find plant feeding caterpillars. *Scient. Amer*, (in press).

Turlings, T.C.J., J.H. Tumlinson & W.J. Lewis (1990). Exploitation of herbivore-induced plant odors by host-seeking parasitic wasps. *Science* **250**: 1251-1253.

Vet, L.E.M., W.J. Lewis, D.R. Papaj & J.C. van Lenteren (1990). A variable-response model for parasitoid behavior. *J. Insect Behav.* **3**: 471-490.

Wäckers, F.L. & W.J. Lewis (1992). Olfactory and visual learning and their interaction in host site location by the parasitoid *Microplitis croceipes. Biol. Control* (in press).

Proc. 8th Int. Symp. Insect-Plant Relationships, Dordrecht: Kluwer Acad. Publ.
S.B.J. Menken, J.H. Visser & P. Harrewijn (eds), 1992

Systemic releases of volatiles by herbivore-damaged plants: what possible functions?

Ted C.J. Turlings, Philip J. McCall, Hans T. Alborn and James Tumlinson
USDA-ARS, Insect Attractants, Behavior, and Basic Biology Res. Lab., Gainesville, Florida, USA

Key words: Attractants, *Cotesia marginiventris*, parasitoids, plant defence, *Zea mays*

Several hours after caterpillars start feeding on corn seedlings the plants respond with the release of relatively large amounts of terpenoids. Parasitic wasps that lay their eggs in caterpillars exploit this plant response and use the volatile terpenoids to guide them into the vicinity of the caterpillars (Turlings *et al.*, 1990, 1991). In this paper we will first present results that indicate that the herbivore-injured plants indeed actively emit chemical signals. Subsequently, we will argue that such a signalling function most likely has evolved secondarily and that the emitted chemicals primarily function in the direct defence against herbivores and pathogens.

Systemic signalling

A specific response to herbivore-damage. The release of terpenoids by corn can not be induced with artificial damage (Turlings *et al.*, 1990). When we allowed beet armyworm (BAW) caterpillars to feed on corn seedlings for two hours, the following day (15 hours later) the seedlings released terpenoids in very significant amounts. While the BAW caterpillars were feeding on their seedlings the damage they inflicted in the plants was mimicked in another group of seedlings with razor blades and micro-scissors. These seedlings released terpenoids only in very minor amounts the following day. If, however, seedlings that underwent such artificial damage were treated with BAW regurgitate *i.e.*, caterpillar oral secretions, by rubbing it over the damaged sites just after damage was inflicted, then the following day they released terpenoids in amounts similar to the caterpillar-damaged seedlings. Caterpillar regurgitant had no effect when rubbed over undamaged leaves. The chemical differences that we observed in the plants which underwent these different treatments was reflected in the attractiveness of these plants to female *Cotesia marginiventris* (Cresson). In flight tunnel bio-assays, the wasps were far more attracted to the terpenoid releasing plants (Turlings *et al.*, 1990).

An active process. The response by the plants is not instantaneous: the terpenoid emissions reach significant amounts only several hours after damage (Turlings *et al.*, 1990). This already indicates that an internal process occurs in the plant that is triggered by the feeding caterpillars. This is also evidenced by the fact that the release of the volatiles is not limited to the damaged sites but occurs throughout the plant. Several hours after we damaged two of the leaves of seedlings, with three leaves significant amounts of terpenoids were released by not only the damaged leaves but also by the undamaged leaf. Control plants, unharmed seedlings that had been standing directly next to the treated seedlings, showed no changes in their volatile emissions. Again the plant responses corresponded with attractiveness to *C. marginiventris*. Several hours after damage, the wasps were more attracted to undamaged leaves from damaged plants than to leaves from control plants (Turlings & Tumlinson, unpubl.). This clearly demonstrates an internal process whereby chemical factors are transported throughout the plant and also to the surface of the leaves.

Caterpillar spit. To characterize the active factor in the regurgitate we developed a bio-assay in which plants could be induced to release terpenoids without actual surface damage to the leaves. Corn seedlings were cut from trays in which they had been growing in a greenhouse and were placed in vials with their cut stem submerged in BAW regurgitate (10 x diluted with water). Control seedlings were placed in vials with water only. The following day the plants in regurgitate released terpenoids in very significant amounts, while control plants showed no chances in volatile emissions. As expected from the above results, parasitoids were strongly attracted to the terpenoid releasing seedlings.

Conclusions

In our studies with corn, BAW and *C. marginiventris* we have found evidence for an elaborate tritrophic interaction. The above shows that when caterpillars feed on corn seedlings they cause the plant to actively and systemically emit volatiles that are highly attractive to the parasitoid. It is tempting to suggest communication between plant and parasitoid especially because the interaction benefits both parties. Such an interaction may have evolved secondarily and parasitoids and predators can be expected to add selection pressures that favour the induced releases of volatiles. However, we suspect that the interaction between herbivores and plants formed the basis for the production of terpenoids by plants. As several studies indicate (Turlings & Tumlinson, 1991), terpenoids are most likely defensive chemicals or by-products of such chemicals aimed at the herbivores themselves or at pathogens that may invade at the vulnerable damaged sites.

References

Turlings, T.C.J. & J.H. Tumlinson (1991). Do parasitoids use herbivore-induced chemical defenses to locate hosts? *Fla. Entomol.* **74**: 42-50.

Turlings, T.C.J., J.H. Tumlinson & W.J. Lewis (1990). Exploitation of herbivore-induced plant odors by host-seeking parasitic wasps. *Science* **250**: 1251-1253.

Turlings, T.C.J., J.H. Tumlinson, R.R. Heath, A.T. Proveaux & R.E. Doolittle (1991). Isolation and identification of allelochemicals that attract the larval parasitoid, *Cotesia marginiventris* (Cresson), to the microhabitat of one of its hosts. *J. Chem. Ecol.* **17**: 2235-2252.

Proc. 8th Int. Symp. Insect-Plant Relationships, Dordrecht: Kluwer Acad. Publ.
S.B.J. Menken, J.H. Visser & P. Harrewijn (eds), 1992

Natural enemy impact varies with host plant genotype

Robert S. Fritz
Dept of Biology, Vassar College, Poughkeepsie, New York, USA

Key words: Genetic variation, sawfly, tritrophic level, willow

The concept of Three Trophic Level interactions (Price *et al.*, 1980) states that interactions between members of two trophic levels will have effects on the third trophic level. Variation among host plants is predicted to affect traits of herbivores in ways that, in turn, influence the susceptibility of the herbivores to their natural enemies. Genetic variation among crop plant cultivars has been shown to affect natural enemy impact on herbivores, but the effect of plant genetic variation on natural enemy impact has seldom been demonstrated for natural populations of host plants, herbivores, and natural enemies. Previous work showed that survival and enemy impact on a gall-forming sawfly (*Phyllocolpa* sp.) varied significantly among potted clones of the willow (*Salix lasiolepis* Benth.) in Arizona (Fritz & Nobel, 1990). I tested the hypothesis that in a different willow-herbivore system, plant genetic variation would also have a significant effect on the susceptibility of two herbivore species to natural enemy impact. I also tested the hypothesis that densities of heterospecific herbivores are correlated with natural enemy impact.

Materials and methods

The hypothesis that plant genetic variation affects enemy impact was tested using potted clones and half sib progeny of silky willow, *Salix sericea* Marsh. & Vill., and two herbivores, a leaf mining moth, *Phyllonorycter salicifoliella* Chambers (Lepidoptera: Gracillariidae) and a *Phyllocolpa* species (Hymenoptera: Tenthredinidae). Clones (16) or half-sib progeny (13 sires) grown in pots for two years were placed in randomized block designs in the field near naturally occurring *S. sericea* plants during 1989, 1990, and 1991. After herbivores oviposited and developed to the pupal stage, densities of the herbivore were recorded and leaf mines or galls were collected and dissected in the lab to determine survivorship and causes of mortality. Data were analyzed using ANOVA or log-linear analyses to determine significant clone or sire effects on density, survival, and causes of mortality. Clone mean correlations were calculated using clone or sire means. Genetic correlations were calculated between survival, enemy impact, and conspecific and heterospecific densities for clone data.

Results and conclusions

Genetic variation in plant resistance was demonstrated by the significant variation in densities of *Phyllonorycter* and *Phyllocolpa* among clones and sires in 1989 and 1991. In 1990 there were no significant variation in density among sires. Densities of *Phyllonorycter* and *Phyllocolpa* varied by at least 3 fold among clones in 1989, and density of

Phyllonorycter varied by 8 fold among sire half-sib families in 1991. Estimates of heritabilities from half-sib analyses for resistance to *Phyllonorycter* ranged from 0.54 in 1989 to 0.14 in 1991. In 1991 heritability of resistance to *Phyllocolpa* was 0.24.

Survival and mortality caused by egg and larval parasitoids also differed significantly among clones and sires for both herbivore species, demonstrating genetic variation in natural enemy impact (Table 1). For *Phyllonorycter* larval parasitism by the wasp *Pholetesor salicifoliellae* Mason (Hymenoptera: Braconidae) varied significantly among clones and sires in 1989 and 1991, and total parasitism, including parasitism by chalcidoid species, varied significantly among sires in 1990. Sire half-sib family means for *Pholetesor* parasitism varied from about 3% to 13% in 1991. For *Phyllocolpa* spp., egg parasitism, presumably by a *Trichogramma* species, varied significantly among clones in 1989. Other causes of egg and larval death did not vary among clones or sires in any of these years.

Table 1. Tests of significance of plant genetic variation on survival and enemy impact for *Phyllocolpa* spp. and *Phyllonorycter salicifoliella* on cloned willow plants (1989) and on half-sib progenies (1990 and 1991). Analysis of variance was performed on angular transformed proportions on clones (df = 15, 164) and log-linear analyses were performed on data in 1990 and 1991. Death due to several other causes did not vary significantly among genotypes and is not shown

| *Phyllocolpa* spp. | | *Phyllonorycter salicifoliella* | | | |
	1989		1989	1990	1991
Source	F	Source	F	G	G
Survival	1.85*	Survival	1.93*	27.52**	23.12*
Total Egg Death	1.96*	Total parasitism	2.48*	24.66**	21.29*
Other Egg Death	1.61	*Pholetesor*	2.76***	17.47	29.99**
Egg Parasitism	1.95*	Chalcidoid	1.86	7.21	15.31

* = P < 0.05; ** = P < 0.01; *** = P < 0.001

Survival of *Phyllonorycter* was inversely density-dependent among individual plants (phenotypically) and clones in 1989 (R = -0.544, P < 0.05), but survival of *Phyllocolpa* was unrelated to its density phenotypically or clonally (R = -0.155, NS). For both, *Phyllonorycter* and *Phyllocolpa* survival was inversely related to parasitism by larval and egg parasitoids (R = -0.817, P < 0.001; R = -0.701, P < 0.01; respectively), but not with other causes of mortality, thus parasitism explained a large amount of the variation in survival among clones. When multiple regressions were performed with conspecific and heterospecific densities entered as independent variables, heterospecific density explained a significant portion of the variation in percentage parasitism for each species. Regression of residual *Phyllonorycter* parasitism among clones (after adjusting for intraspecific density-dependence) on density of *Phyllocolpa* among clones was highly significant (R = -0.733, P = 0.0012). Regression of residual *Phyllocolpa* parasitism on density of *Phyllonorycter* among clones was also significant (R = -0.526, P = 0.036). These data may suggest that either: (1) the presence of heterospecifics interferes with host finding by parasitoids of the other species, or (2) that genetic variation among clones confers susceptibility to one species but also confers some degree of protection from enemy impact for the other herbivore species (pleiotropy). One bit of evidence suggests that the latter explanation is more likely. Oviposition of *Phyllocolpa* and probably therefore egg parasitism occurs prior to oviposition by *Phyllonorycter*, so that leaf mines are not likely present when egg parasitism is taking place and thus no direct effect is possible.

These data strongly support the idea that host plant genetic variation can have a substantial effect on herbivore-natural enemy interactions in natural systems. The prediction from the Three Trophic Level concept that there is plant genetic variation in susceptibility of herbivores to their natural enemies was strongly supported for two herbivores on willow. There were apparent interactions between the two herbivore species that appeared to be mediated by natural enemies, but it is suggested that these are due to clonal correlations not to interactions between the herbivore and parasitoid species.

References

Fritz, R.S. & J. Nobel (1990). Host plant variation in mortality of the leaf-folding sawfly on the arroyo willow. *Ecol. Entomol.* **15**: 25-35.
Price, P.W., C.E. Bouton, P. Gross, B.A. McPheron, J.N. Thompson & A.E. Weis (1980). Interactions among three trophic levels: Influence of plants on interactions between insect herbivores and natural enemies. *Annu. Rev. Ecol. Syst.* **11**: 41-65.

ESF Workshops

Proc. 8th Int. Symp. Insect-Plant Relationships, Dordrecht: Kluwer Acad. Publ.
S.B.J. Menken, J.H. Visser & P. Harrewijn (eds), 1992

Chemistry of insect-plant interactions

Wittko Francke
Institut Organische Chemie der Universität, Hamburg, Germany

Chemistry plays a major role in insect-plant interactions. A workshop on the subject was held by the European Science Foundation (ESF) Network on Insect-Plant Interactions in Boldern (Switzerland), 4-7th November 1990. Since progress in this area of research needs a multidisciplinary approach, contributions in the forum of lectures and discussions came from both chemists and biologists. During the workshop, general questions concerning the contribution of chemistry were discussed, both with respect to the impact of independent research projects and service activities within a team.
- *Modern analytical techniques*:
Progress and limitations in the analysis of volatile and non-volatile compounds from insects and plants; behaviour-mediating substances like attractants and repellents as well as chemicals acting as stimulants or deterrents for oviposition and/or feeding; substances showing physiological effects.
- *Expectations and possibilities*:
Requirements for chemical support in the research on biological phenomena; synthesis of (enantiomerically) pure natural compounds and corresponding analogues.
- *Structures of interdisciplinary cooperation*:
Approaches of independent classical disciplines and overlapping areas in a field of mutual interest; studies of biosynthetic pathways in plants and sequestration or metabolic transformation in herbivores; formulation and application of plant and insect derived compounds in systems of integrated pest management.

Preliminaries

A prerequisite of the isolation or identification of any active compound involved in a biological phenomenon is a clear definition of the project and a reliable and sensitive bioassay. Electrophysiological experiments, including on-line techniques like EAD (*i.e.*, the use of insect antenna as a highly selective detector in gas chromatography) provide excellent tools to trace candidate components during the fractionation of crude natural material (Pickett), however, *in vivo* bioassays (behaviour tests) are indispensable (Simmonds). Finally, the activity of the candidate compounds needs always to be confirmed using synthetic compounds. Problems arise in testing thermodynamically unstable substances or mixtures which may change their proportions due to different volatilities of the components (Bergström).

Many active substances which play a role as chemical messengers are produced in particularly small amounts and are embedded as trace components in complex mixtures. The biological significance and ecological impact of mixtures of behaviourally active compounds is recognized, but only in a few examples of insect-plant relationships are the crucial mechanisms understood; the physiological background of synergism is almost unknown. Unfortunately many research projects concerning the influence of non-volatile compounds on host selection or on plant defense deal only with single (commercially available) compounds or with crude "natural" extracts which represent chemically less defined mixtures. The isolation and identification of semiochemicals represents a challenge even to laboratories of highest analytical standard. Sampling and separation techniques will largely depend on physicochemical properties of the target compounds, while identification procedures are governed by the amounts available and the data required for structure elucidation.

Isolation

Biologically active material from natural sources such as plants may be obtained by total extraction or from surface washings with organic solvents. Drawbacks of this rather crude method lie in the tremendous amount of extracted matrix components which may seriously hamper further separation steps. Biologically active compounds from insects can be obtained by direct extraction of specific tissues or glands at the production site. The use of selective solvents may largely facilitate the isolation procedure. Volatile compounds may be advantageously collected "on-site" by a variety of head space techniques (Bicchi). Thermodynamically stable, volatile substances represent the least problematic group of semiochemicals, since gas chromatographic methods may be applied. Gas chromatography (GC) is by far the most versatile and advanced separation technique. Using capillary columns, modern stationary phases provide extreme selectivity (including enantiomeric separation) which may be even increased by two-dimensional GC (Bicchi). Mild injection techniques like on-column, thermo-desorption or solid-sample treatment as well as specific detectors have been developed; on-line coupling with spectroscopic methods is well-established. Non-volatile compounds require purification conditions like high pressure liquid chromatography (HPLC), super critical fluid chromatography (SFC) and capillary electrophoresis. The latter technique is not only suitable for the separation of charged compounds of high molecular weight, but also for neutral low molecular weight substances (Van Beek). As compared to GC, methods using liquid mobile phases often suffer from limited efficiency in separation, less sensitivity at the detector site and difficulties in on-line coupling techniques. Methods for the handling and characterisation of minute amounts of mixtures of non-volatile compounds have been largely facilitated during recent years, but still need a lot of improvement (Nielsen).

Dealing with unstable compounds will always cause trouble, and techniques will have to be adjusted to the given problem: thiosulfinates which play an important role in the relations between *Allium* and their specialist insects, *Acrolepiopsis assectella* (Zell.) and its parasitoid wasp, *Diadromus pulchellus* (Wesm.), tend to degrade to a multitude of

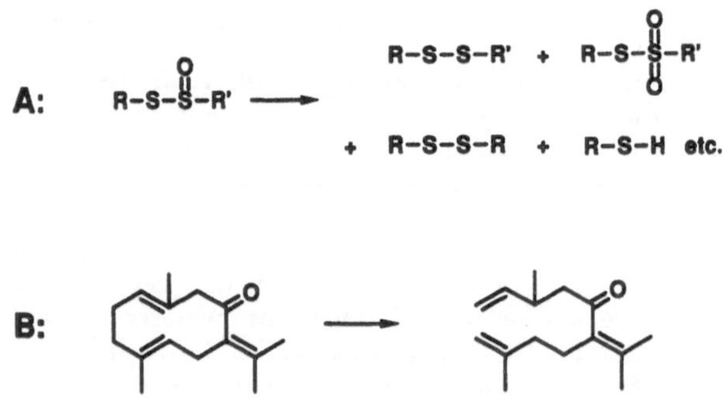

Figure 1. A: Thiosulfinates and their degradation products. B: Germacrone and its product or rearrangement, elemone.

374

compounds (Fig. 1A), and in pure state easily disproportionate into the stable disulfides and thiosulfonates (Auger). Another example is the formation of the terpene ketone, elemone, from germacrone (Fig. 1B) similar Cope-rearrangements are also known from other germacrene derivatives which has been identified as an antifungal component from the volatile oil of *Myrica gale* L., the production of which proved to be enhanced after attack by the sap-sucking capsid bug, *Amblystylus nasutus* (Kirschbaum) [Waterman & Carlton].

Identification

Detection of a target compound is usually followed by identification. Most significant data are provided by NMR-techniques, however, their application is limited because of their relatively low sensitivity. Even modern instruments require *ca* 1 µg for an almost pure sample for ^{1}H-NMR and 10-15 µg for a C-H correlation analysis. In contrast, on-line methods like GC/MS (including high resolution MS) or GC/FT-IR as well as two-dimensional gas chromatography coupled to tandem mass spectrometry (GC-GC/MS-MS) combine high selectivity at the separation site and high sensitivity (picogram to nanogram region) at the detector site (Tabacchi). Unfortunately the data obtained with these methods are often less definite in case entirely new compounds are involved.

It may be generalized that: The more significant the *a priori* available spectroscopic or chromatographic data are, which characterize a target compound (GC-retention index, MS-library, pronounced mass ion, FT-IR-library etc.) the smaller the amount needed for structure assignment. In contrast, the more specific the data which needs to be obtained for structure elucidation, the greater the amount of target compound required. In other words, components with registered mass spectra like 2-nonanol may be easily identified (including absolute configuration) even in a complex mixture at the picogram level, while structure elucidation of a new compound, especially if it has several chiral centers, may need mg-amounts of pure crystals for x-ray analysis.

Monitoring of already known components which are involved in plant physiology (biosyntheses, genetics, transportation within the plant, compartmentalisation, stress induced changes, daily and seasonal dynamics of secondary plant metabolism etc.) or in insect physiology and behaviour (sequestration and processing of plant constituents, storage and release of volatile and non-volatile compounds) may become routine (even at a subcellular level! - laser-TOF-mass spectrometry being a most promising method). The availability of rapid and reliable techniques for the analysis of large numbers of samples is still lacking, however, and would be very useful to many biologists who need to carry out statistical analyses. Autosamplers connected to conventional analytical systems will only partially solve the problem, however, radio immune assays and other biochemical methods may help.

Case studies

Besides technical developments and principles of interdisciplinary cooperation, the ESF-workshop covered recent results in the identification of substances from plants and insects, focusing on behaviour-mediating compounds, on aspects of defence, and on other physiologically active substances.

Due to the well-established methods in GC/MS, much more is known about the chemical structures of volatile compounds as compared to non-volatile ones. At least in part this is due to the tremendous input of the flavour industry (Bicchi). Advanced GC/MS-techniques provide an ideal tool for investigations on mechanisms in host selection and pollination.

Females of the European corn borer, *Ostrinia nubilalis* (Hübner) can distinguish between different corn hybrids by means of volatiles emitted by the leaves. It can be shown that the compounds act at a short distance and influence selection of oviposition sites. Effects of pattern recognition may be involved (Thiery & Marion-Poll). The same phenomenon plays a major role in pollination. Careful studies of various plant families revealed clusters of stereotypic volatiles which can be attributed to distinct chemical classes. Interesting differentiations of volatiles from different parts of the plants were observed (Bergström).

A. assectella and *Plutella xylostella* (L.), when feeding on *Allium* or other sources, metabolise sulfur containing compounds to the same three volatile disulfides, namely to the dimethyl-, dipropyl-, and methyl-propyl-derivative which stimulate the parasitic wasp, *D. pulchellus* [Auger]. Although the orientation of bark beetles towards monoterpene hydrocarbons which evaporate from host plants and the mechanisms of their conversion to corresponding oxygenated behaviourally-active compounds, is at least partly understood, almost nothing is known about the role of non-volatile compounds in host selection. Chemical communication among bark beetles appears to originate from detoxification mechanisms of plant compounds. Allylic oxidation of (toxic) monoterpene hydrocarbons is a widespread basic principle (Fig. 2A). Since autoxidation of monoterpene

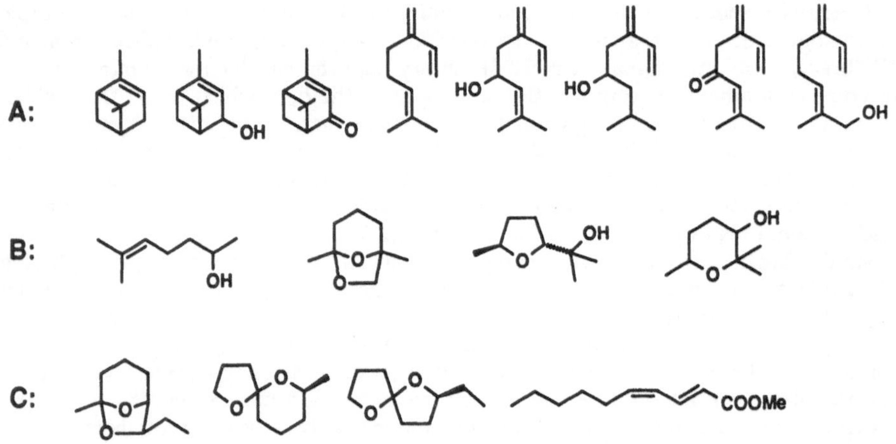

Figure 2. A: α-Pinene and myrcene and their products of allylic oxidation known as bark beetle semiochemicals. B: Oxygenated bis-nor-terpenoids known as bark beetle pheromones. C: Acetogenines known as bark beetle pheromones.

hydrocarbons also involves allylic oxidation, such communication channels are prone to jamming. More complex and intriguing systems rely on terpenoids which are produced in a sequence of several steps, thus offering more specificity (Fig. 2B). Some, perhaps highly developed systems largely avoid terpenoids and use acetogenines instead (Fig. 2C) which may even be superior (Francke).

High boiling, polar compounds like higher terpenoids, lipids and aromatics may play an important role in host selection either as attractants or repellents, as phagostimulants or antifeedants. Lignans consist of two phenylpropanoid units linked at C-2 of the side chain. The three basic structures, dimethyl-1,4-diphenylbutane, dimethyl-1-phenyltetralene and dimethyldibenzocyclooctene may be differently oxygenated yielding a variety of

376

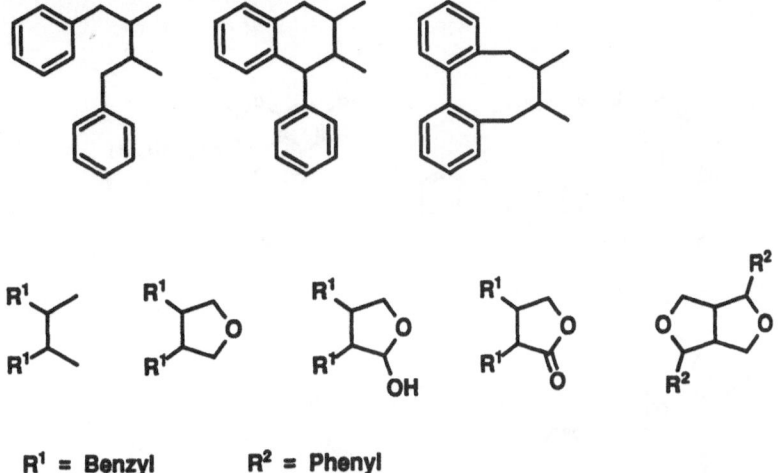

R¹ = Benzyl   R² = Phenyl

Figure 3. Principle types and structures of lignans.

tetrahydrofurane derivatives (Fig. 3). Some of these compounds show striking antifeedant activity either per se or as synergists. Most of the more simple lignans are easily prepared by chemical synthesis or through biotechnical methods. They comprise a most promising class of candidates for pest management (Harmatha).

Chemical aspects of the interactions between grape vine and the pathogenic fungus *Eutypa lata* involve a series of disubstituted phenols or phenol ethers as fungal metabolites among which 3-(3-methyl-1-butyn-3-enyl)-4-hydroxy benzaldehyde (Eutypine) shows phytotoxic activity (Tabacchi). The previously mentioned feeding of the sap-sucking capsid bug *A. nasutus* on sweet gale causes a local necrosis in the leaf and enhances the production of the volatile antifungal germacrone and induces *de novo* synthesis of the non-volatile flavonoid, kaempferol-3-rhamnopyranoside shown in Fig. 4; 1. These changes do not appear to deter the herbivore, but they do enhance the chemical defence of the leaf against secondary invasion by fungi (Waterman & Carlton).

Flavonoid glycosides like kaempferol-3-lathyroside (Fig. 4; 2) from horseradish stimulate feeding in the monophagous horseradish flea beetle, *Phyllotreta armoraciae* (Koch), but not in the related oligophagous *P. nemorum* L. It is suggested that a chemical profile containing a limited number of compounds from the host plant may allow crucifer feeding insects to discriminate between host and non-host plants (Nielsen). Glycosinolates and their isothiocyanate metabolites present in crucifers (Fig. 4; 3) may be decisive components of such profiles. According to Pickett, the modification of such profiles may lead to selective pest management systems: the aim would be to reduce glucosinolates (and corresponding isothiocyanates) in the vegetative parts of the plants which cause attraction of the adapted pests and to increase those compounds which reduce disease, development or have an antifeedant effect. The overall content of glycosinolates should, however, be kept low in the seed for nutritional and commercial purposes. Reducing the attractiveness or acceptability of a crop may cause aggregation of the pest on discard or trap crop areas where they can be destroyed: within single plants the growing tip and seed-producing part of the plant can be protected with antifeedants or inhibitors, while

377

attractants applied to the lower leaves lure insects down to places where they can be

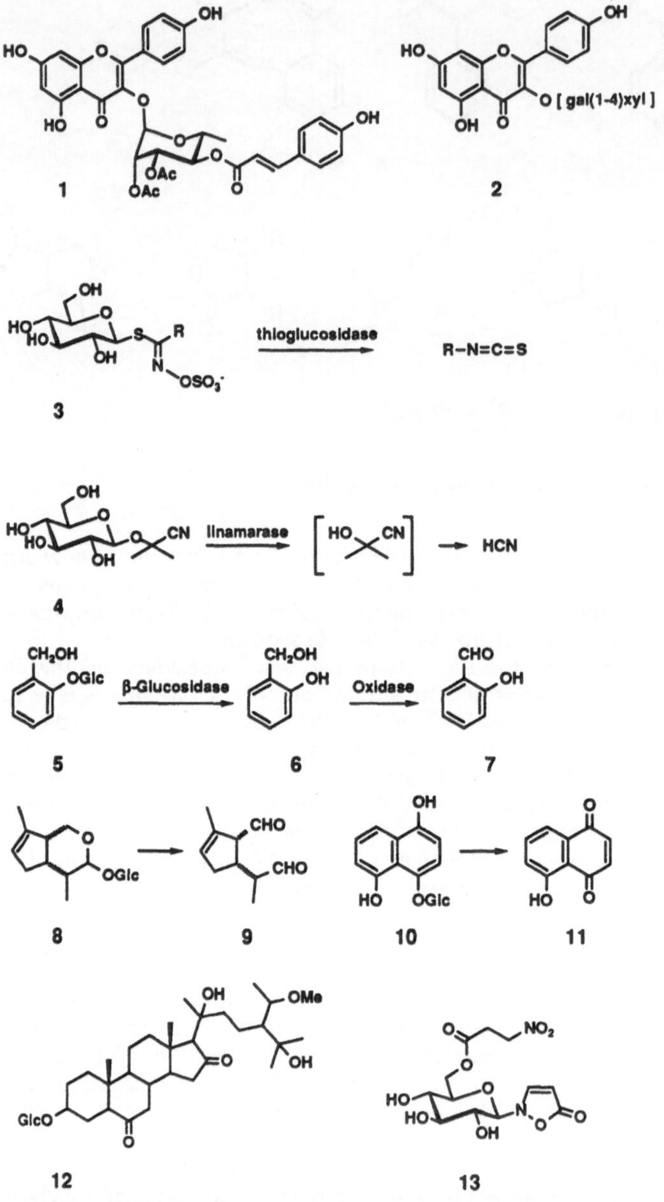

Figure 4. Glycosides and their degradation products. 1: Kaempferol-3-rhamnopyranoside, 2: Kaempferol-3-latyroside, 3: Glucosinolates, 4: Linamarin, 5: Salicin, 6: Saligenin, 7: Salicylic aldehyde, 8: Iridoid glucoside, 9: Iridodial, 10: Naphthol glucoside, 11: Juglone, 12: Cardenolide, 13: 6-Acyl-isoxazolinone glucoside.

successfully attacked by biological control agents like fungal pathogens (Pickett).

Though antifeedants certainly will be valuable components in pest management systems, careful investigations are required with respect to proper application. Salicin (Fig. 4; 5) which is a potent antifeedant against larvae of *P. xylostella* was much more effective when applied to sucrose-treated glass-fiber discs and discs from turnip leaves than when applied to discs from cabbage leaves. This effect did not result simply from a preference for cabbages, but was due to changes in the salicin mediated response to phagostimulants present in the surface of leaves (most humans do not taste the bitter acetylsalicylic acid when it is taken together with vitamin C!). This experiment illustrates the fact that the mode of action of behaviour modifying compounds is, unfortunately, still largely unknown (Simmonds & Blaney).

Glycosides represent a widespread type of storage form of lower boiling semiochemicals. While salicin may act as a plant antifeedant (see above), after sequestration by herbivores, it may be enzymatically cleaved to yield salicylic alcohol, which subsequently may be oxidized to the volatile insect defence compound, salicylic aldehyde (Fig. 4; 7). Other glucosides may as well represent cryptic volatile repellents which keep enemies away. Examples are found both among iridoids and quinones which were found in leaf beetles. *De novo* produced cardenolides (Fig. 4; 12) or isoxazolinone (Fig. 4; 13) glycosides (the agyclon being formed from aspartate) are among the "chemical defences" of several leaf beetle species, Chrysolinina and Chrysomelina, respectively (Fig. 4). In only one species of Chrysolinina, *Oreina cacaliae* (Shrank) an influence of the host plant on the chemistry of the defence gland has been demonstrated. This species sequesters pyrrolizidine alkaloids from its host plants (Asteraceae) and has apparently lost the ability to produce cardenolides (Daloze).

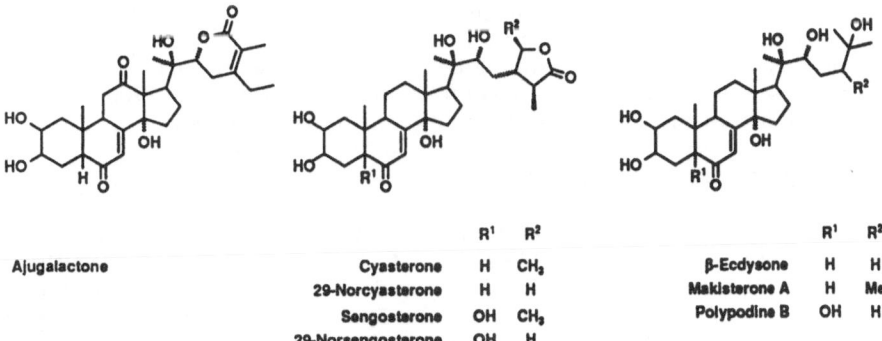

Figure 5. Clerodan diterpenoids and steroids showing antifeedant activity.

379

Glycosides as biologically active storage forms of defence compounds also occur as cyano-glucosides. Cyanogenesis (release of HCN under certain conditions) is typical for some higher plants, however, it was also observed in some Coleoptera, Heteroptera and Lepidoptera. For example *Zygaena trifolii* (Esper) is able to sequester and synthesize *de novo* linamarin (Fig. 4; 4) and lotaustralin. These compounds are the glucosides of acetone cyanhydrine and butanone cyanhydrine, respectively. The aglycons of these compounds clearly originate from the amino acids, valine and isoleucine. These amino acids may also be precursors of 2-methyl propionitril and its homologue, 2-methyl butyronitril, which upon the action of a multifunctional oxidase would produce the corresponding α-hydroxynitrils, *i.e.*, cyanohydrins, which spontaneously decompose to yield HCN (Nahrstedt).

Compounds which show antifeedant activities could be isolated from *Ajuga* species (Camps). Some members of the large group of clerodan diterpenoids (Fig. 5) exhibited activities at 0.3 ppm. Several phytoectosteroids have been isolated from *Ajuga reptans* L. including the toxic compounds ajugalactone and 29-norsengosterone (Fig. 5).

Pyrrolizidine alkaloids (PAs) are generally regarded to serve as protective chemicals in plants. In *Senecio* (Asteraceae), the roots were found to be exclusive sites of PA syntheses. They generally are produced as N-oxides which are subsequently transported to the shoots. Larvae of the Arctiid moth *Tyria jacobaeae* (L.) sequester PAs from their host plant, *Senecio jacobaea* (L.). Interestingly, *Tyria* was shown to produce the "insect PA" callimorphin from retronecine by reesterification with a necic acid of insect origin. Similarly, adults of the arctiid *Creatonotos transiens* (Walker) contain two PAs which may represent products of reesterification of retronecine and an acid which most likely is derived from iso-leucine (Fig. 6). These compounds which are present as N-oxides are the

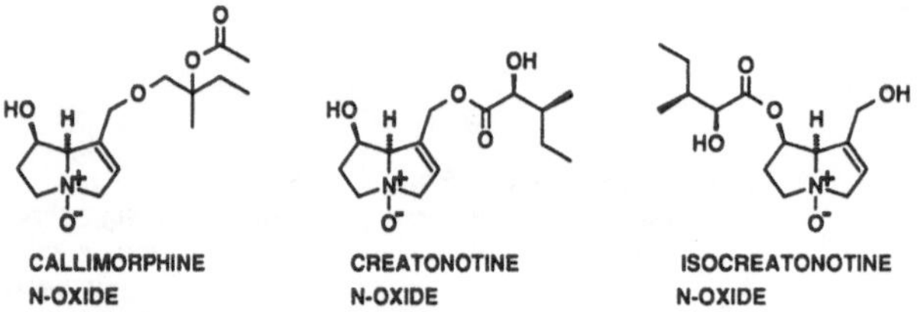

CALLIMORPHINE N-OXIDE   CREATONOTINE N-OXIDE   ISOCREATONOTINE N-OXIDE

Figure 6. "Insect pyrrolizidine alkaloids" made up by "plant amines" and "insect acids".

first examples of such processing of PAs by an insect: a necine base originating from the plant is esterified with a necic acid produced by the insect (Hartmann). While the N-oxides of senecionen and seneciphyllin are sequestered by the specialist, *O. cacaliae* (see above) - the compounds show the same proportions both in the beetle and in the plant - the generalist snail *Arianta arbustorum* L.) is strongly deterred by the N-oxides. Differing widely in their response to one specific group of plant secondary compounds, the snail may exert a selective pressure on the plant to maintain the presence of PAs, whereas the specialist leaf beetles might act in the reverse direction (Rowell-Rahier).

General conclusions and recommendations

The workshop showed that efforts in the identification of polar, non-volatile substances should be given special attention. More needs to be known about the chemistry and physics of plant surfaces and the role they play in host selection behaviour. Structure elucidation of secondary plant metabolites like lipids, alkaloids or phenols, flavones, terpenes and corresponding glycosides is prerequisite to understand the impact of such compounds and mixtures on feeding behaviour, habitat selection, development and reproductive success of herbivores. Only interdisciplinary cooperation will be succesful in advancing our knowledge about insect-plant interactions. Sophisticated methods in structure elucidation require expensive instrumentation and specific expertise which underlines the importance of specialisation and cooperation, especially between chemists and biologists. In a joint approach the biologically orientated laboratory will take the lead in carrying out bioassays and often also in isolation procedures, whereas gathering qualitative and quantitative data on candidate compounds from insects and plants may represent an overlapping area. Final purification, structure elucidation and synthesis of new structures is typically a matter of chemists. Supply with sufficient amounts of pure (optically active) test samples requires substantial input of preparative chemistry. More chemists should be encouraged to be active in this field - and at the universities more biology should be taught to chemistry students during their basic studies. Biologists and chemists will have to learn to understand each other better and join to continue in problem-solving - even if the solution turns out to be the pH-value or glucose.

Papers presented at Workshop

J. Auger: Artefacts during analysis of substances perceived by insects: the case of *Allium* volatiles.

T. van Beek: Scope and limitations of currently available chromatographic and spectroscopic methods for isolation and identification of biologically active substances.

G. Bergström: Chemistry of pollination.

C. Bicchi: Direct capture of volatiles from living plants.

F. Camps: Allelochemicals from *Ajuga* plants.

D. Daloze: Host plant influence on leaf beetles glandular defences.

W. Francke: Insect-host relationships in the chemistry of bark beetle pheromones.

J. Harmatha: Possible role of lignans in the chemistry of insect-plant interactions.

Th. Hartmann: The pyrrolizidine alkaloids: chemical and biological aspects.

A. Nahrstedt: Cyanogenesis in Lepidoptera.

G.K. Nielsen: Separation of plant compounds for subsequent use in insect feeding assays.

J.A. Pickett: Exploiting secondary plant metabolism by strategy chemical identification.

M. Rowell-Rahier: Plant PAs and their interactions with a specialist and a generalist herbivore.

M. Simmonds: Surface extracts of plants influence the potency of insect antifeedants.

R. Tabacchi: Chemistry of interactions between grape vines and *Eutypa lata*.

D. Thiéry: Are corn hybrids volatiles discriminated by ovipositing European corn borer females?

P.E. Waterman: Insect herbivory includes production of antifungal volatile oil and flavonoid compounds in sweet gale (*Myrica gale*).

Proc. 8th Int. Symp. Insect-Plant Relationships, Dordrecht: Kluwer Acad. Publ.
S.B.J. Menken, J.H. Visser & P. Harrewijn (eds), 1992

Insect behaviour

M.S.J. Simmonds [1] and F. Camps[2]
[1] *Jodrell Laboratory, Royal Botanic Gardens, Kew, UK*
[2] *Dept Biological Organic Chemistry, CID, CSIC, Barcelona, Spain*

The European Science Foundation Network on Insect-Plant Interactions held a workshop on Insect Behaviour at Sitges (2-5th March 1991). The talks and resulting discussions at this workshop provided a review of our knowledge about the basic principles governing the behaviour of phytophagous insects in relation to physiology, ethology and ecology. Areas of research that would benefit from further study were then identified and the participants discussed how collaborative projects utilising the diverse skills at various laboratories within Europe might be optimised.

Physiology

Volatile compounds are thought to be of primary importance in habitat and host selection behaviour for many mobile insects. The early electrophysiological studies on olfactory sensilla made use of commercially available volatile compounds known to be present in plants. The results from these studies indicated that the antennae of phytophagous insects have chemoreceptors, each of which responds to many plant odours. These chemoreceptors became known as "generalist plant odour receptors" (Visser). However, recent studies have shown that some of these chemoreceptors are responsive only to a narrow range of compounds. Often these compounds occur at very low concentrations and it is only now, with the use of single cell electrophysiological recordings coupled to a gas chromatograph, that the isolation and identification of these minor volatile components has been achieved and their importance recognised (Wadhams; Wibe & Mustaparta). This electrophysiological-chemical bioassay has been used successfully to isolate compounds involved in the host selection behaviour of aphids (Wadhams) and bark beetles (Wibe & Mustaparta). However, although this technique might be of use in identifying which compounds stimulate receptors it does not enable us to take into account the possible interactions of compounds as they occur in complex mixtures.

Normally more than one compound is involved in initiating a behavioural response from an insect, therefore it is important to know not only the chemical composition of a mixture but how the insects perceive mixtures. Difficulties in isolating compounds from mixtures were discussed. During the chromatographical fractionation of a plant extract the mixture of compounds in each fraction becomes simpler, and compounds that interact synergistically or additively can be separated. This separation can result in the total loss of biological activity. The removal of deterrent substances from a mixture could also obscure the loss of a stimulating substance and, if the behavioural bioassay being used to identify the active compounds is not able to detect both deterrent and stimulatory activity, it can be difficult to quantify the importance of the different compounds in the original plant extract (Roessingh).

When neurones in the peripheral taste sensilla are stimulated by plant compounds they send quantitative and qualitative information to the insect's central nervous system. This information can influence the initiation of a specific behavioural activity. The nature of this activity has been shown to depend on which neurones have been stimulated, on the developmental stage of the insect and on the compounds involved (Simmonds & Blaney). Electrophysiological investigations of chemosensory sensilla have shown that some compounds that deter feeding stimulate a neurone that has become known as the "deterrent neurone", whereas other compounds that promote feeding, stimulate other neurones. By differential adaptation experiments it has been possible to show that many deterrent compounds, although differing greatly in chemical structure, can stimulate responses from the same neurone. These experiments have also shown that in some cases deterrent compounds inhibit neurones that respond to sugars and salts (White et al.; Simmonds & Blaney). Knowing the specificity of neurones in the taste sensilla of insects, can be of use when attempting to ascribe a behavioural activity to compounds present in a mixture. Such information can also be used to investigate whether compounds interact and whether such interactions affect the behaviour of insects (Roessingh; White et al.; Simmonds & Blaney).

In order to understand more about the relationship between the neural input to the central nervous system from these sensilla and a specific behaviour, neural responses are often correlated with various aspects of that behaviour. For example, with feeding behaviour, the neural response from the taste sensilla on the mouthparts of larvae to a specific compound or mixture have been correlated with the probability of starting a meal, the duration of a meal, the percentage of time spent feeding and the number of faecal pellets produced during a known period of time. The neural responses used in this type of correlation have often been obtained from either one neurone (labelled line) or from more than one neurone in one or two sensilla (across-fibre patterning).

Although this neural input can be statistically correlated with a specific behaviour we cannot infer that the relationship is causal. Neither, should we conclude that the relationship is meaningless if there is no statistical correlation (Simmonds & Blaney; Visser). Despite these problems, correlating various aspects of neural activity with behaviour has enabled, not only the specificity of various neurones to plant compounds to be established but also the characterisation of different neural codes. For example, one particular neural code which is usually associated with rejection behaviour involves the neural interaction between neurones responding to stimulants and neurones responding to deterrents. So far the neural codes associated with acceptance behaviour have been found to be simpler than those associated with rejection behaviour (Simmonds & Blaney; White et al.; Roessingh).

Another level of complexity that has to be taken into consideration when studying the relationships between neural input and behaviour is that these relationships are not static but appear to be modulated by the insect's physiological condition and past experience (Simmonds & Blaney). Changes in neural responsiveness to nutrients can be mediated by depriving the insect of specific nutrients, whereas previous exposure to a deterrent compound such as an antifeedant can result in either a decrease or increase in neural responsiveness to that compound, which in turn relates to changes in behaviour. Thus, while it is possible to explain some changes in insect behaviour by changes in neural input, it still remains very difficult to predict a behavioural response from a set of electrophysiological data. In order to clarify the relationship between the neural input and behaviour it is imperative that we known more about the processing of neural information by the central nervous system (Visser; Mustaparta).

Ethology

Observations in the laboratory and field indicate that polyphagous insects are able to oviposit on a greater range of plants than oligophagous or monophagous species. Generally, the more specific the insect, the greater the range of plant compounds that will deter it from ovipositing or feeding. However, some polyphagous species including *Ostrinia nubilalis* (Hübner) show very precise preference behaviour but the exact compounds involved in mediating this preference behaviour are unknown (Derridj). The adults appear to land randomly on plants, which suggests that volatile compounds do not play an important part in the preference behaviour. The chemical information obtained by the adult when in contact with the phylloplane appears to be important in influencing its behaviour. Studies have shown that nutrients in the form of soluble carbohydrates and free amino acids are present on the phylloplane of plants. The proportion of these nutrients on the phylloplane varies between species of plants and between leaves on a plant. This profile of nutrients as well as of organic acids could characterise the physiological state of a leaf and be used in the host selection behaviour of *O. nubilalis* (Derridj). The sensilla on the tarsi and ovipositor of *O. nubilalis* are very responsive to soluble carbohydrates, such as sucrose, fructose and glucose, which occur in high concentrations on chosen leaves (Derridj).

Compounds on the phylloplane are also important in the oviposition behaviour of the leek moth, *Acrolepiopsis assectella* (Zell.) which lays eggs on leeks and onions (Thibout). The volatile sulphur compounds which are characteristic of species of *Allium* L. are weak stimulants for this insect. The precursor sulphur amino acid, S-propyl-cysteine sulphoxide occurs on the surface of leaves and stems and this amino acid, along with other compounds in the epicuticular wax, could be important in the modulation of the oviposition behaviour of *A. assectella* [Thibout].

Another complexity that needs to be taken into account when studying oviposition behaviour is that many insect species deposit compounds which can deter conspecifics from oviposition on the same plant (Thiery). These compounds, known as oviposition deterrent pheromones (ODP), are thought to play a major role in the dispersal of eggs by conspecifics. Many species of insects produce ODP but there is little electrophysiological information available to show whether these compounds modulate the neural input associated with the plant compounds present on the host plant.

The secondary compounds in plants can also provide the insect with the building blocks with which to make oviposition deterrent pheromones, intraspecific pheromones or toxic defensive compounds (Boppré). In some cases an insect's fitness can depend on the ingestion of specific secondary compounds and the insect will select plants that contain these compounds. In some "pharmacophagous" insects the intake of these secondary compounds, for example, pyrrolizidine alkaloids, is thought to be independent of the uptake of nutrients (Boppré).

Ecology

The continua of interactions between insects and plants at the ecological level and the diverse effects these interactions can have on insect behaviour in natural ecosystems and agro-ecosystems are poorly documented (Visser). The selective pressures that influence insect behaviour in these two systems may be partially determined by the stability of the interactions between the phytophagous insects and plants within each ecosystem and by the role of parasitoids and predators. The influence these pressures have on an insect species will be determined both by the genetic variation within the population and by the plasticity of this variation.

The elucidation of the role of secondary compounds in the evolution and ecology of insect-plant interactions often benefits from an in-depth study of the relationship between endemic plants and their adapted insects. The Canary Islands have a Tertiary flora which includes plants that are known to be resistant to many species of insect (González Coloma). The Canarian laurel forest consists of plants that contain diterpenes that are toxic to both endemic insects (*Macaronesia fortunata*) and crop pests (*Helicoverpa armigera* (Hübner)). However, some insects are able to feed on these plants but nothing is known about their behaviour and ecology (González Coloma).

Very little is known about the genetics involved in the evolution of host specialisation. Minor genetic changes are known to have a profound effect on behaviour and a change in a gene that determines the receptor proteins involved in the perception of host chemicals may result in a change in host perception and thus influence the ecology of the insect.

The behaviour of an insect is also influenced by its fitness, which in turn is influenced by the quality of the resources obtained from the plants that the insect is associated with in a specific habitat. Plant resource quality has been shown to play a role in the population dynamics of insects such as aphids. Aphids, like many other insects, respond not only to the nutrients within plants but also to the levels of secondary plant compounds. For example, hydroxamic acids, specifically DIMBOA could be involved in the resistance of winter cereals to attack by aphids in the autumn (Castañera). However, levels of DIMBOA decline rapidly as the plant matures. Therefore, other mechanisms must be involved in the resistance exhibited by older plants. Studies have shown that alpha-amylase inhibitors could be involved in the resistance of mature plants to aphids and different classes of inhibitors have been isolated from *Triticum aestivum* L. The isolation of proteins that confer resistance to a plant has far reaching implications for the development of transgenic plants. Some of these proteins have already been introduced into plants but their influences on insect behaviour at the physiological and ecological level are not yet known.

Conclusions

The patterns of behaviour exhibited by insects in the various stages of insect-plant interactions (habitat location, host location, host selection and host acceptance) are diverse and vary in complexity depending on the life style of the insect. At the moment there is very little information about the mechanisms which determine when one type of behaviour ends and another begins or about which endogenous factors modulate insect behaviour. Not only is it important to identify which factors influence behaviour but, in order to discuss the relevance of a change in behaviour, it must be possible to define the behaviour clearly and accurately, and to quantify how it has changed. Thus it is important that more care is taken in defining the different behaviours involved in insect-plant interactions. Clarity in describing a specific behaviour becomes very critical when comparisons are made between species.

Many compounds in plants are thought to play a crucial role in influencing insect behaviour. Volatile plant compounds provide insects with information not only about which plants are present in a habitat but whether a plant is in flower, pollinated or damaged. Insects are able to detect slight changes in the composition and concentration of compounds in the vicinity of a plant and this information enables them to assess the suitability of the plant. Despite our knowledge about the volatile compounds that have been isolated from plants, their precise role in the hierarchy of behavioural steps associated with host selection remains in most cases unsubstantiated. For example, many volatile plant compounds have been cited as being important in host selection, although relatively little is known about the role of these compounds once the insect has alighted on the plant and is exposed to the non-volatile "contact" compounds on the plant surface.

Recommendations

Overall, it was felt that priority should be given to increasing our basic knowledge about how insects perceive plant compounds, how the insect's central nervous system uses the neural information obtained when the peripheral chemoreceptors are stimulated with plant-derived compounds or mixtures from host and non-host plants. This would enable us to determine how specific the host-selection behaviour is and whether variability in plant chemistry is important in influencing insect behaviour. Insect-plant interactions are not static and insects can modulate their behaviour, they can on the one hand learn to avoid plants or on the other hand selectively seek out plants. These phenomena need further study. Other specific questions were identified:

a) What is the role of the phylloplane in insect host selection?
b) What genes are implicated in insect behaviour?
c) What is the effect of the plant genotype on the expression of behaviour?
d) What are the direct and indirect effects of parasitoids and predators on the behaviour of phytophagous insects?

It was felt that no one insect could serve as a model for elucidating the behavioural mechanisms involved in insect-plant interactions. Instead a series of "models" were proposed and it was suggested that comparative studies involving specialist and generalist insects from different families might help identify the importance of factors that influence insect behaviour.

Papers presented at Workshop

M. Boppré: Pharmacophagous insect behaviour and non-nutritional herbivory.

P. Castañera, C. Gutierrez, R. Sánchez-Monge & G. Salcedo: Evidence of a successive chemical defense strategy of certain cereals to insect pests.

S. Derridj: Role of leachates on insect behaviour, with an example concerning oviposition preference.

A. González Coloma: Chemical ecology in the Canary Islands: importance, problems and solutions.

H. Mustaparta: Receptor neurons: specificity, sensitivity and responses to mixtures in interpreting biological significance of odour cues.

P. Roessingh: Identification of bioactive plant compounds: problems and possibilities of behavioural and electrophysiological assays.

M.S.J. Simmonds & W.M. Blaney: Gustatory codes: does neural input explain feeding behaviour?

E. Thibout: Oviposition behaviour of moths and plant allelochemicals.

D. Thiéry: Some semiochemicals that modify the oviposition behaviour in a phytophagous insect.

L. Wadhams: The role of electrophysiology in understanding chemically-mediated behaviour in insects.

P. White, R.F. Chapman & A. Ascoli-Christensen: Tarsal responses to deterrents: relating behaviour to electrophysiology.

A. Wibe & H. Mustaparta: Linked GC-single cell recordings in identification of host odours for the pine weevil (*Hylobius abietis*).

J.H. Visser: How to interpret insect behaviour.

Proc. 8th Int. Symp. Insect-Plant Relationships, Dordrecht: Kluwer Acad. Publ.
S.B.J. Menken, J.H. Visser & P. Harrewijn (eds), 1992

Variability in insect-plant interactions

W.M. Blaney[1] and M.S.J. Simmonds[2]
[1] *Dept of Biology, Birkbeck College, London, UK*
[2] *Jodrell Laboratory, Royal Botanic Gardens, Kew, Richmond, Surrey, UK*

A workshop on this topic was held by the European Science Foundation (ESF) Network on Insect-Plant Interactions in Budapest, 25-26th August 1991. Discussions addressed the causes and effects of differences occurring between individual organisms, or within the same organism at different times, whether attributable to known external or internal causative factors or not.

It is widely accepted that the majority of phytophagous insects have a very restricted host range, that is they are specialists. This specialisation persists despite the fact that the attributes involved in the maintenance of insect-plant relationships are very variable, both in the insects and in their host plants (Jermy). How then do so many associations between plants and their specialised herbivores remain so constant over time and space in nature?

Two processes which are critically important in determining specific interactions, namely plant recognition and plant acceptance, depend largely on the chemical profile of the plant and the sensory system of the insect. Neither of these aspects of the relationship is reliably constant and this prompts us to ask the following questions.

1. How variable is the chemical profile of the plant and for what reasons does it vary?
2. How constant is the sensory system of the insect and what causes it to change?
3. How do these two factors interact to keep the plant profile within the range of acceptability to the insect?
4. How does the insect compensate for the imperfect plant and how does the plant compensate for herbivory?
5. To what extent is the variability of both partners genetically determined?

Most laboratory investigations of insect-plant interactions have been designed to eliminate variability and to minimise its consequences. The aim has been to standardise the biological material and to dissect its complexity by pursuing reductionist models which address one aspect of the relationship at a time. This approach has yielded much information, for example, on the sensory and behavioural responses of insects to relatively simple solutions, and has allowed the identification of key allelochemicals, and in some cases the characterisation of receptor types. Between these sharply focused studies on the one hand, and our understanding of the ecology and evolution of insect-plant relationships on the other hand, lies an imperfectly understood area characterised by variability and complexity of its elements and of the interactions between them. The mechanisms involved and the principles governing their operation are quite well known, as a consequence of the reductionist approach, but the limits within which they operate and interact are not well known. During this workshop some examination was made of the key aspects of the insect-plant relationship, with particular emphasis on variability and its consequences.

Plant variability

The distribution and concentration of allelochemicals in plants is of outstanding ecological significance, but information about this is known mainly from cultivated plants. Little is known about variability in the occurrence of allelochemicals in wild plants, and especially little is known about the effect of that variability on insect-plant interactions. However, such variability does occur, due to both genetic and environmental factors (Van der Meijden). Investigations of the chemistry of *Solanum dulcamara* L. in Europe have revealed that the three dominant alkaloids tomatidentol, soladulcidine and solasodine, vary in concentration in plants from different geographical and genetic backgrounds (Máthé). In the vegetative parts of the plant, tomatidenol is the principle alkaloid in plants from Western Europe while soladulicidine predominates in plants from Eastern Europe. The significance of these differences for phytophagous insects associated with the plants is not known. Similarly, investigations of floral scent in the orchid genus *Platanthera* Rich. has revealed variation between individuals and populations (Tollsten). The main source of variation is differences in the occurrence of linalool or lilac aldehydes and alcohols. This scent variation could result from selection due to pollination preferences.

Aspects of the physical environment can influence the relationship between a plant and its insect predators, as in the case of the fruit fly *Urophora cardui* (L.) which produces large multilocular galls on the creeping thistle *Cirsium arvense* (L.) Scop. [Zwölfer]. Shade and moisture are particularly important in influencing "gall quality" and survival of the fly. The Labiate *Ajuga reptans* L. produces phytoecdysterones which are deterrent to insects. A study which looked at the ratio of C28 to C29 phytoecdysterones, representing over 90% of the total content, found that the ratio was less than 1 in leaves of wild plants but ranged from 3 to 5 in greenhouse plants (Camps). In micro-propagated plants of the same species the levels of phytoecdysteroids in leaves were very low, whereas the levels in roots were higher than in any other plants tested. Callus cultures from leaves lost their ability to produce phytoecdysteroids. The effect of these chemical variations on potential predatory insects is unknown.

Insect variability

The reductionist approach referred to earlier has acted to minimise the expression of variability in studies of chemosensory physiology in phytophagous insects. Such variability was implicitly regarded as a manifestation of some unknown and undesirable malfunction and it is only in recent years that systematic studies have been undertaken to quantify this variability and assess its significance for the sensory coding, especially of taste quality (Schoonhoven). Most studies have involved lepidopteran larvae and have focused on the maxillary sensilla styloconica which have been shown to be critically important in the host plant selection process.

The two closely related species of Lepidoptera *Papilio machaon* L. and *P. hospiton* Guenée differ in that the former is oligophagous and the latter monophagous. Electrophysiological recordings of the responses of the maxillary sensilla styloconica to saps of various host plants reveal differences in the response of comparable neurones in the two species (Crnjar). These differences are greater when larvae of the two species have been reared on different food plants than when they are reared on the same plants. Similar differences were found between larvae of *P. machaon* when they had been fed on different host plants. Thus, variability in host plant chemistry can influence the way in which insects reared on these plants assess the chemical characteristics of the plants.

Similar, more detailed studies on larvae of *Spodoptera littoralis* (Boisduval) reveal that feeding history and diet influence the responsiveness of peripheral chemosensory neurones both to nutrients and to allelochemicals (Blaney & Simmonds). The response to nicotine in a neurone coding for deterrence is higher in last stadium larvae which have never encountered nicotine than in similar larvae fed throughout the larval period on a diet containing small quantities of nicotine. The overall responsiveness of the maxillary sensilla styloconica is low at the beginning and end of a stadium and high in the middle. This pattern of responsiveness corresponds closely with the pattern of food intake during the stadium. Thus differences in behavioural responses of insects to plants are influenced by changes in the insect's perception of the plant as well as by changes in the insect's central neural processing of information.

Insect-plant interactions

Many of the ecological theories about insect-plant interactions are short lived (Larsson). This is because they are empirical generalizations based on data from only a few studies and do not take into account the variability of either the plant or the insect. For example, the effects of variability in plant chemistry on phytophagous insect behaviour has not received much attention. The antifeedant effects of extracts from the leaves of *Teucrium polium* L. (Labiatae) were observed to vary from month to month and this variability in activity was investigated using larvae of the oligophagous *Spodoptera exempta* (Walker) and the polyphagous *S. littoralis* [Blaney & Simmonds]. Plants were harvested each month from March to October and ethanol extracts applied to glass-fibre discs. These discs were offered to the larvae in no-choice bioassays so that the effects of any month-by-month changes in the plants could be monitored. The antifeedant effect of the plants varied significantly over the eight month period, being greatest in July and August and lowest in April. The amount of terpenoids present in each of the monthly extracts was measured and compared with the responses of terpenoid-sensitive taste neurones in sensilla on the mouthparts of the larvae. In *S. exempta* the magnitude of the neural response was greater, and its variability was less, than with *S. littoralis*. Despite this variability, and variability in levels of terpenoids in the plants, there were good correlations between behaviour, neural response and plant chemistry with *S. exempta*, but much less so with *S. littoralis*. The variability in terpenoid levels between individual plants was much less than the variation in levels that occurred from month to month. It would appear that seasonal variation was the key determinant of insect behaviour, overriding the effect of inter-individual variability in the insects.

Many insects derive protection by sequestering toxins from their food plants and this may have led to some insect species developing specialist associations with particular toxic plants and so escaping from generalist predators. However, the occurrence of allelochemicals in plants is very variable, suggesting that host-derived defence may be unreliable. An investigation of this relationship with leaf beetles of the Chrysomelini has concluded that *de novo* synthesis of protective chemicals is the normal occurrence and host-derived defence is an exception (Pasteels). In contrast, the Lepidoptera more often rely on sequestration of host chemicals. The effects of plant variability in both situations has yet to be thoroughly investigated.

Compensation

Both insects and plants are able to compensate for the effects of the relationship between them, the plant by changes in its growth patterns, the insect by changes in its behaviour and physiology.

An investigation has been made of the effect of two root herbivores, the moth *Agapeta zoegana* (L.) and the weevil *Cyphocleonus achates* Fåhrs. on the growth and development of individual *Centaurea maculosa* Lam. plants, stressed in various ways (Müller-Schärer). Competition with the grass *Festuca pratensis* Hudson had much more effect than herbivory in reducing survival, growth and reproduction of *C. maculosa*. The reaction of *C. maculosa* to root herbivory was highly plastic and consisted mainly in compensatory growth involving increased allocation of nitrogen and biomass to the roots.

Phytophagous insects are generally sufficiently mobile to allow them to exercise a choice between potential host plants, or between different regions of the same plant. The choice behaviour of insects experimentally deprived of selected nutrients has been investigated by the use of carefully defined artificial diets (Simpson). There is no reason to argue that similar compensatory selection cannot occur in the more complex context of host plants. When final stadium nymphs of *Locusta migratoria* L., conditioned for eight hours on a diet containing digestible carbohydrates but deficient in protein, are offered a choice of protein-containing and carbohydrate-containing diets, they will select predominantly the protein diet, substantially redressing the nutritional imbalance during the first hour of choice. Final stadium larvae of *Spodoptera littoralis* have a similar capacity but take longer to redress the imbalance. This ability to compensate is at least partly determined by changes in peripheral responsiveness which is related to the amounts of nutrients consumed. When the conditioning diets offered to the locusts do not allow their nutritional requirements to be met by adjusting the food intake, they may be able to reach their "nutritional target" by differential utilisation of the ingested food. Thus it seems that phytophagous insects have sufficient plasticity in their behavioural repertoires and metabolic capabilities to allow them to compensate for wide variation in the nutrient quality of their host plants.

Genetics

Many insect species live in discrete populations spatially separated by unsuitable habitats, which prevent or restrict movement of individuals between the populations. This situation can result in the occurrence of genetic differentiation between the populations, either by random genetic drift or by the influence of different selective pressures acting on the populations. This segregation can be affected by variability in host plants, and genetic differentiation in the insects can affect their ability to adapt to local environmental conditions, including plant chemistry.

Two central European species of leaf beetles *Oreina cacaliae* (Schrank) and *Oreina speciosissima* (Scopoli) are oligophagous on the same set of host plants belonging to the Asteraceae (Rowell-Rahier). In the populations of both insect species genetic drift and inbreeding are high, resulting in large genetic divergence between local populations. Since these beetles are fairly immobile it is likely that speciation can be influenced by variability in their host plants.

In the Seychelles, *Drosophila sechellia* breeds only on the fruits of *Morinda citrifolia*, which contain highly toxic compounds to which *D. sechellia* is tolerant (Carton). Its close sympatric relative *Drosophila simulans* Sturtevant breeds on a wide variety of resources. *D. sechellia* is attracted to oviposit on *M. citrifolia*, whereas *D. simulans* is repelled. F1 hybrids and reciprocal backcross individuals exhibit intermediate, approximately additive behaviour and the tolerance of *D. sechellia* to *M. citrifolia* is fully dominant in F1 hybrids. Thus in this case several characters of the insects influencing host plant specialisation are genetically determined.

Recommendations

Following discussion of the papers presented at the workshop the participants were able to identify aspects of the subject which are inadequately understood and where further study is urgently needed. In order to identify fundamental features of insect-plant relationships it is imperative that the variability of insect-plant systems should be analysed. This analysis should be made at the level of individuals as well as of populations and attention should be given to the third trophic level. In view of the very limited knowledge of the genetical component in the variability of insect-plant systems, genetic analyses of such systems are badly needed. Lack of knowledge of the dynamics of a plant's chemical profile is a potential limitation to the interpretation of most studies on insect-plant interactions. It would be helpful if more phytochemists and plant scientists could elucidate chemical and morphological variations in plants, but in order to maximise the impact of this information on studies of insect-plant systems, close interactions with entomologists are essential. The extent of variation of plant chemistry and insect feeding behaviour should be analysed in crop systems and compared with that of natural systems. Studies on insect-plant interactions should not focus only on the effect of the plant on the insect but also investigate the phenomenon of responses induced in the plant by insect attack. This phenomenon should be studied in well defined systems in which the changing profiles of chemicals in the plant are elucidated, and their effect on insect behaviour investigated. The workshop recognised the crucial role of taxonomy and evolutionary biology as a basis for the wider interpretation of studies on insect-plant interactions.

Papers presented at Workshop

W.M. Blaney & M.S.J. Simmonds: The effect of variability in plant chemistry on insect behaviour.

F. Camps: Phytoecdysteroids production and accumulation in *Ajuga reptans* L. *in vitro* and *in vivo* cultures.

Y. Carton: Genetic analysis of host-plant specialization in *Drosophila* species.

R. Crnjar: Diet-dependent changes of gustatory discrimination patterns in Lepidoptera.

T. Jermy: Variability and its limits.

S. Larsson: Why theories in plant-insect ecology are short lived.

I. Máthé, Jr.: Variable steroid production of *Solanum* species as manifestation of internal and external factors.

E. van der Meijden: Variability in plant defence against herbivores.

H. Müller-Schärer: Plant responses to root-feeding insects: variations, causes and mechanisms.

J.M. Pasteels: The consequence of host-plant chemical heterogeneity on the evolution of chemical defence in leaf beetles.

M. Rowell-Rahier: Genetic variation and differentiation between populations and sexes of two *Oreina* species (*oreina cacaliae* and *O. speciosissima*, Chrysomelidae).

L.M. Schoonhoven: Inconstancies of chemoreceptor sensitivity.

S.J. Simpson: Rails and arcs: the geometry of compensatory feeding.

L. Tollsten: Some aspects of variation in floral fragrance chemistry.

H. Zwölfer: Host plant and environmental variability: key factors of the *Urophora cardui* food web on *Cirsium arvense*.

Proc. 8th Int. Symp. Insect-Plant Relationships, Dordrecht: Kluwer Acad. Publ.
S.B.J. Menken, J.H. Visser & P. Harrewijn (eds), 1992

Specialization in herbivorous insects

Hanna Mustaparta
Dept of Zoology, University of Trondheim-AVH, Norway

The workshop on specialization in herbivorous insects, held by the European Science Foundation Network on Insect-Plant Interactions at Oslo in May 1991, brought together scientists working on ecology, population genetics, behaviour and physiology. The presentations provided a forum to review and discuss our knowledge about specialization in herbivorous insects. Areas within this topic that need further study were identified. It was recognized that although many European laboratories have an experience in the study of specialization, the subject would benefit from more collaborative projects between these laboratories and colleagues in other parts of the world. The topics covered included: A. factors influencing the driving force in the evolution of insect host plant specificity ("Why" questions), and B. mechanisms underlying specialization *vs* polyphagy ("How" questions). The discussions centered mainly on topic B.

A. Evolution of insect host plant specificity: Why does an insect species usually utilize a narrow range of plant species?

There are many hypotheses on the evolution of host specificity and there is some debate as to whether host-race formation and host specialization exist and what conditions lead to speciation. Plant chemistry is a major determinant of diet breadth and insects usually feed on chemically related plants which are not always taxonomically related. Thus, cladograms of insects and their host plants are often not congruent. Factors which would bring about dietary changes include oviposition mistakes, intra- and interspecific competition and natural enemies. These factors might either lead to specialization or diet broadening, depending on the following conditions; extent to which environments are different, genetic variability in response to these environments, and gene flow among the parts of a population which experience selection by each environment. For instance, without gene flow locally specialized populations are expected, whereas generalized genotypes would evolve if migration occurs. Studies on *Yponomeuta* Latreille (Menken) and aphids (Guldemond) are providing evidence that sympatric speciation could be an important mode of speciation, a hypothesis that has not yet become widely accepted. Investigations into the population genetics of *Yponomeuta* suggest that host races occur (Raymann & Menken, this vol., pp. 209-211). Another interesting insect group for studying host races, is aphids. A high proportion of aphids are host specific on different plants with sympatric distribution and morphologically similar characteristics. A high rate of populational increase, along with the ability to shift to a new host with loss of fitness on the previous host, and reduced gene flow, are characteristics which preadapt aphids to host-race formation and subsequent sympatric speciation (Guldemond). Biosystematic research in many insect-plant systems are in progress and may further enlighten the evolutionary aspects of host specialization (Drosopoulos).

    The importance of natural enemies as driving forces for host specialization is clearly demonstrated in herbivorous caterpillars (Bernays). For example, by exposing vespid wasps or predatory ants to pairs of caterpillar species of similar size, but with different host range (narrow *vs* broad diet) it was found that generalist caterpillars were significantly more vulnerable to attack by the wasps or ants than specialists. The predators tended to catch species with a broad host range more readily than those with a narrow host range. These and other experiments suggest that generalist predators may drive and maintain specialization in their prey, whereas specialist enemies may lead to host switches or polyphagy.

Factors that could be important in the evolution of host specificity in aphids include the chemical and morphological structure, phenology and abundance of plants. Here, it is not nessesary to invoke concepts like "enemy free space" or "rendevous hosts" to explain the host specificty of aphids. These insects have a high rate of population increase, and a very small difference of the intrinsic growth rate (r_m) renders it advantageous to change host even if the risk associated with migration is high. An increase in the difference in the perceived quality of a host will result in more aphids colonizing the better quality host. This in turn increases the probability of chance mutations, thus improving the adaptation to that particular host.

Studies of the interactions of specialized bark beetles and their host trees have identified two different strategies by which the beetles can overcome the defences of the host tree; mass-attack and detoxification (Gregoire). In the mass-attack strategy, the bark beetles aggregate on the tree, inoculate pathogenic fungi that results in the death of the host tree. However, the phloem in these trees remains functional for several weeks, allowing the beetles time to breed. The disadvantage of this strategy to the beetle is high intra-specific competition which is the major cause of brood death in the bark beetle *Ips typographus* (L.). The benefit is, however, low investment in detoxification processes. In contrast, the bark beetle *Dendroctonus micans* exhibits the detoxification strategy and solitary females attack healthy trees. These beetles are remarkably tolerant to the resin of the host tree and the larvae feed gregariously in the phloem of the living tree. Thus, the cost for this bark beetle is investment in detoxification processes, whereas the benefit is low intra-specific competition and protection against generalist natural enemies associated with other bark beetle species. Even though fewer parasitoides and predators attack *D. micans*, compared to the numbers attacking *I. typographus*, there is one specialist predator, *Rhizophagus grandis* which seems to be the major factor in regulating the population of *D. micans* [Gregoire].

These examples clearly demonstrate the complexity and diversity of insect-plant interactions and the variability of factors that are important in host specialization.

B. Mechanisms underlying specialization: What mechanisms are involved in host specialization and how do insects select their host plants?

The presentations concerning these "How" questions were linked to the three following aspects of insect-plant interactions.

Behaviour associated with habitat selection

The distinction between habitat location and host location is difficult in phytophagous insects. In parasitoides, however, distinctions can be made on the basis of the source of the cues; the host itself, or its host plant. In micro-habitat location, specialist parasitoids that parasitize monophagous or oligophagous aphids use plant volatiles as cues for finding their hosts, whereas it is unlikely that parasitoids of polyphagous aphids, use such cues (Hofsvang). The parasitoid use of plant volatiles for attraction to Diptera and Lepidoptera larvae is well documented. For example, the specialist parasitoides of the leek moth (*Acrolepiopsis assectella* Zeller) use volatile sulphur compounds of leek for finding their host (Thibout). It was recognized that more information about several factors such as host diversity, phenology and density is needed for understanding the mechanisms for habitat selection.

Host location

Role of volatile compounds. The importance of host-plant volatiles has been shown to be of primary importance in the host-selection behaviour of many phytophagous insects and a large part of their olfactory system is thought to be specifically involved in conveying and processing information about plant volatiles. However, little is known as to which compounds, blends or profiles of plant volatiles are used by the insects to locate their host plants. Our knowledge in this respect is restricted to some single compounds in various species. In this workshop the presentations concerned plant volatiles used as host attractants in Coleoptera and Lepidoptera in general (Mustaparta), bark beetles (Tømmerås), the pine weevil (Wibe & Mustaparta, this vol., pp. 127-128), the apple blossom weevil (Kalinova) and the leek moth (Thibout). In the last mentioned case specific sulfur volatiles from the host plant attract the female from a distance to its host plant for egg-laying. Other leek-specific compounds elicite egg-laying and non-volatile sulphur compounds act through gustation. The other studies, were aimed at identifying more completely the blend of host-plant compounds used by the insect species. Here, it was chosen to start with trapping volatiles from the plant's "head space". The gas-chromatographic separation of these volatiles was linked to electrophysiological recordings of receptor responses (either EAGs or single cell responses) in order to identify which fractions of the volatile mixture actually influence the receptor cell activities. Both in bark beetles and the pine weevil such experiments demonstrated that minor components are important constituents in the host odour blend. Most receptor cells of bark beetles respond to only minor components in this blend, the pine weevil possesses receptor neurons that respond to major and minor components. In general, both in the bark beetles and the pine weevil, the olfactory system for detecting plant volatiles seems to possess specialized neurons rather than neurons broadly tuned to many components. However, one should be careful in generalizing from one species to another. It is possible that both labelled-lines and across-fiber patterns exist as neural coding of plant odours, and that these mechanisms may vary between species. It is also possible that one species may use both systems, a labelled line system for important specific compounds and an across-fiber mechanism for more general green odours. Furthermore, receptor neurons labelled for one compound may become labelled to another (analogue) when the "key compound" is not present (Wibe & Mustaparta, this vol., pp. 127-128). Clearly, our knowledge in this area is scarce. The need was stressed to focus on perception of odours (host and non-host) in order to find out how insects select plants and if token stimuli are more important than the presence or absence of deterrents. The pheromone system can serve as a model for the host odour olfactory system, since the same questions and methods are employed. The olfactory mechanisms for pheromone perception are better elucidated mainly because it has been easier to identify the biological signals produced by the insects. Therefore, in studies of insect-plant interactions there is a need to identify more of the plant volatiles detected by the insects, before investigating how receptors respond to these volatiles and how the information contained in the response from the olfactory neurons is further integrated in the central nervous system. Attraction and avoidance are important characteristics of olfaction, in general. It is therefore important that signals from both host and non-host plants are studied and that these signals are analysed as regards the effect of single compounds and mixtures.

Role of non-volatile compounds. In many host-plant groups, data exist as to which non-volatile compounds act as phagostimulants or feeding deterrents to insects. There is a need, however, to identify the more complete "chemical profiles" or "gestalts" that are important in host recognition. For example, studies on crucifers have shown that in addition to the glucosinolates other compounds act as phagostimulants for Crucifer-eating insects (Nielsen). The other presentations dealing with host recognition concerned the relation between host specificity and the responsiveness of insects to allelochemicals. Evidence was provided that

397

insects which are more restricted in their diet have gustatory receptors that are responsive to a wide range of allelochemicals, the deterrents receptors. As an extrapolation from this, it was suggested that diet breadth could be changed simply by varying the sensitivity of peripheral receptors (Chapman). Differences in the responsiveness of gustatory sensilla of taxonomically related insects could result in specializing on different host plants. For example, two species of *Spodoptera*, *S. exempta* Walker (oligophagous) and *S. littoralis* (Boisd.) (polyphagous) were screened for receptor sensitivity and behavioural responses to 250 allelochemicals. It was found that 58% of the compounds acted as antifeedants against the polyphagous species *S. littoralis*, whereas 75% were antifeedants against *S. exempta* (Simmonds). On the basis of these and other studies, Chapman compared sensitivity thresholds in monophagous, oligophagous and polyphagous insects and showed that sensitivity to many deterrents is "lost" in polyphagous species.

The evolution of specificity would necessitate the loss of sensitivity to secondary compounds in a potential host. It is suggested that a shift in the responsiveness of deterrent receptors could result in a subsequent change in the insect's response to a plant leading to specialist or generalist insects. This indicates that sensitivity is not a separately evolved adaptation, but an inherited characteristic. Studying the phylogeny of a specific insect-plant relationship would show whether insects change their responsiveness to compounds associated with a specific plant. .

Such a study has been undertaken on *Yponomeuta*. Here the inheritance of gustatory sensitivity was studied in reciprocal interspecific crosses of taxonomically closely related species that use taxonomically and phytochemically different host plants. *Y. cagnagellus* Hübner is monophagous on *Euonymus europaeus* L., whereas *Y. malinellus* Zeller feeds mainly on *Malus*. Dulcitol acts as a phagostimulant for *Y.cagnagellus* and the stereoisomer, sorbitol, stimulates *Y. malinellus* to feed. Phloridzin (*Malus*) and prunasin (*Prunus*) act as antifeedants for *Y. cagnagellus*; both compounds are absent in the host of the species. The progenies (F1) of the two moth species have been tested for sensitivity to dulcitol, phloridzin and prunasin. All F1's were sensitive to phloridzin and prunasin (Van Loon). Thus, sensitivity is dominant over non-sensitivity. Monofactorial autosomal inheritance of sensitivity seems likely in the case of responsiveness to dulcitol and prunasin, whereas sensitivity to phloridzin suggests additive inheritance.

A basic knowledge of the peripheral sensory perception of single compounds exists for the gustatory system. The specificity of receptor neurons for a number of phagostimulants and deterrents have been determined in various species, demonstrating a coding system which is based on a combination of "labelled-lines" and "across-fiber-patterns" (Schoonhoven). This information has enabled simple sensory models to be developed which provide clues to the understanding of host-recognition behaviour. However, it was stressed that there is a need for extended studies on the neural responses to mixtures as well as the central nervous integration of peripheral information.

Host selection and host acceptance

This topic has been investigated in many species, *e.g.*, aphids, Coleoptera, Lepidoptera, Diptera, and grasshoppers. There is information on the physical and chemical factors that are important in host-selection behaviour. However, it was agreed that further behavioural studies are needed to help identify key stimuli in a range of specialist and generalist species (though avoiding "stamp collection") in order to find out whether this behaviour is modulated by a range of similar compounds or specific plants, depending on specific insect-plant associations. It was recognized that feedback mechanisms associated with the ratio of nutrients and allelochemicals in plants can be very important in host acceptance and the resulting fitness of individuals developing on a plant.

Conclusions and recommendations

The presentations and discussions, examplifying the diversity of factors and mechanisms involved in interactions between phytophagaous insects and their hosts, led to the conclusion that it is by no means sufficient to study simply one insect-host-plant system in order to increase the knowledge and understanding of mechanisms underlying the interactions. Depending on the questions being asked, an appropriate insect-host-plant model should be selected.

The importance of understanding how various factors influence the evolution of host plant specialization and polyphagy was recognized, and it was stressed that it is important to link the research on mechanisms (physiology-behaviour) to the ecological-evolutionary questions in multidisiplinary ways (e.g., combined genetical studies as in Yponomeuta). It was decided that the Network ought to focus on the elucidation of mechanisms but also encourage collaborative projects with links to ecology and evolution.

It was recommended that priority should be given to studies on the functioning of the central nervous system and to studies on the phenotypic and genotypic bases of variability in host-selection behaviour. Such studies would enable us to elucidate the importance of behavioural plasticity in host-plant interactions, its consequences for host shifts, and the development of new crop pests as well as its importance in the ability of insects to overcome resistance in insect-resistant cultivars of economically important crops.

Papers presented at Workshop

Bernays, E.A.: Evolution of dietary specialization in insect herbivorous.

Chapman, R.F.: Chemosensory aspects of the evolution od host plant specificity.

Dixon, T.: Why are aphids highly host specific?

Drosopoulos, S.: Insect-plant assocations: an evolutionary approach based on biosystematic research.

Gregoire, J.-Cl.: Specialization in herbivorous isnects: How does it extend to the third trophic level? Examples from conifer bark beetles.

Guldemond, J.A.: Host shift and sympatric speciation in aphids.

Hofsvang, T.: Host selection in aphid parasitoids.

Jermy, T.: Unanswered questions of host plant specialization in phytophagous insects.
Kalinova, B.: What is the role of olfaction in host plant discrimination in apple blossom weevil, Anthonomus pomorum.

Menken, S.B.J.: Host race formation in phytophagous insects: Does specialization lead to speciation?

Mustaparta, H.: Specialization of receptor neurons for host plant volatiles in forest insects.

Nielsen, J.K.: Is it possible to describe chemical profiles which would allow specialized insects to distinguish between host and nonhost plants?

Schoonhoven, L.M.: Neural restraints upon diet breadth.

Simmonds, M.S.J.: A comparison of neural and behavioural responses in oligophagous and polyphagous caterpillars.

Thibout, E.: The various levels on which the sulfur allelochemicals are active on the leek moth.

Tømmerås, B.Å.: Specialization of receptor neurons to host volatiles in forest insects: bark beetles.

Van Loon, J.J.A.: Could chemosensory correlaytes of specialized selection behaviour be more than only correlates.

Wibe, A. and H. Mustaparta: Specialization of receptor neurons for host plant volatiles in forest insects: the pine weevil, Hylobius abietis.

Synopsis

Proc. 8th Int. Symp. Insect-Plant Relationships, Dordrecht: Kluwer Acad. Publ.
S.B.J. Menken, J.H. Visser & P. Harrewijn (eds), 1992

Impressions of the symposium, thoughts for the future

Erich Städler
Eidgenössische Forschungsanstalt, Wädenswil, Switzerland

In general

It is a very difficult, if not an impossible task, to summarise 74 oral presentations, 54 posters and the discussions among 180 participants. This has been mentioned earlier by Bernays (1991) and Dethier (1987). Thus, nothing more can be expected from this contribution than a personal evaluation of the present knowledge and an attempt to identify the problems that should be tackled in the future. The contributions of this 8th symposium covered, in accordance with its tradition, many aspects of insect-plant relationships (IPRs). New attractive facets have been added so that the range of topics can be regarded as a representative cross-section through the IPRs.

The level of sophistication and the complexity of experimental techniques presented in the contributions seem to increase continuously. Techniques presented exemplifying this trend were numerous. Electropenetrationgraphs (EPGs) are now used in many studies involving the host plant selection of aphids and other sap sucking insects. The olfactory responses of insects are often studied using plant volatiles which were identified by gas chromatography (GC) and mass spectrometry (MS). The volatiles are assayed with behavioural observations in wind tunnels, testing for anemotaxis, and with electro-physiological techniques like electro-antennograms (EAGs) and single sensilla recordings which are often linked with simultaneous GC plots. In the case of non-volatile, polar compounds, less reliance is now made on commercially available "model compounds" known to occur in specific plants. Instead, analytical techniques such as high performance liquid chromatography (HPLC) are applied to isolate, identify, and quantify the plant compounds relevant for particular insects. The majority of such investigations are based on the intense collaboration between phytochemists and biologists. Despite these advances, the isolation and identification of active phytochemicals still requires a lot of time and effort. It can be hoped that in a time of constraints and limited support for research, granting agencies around the world will appreciate the importance of the identification and quantification of the plant compounds which are important elements of IPRs.

Special developments and considerations

1. Insect-plant communities and evolution. The recent review by Feeny (1992) revealed that there is no shortage of ecological and evolutionary hypotheses explaining different types of IPRs. Therefore, it seems important that future investigations will provide more experimental and observational data testing the proposed theories. The approach of comparing metabolic costs with benefits of plant defence (plant fitness) underlines that in the future new answers have to be found to explain the principle how plant resistance works. Thus the test of an ecological and evolutionary hypothesis may raise important and new physiological questions which should be tackled by specialists, like plant ecologists and physiologists. The present series of symposia would be the ideal forum to initiate such multidisciplinary co-operations. But unfortunately, the field has not yet attracted the interest of plant scientists which, as Bernays (1991) has already emphasised, seems necessary to advance in the many aspects underlying IPRs.

It has long been known that models of interactions between insect species and host plants allowing comparative investigations offer powerful approaches to the study of the biology of IPRs. One of the first, and most ambitious projects in terms of different types of disciplines involved, is the "*Yponomeuta* model" developed by the late W.M. Herrebout (reviewed in Herrebout, 1991; Menken *et al.*, 1992). This model relationship and additional new model systems will be developed further and will certainly bear fruit in the future. Wim Herrebout and his many colleagues demonstrated that the collaboration between different disciplines in such endeavours is extremely valuable.

Studies of the multitrophic relationships of IPRs have also highlighted the value of a comparative approach in two completely different groups of insects. In one case, termites of different evolutionary groups were shown to differ in the way the cellulases less for the digestion of cellulose are acquired, essential for the conversion of this major constituent of the collected food. Evolutionary more primitive species were shown to produce their own digestive cellulases, whereas more evolved species ingest and supplement the necessary enzymes with fungal mycelium and fruiting bodies. The most specialised termite species have lost their own cellulases and depend completely on the enzymes produced by fungi growing in the fungal gardens (Campbell, 1989; Douglas, this volume). A similar evolutionary process was presented for the cardenolides as defensive substances of some Chrysomelid beetles (Rowell-Rahier *et al.*, 1991; Rowell-Rahier & Pasteels, this vol., pp. 341-342). The most ancestral species were shown to produce their own cardenolides. This biosynthetic capacity seems to have either partially or completely disappeared in the more evolved beetles which sequester cardenolides present in food plants and depend on these allelochemicals for their defence.

2. Host-plant selection. In contrast to earlier symposia, more reports were presented dealing with polar, non-volatile substances, which can play as an important role in host selection as the better investigated volatiles. Further evidence was provided that the chemicals influencing herbivore and parasitoid behaviour are, without exception, compound mixtures which are perceived together with visual and physical characteristics of the host plant. These mixtures have now been analysed in a few insects, not only from one single host plant, but also from different host and non-host plant species. Future progress along this line may reveal whether the insect herbivores respond to unique, or alternatively, to different host-specific plant and non-host plant templates ("host plant images" composed of visual and chemical characteristics termed "Gestalt"; see also Dethier, 1987). In addition, these data will provide a sound basis for our understanding of the decoding of the complex environmental stimuli by the insect nervous system.

The plant surface and the compounds present in it has long been recognised to be important for different organisms (reviewed by Städler & Roessingh, 1991). With some notable exceptions, little progress has been made in the identification of the compounds involved. We know close to nothing about the plant physiological mechanisms regulating the transport of the compounds to the leaf surface (phylloplane) and their distribution within the wax layer. Only a few attempts have been made so far to identify the active compounds on the surface that affect feeding or oviposition behaviour. We assume, but do not know for sure, that volatiles accumulated in the boundary layer of the leaf surface are perceived by olfactory receptors. The perception process of non-volatile, dry compounds, which in the case of the *Brassica* leaves, can be obtained only through extracting the wax layer first (Städler & Roessingh, 1991) is still more mysterious. A possible link to the perception of theses compounds may be the composition and function of the dendritic fluid of the contact chemoreceptor sensilla. This was realised by Bernays *et al.* (1975), but since then, no further investigations have addressed this unsolved question. Thus, the plant surface compounds and the insect chemoreceptors involved, need to be studied in much more detail until this important interface between insect and plant can be better understood.

No new electrophysiological recording techniques for the analyses of receptors sensitive to plant "chemical images" have been developed in recent years. But the availability of laboratory computers and dedicated data management programs has reduced the amount of expensive equipment and work involved in the examination of the recordings. A new analytical tool that is "neuronal network analysis" has been developed to allow for the identification and classification of both individual nerve impulses in the extracellular recordings and total spike train patterns of different cells over time (Hanson *et al.*, this vol., pp. 173-175). This is an approach which opens new possibilities to analyse electrophysiological recordings and to test hypothesis about neural (across-fibre) coding and information theory already proposed by Dethier (1987).

The peripheral chemoreceptors got more attention in recent years as a result of the increasing knowledge about behaviourally active compounds. In contrast, the exploration of the integration by the central nervous system neurones, which Dethier (1987) expected to be a new research focus, does still lag behind. The investigation of the olfactory brain of the Colorado potato beetle (De Jong & Visser, 1988) was the first successful attempt, but the recordings from the central neurones proved to be technically very difficult. In contrast, the investigation of the central neurones involved in the decoding of the responses of the antennal pheromone receptors has in recent years, been much more successful (*e.g.* Christensen *et al.*, 1991). Therefore it is hard to understand why so few investigations were initiated to study the brain parts that process information from receptor cells sensitive to host and non-host compounds. Apart from the difficulties experienced by De Jong & Visser (1988), which may be related to the size and the type of preparation, there must be other reasons. It seems that we lack the necessary knowledge of the feeding or oviposition behaviour and the receptor physiology of herbivore insects with sufficiently large brains as well as the composition of relevant plant compounds. These obstacles cannot be invincible, and hopefully, thanks to progress in the pheromone field, the recording techniques will advance to the point that smaller and more difficult parts of brains can be successfully investigated too.

Feedback mechanisms influencing host plant selection behaviour and chemosensory sensitivity have been shown to be important already in earlier symposia (Blaney *et al.*, 1991; Simpson *et al.*, 1991) and also in more recent investigations (Simmonds *et al.*, 1992). Previous exposure to plant nutrients and allelochemicals do play a role in subsequent food choices, and associative learning now seems so well documented that it has to be considered in any investigation of foraging or host plant selection. A special type of feedback is operating in the diet mixing (switching) behaviour in a generalist grasshoppers (Bernays, this vol., pp. 146-148). The evidence presented showed that this behaviour is apparently an innate characteristic, influenced by nutritional imbalances and distinct flavours in foods. Since similar effects have been observed also in molluscs (refs. in Bernays & Raubenheimer, 1991; Rueda *et al.*, 1991) it may well be a phenomenon common to different generalist herbivores.

3. Host-plant resistance - plant genetics. The variability of plants and insects and the impact of environmental and genetic factors have not attracted sufficient attention in earlier years (Bernays, 1991; Dethier, 1987; Schoonhoven, 1991). But now, progress has been made in many studies to record at least the variability and to analyse in a few cases also its causes. However, there is still a great need in almost any aspects of IPR for genetics (Mendelian, quantitative and molecular). It is also becoming apparent that advances in this field are not only necessary but also ambitious in the sense that the amount of data can be overwhelming for the constraints in time and available resources. Thus, careful planning, simple experimental designs and co-operation between different investigators appear to be more than ever of paramount importance.

Studies of host-plant resistance to insects have been expanded to many pests of temperate and tropical agriculture and forestry. The prospects of the application of IPR seem almost unlimited and fulfil earlier hopes of the pioneers in this field (reviewed by Kogan, 1986) that host-plant resistance will be a crucial element of integrated pest management. The reported progress in the analysis of plant genetics and molecular biology revealed that much more development in this field can be expected, as anticipated already by Bernays (1991). Furthermore, any pest control strategy, including the development and deployment of resistant crop varieties, needs to consider the selection of new biotypes of insects that are able to overcome the new barriers. Insect populations are being monitored successfully using new tools such as alloenzyme and DNA sequence analysis that have been developed specifically for this purpose.

4. *Multitrophic interactions.* The ecology and evolution of IPRs cannot be understood without considering additional trophic levels, as pointed out by Bernays (1991). In this regard, symbionts and defensive secretions have already been mentioned. Significant progress has been reported in the identification of compounds originating from plants and herbivores which affect the foraging behaviour of parasitoids. This progress was accomplished by collaboration between different disciplines and emphasised the value of such an approach to complex investigations in this field (see, for example, in this volume: Dicke, pp. 355-356; Turlings *et al.*, pp. 365-366). Far less seems to be known about non-volatile compounds originating from host insects determining the acceptance by parasitoids. This parallels the relative lack of knowledge of the non-volatiles in the herbivore-plant relationship. The recent comparative analyses of the "surface chemistry" of both herbivore insects and their host plants by Espelie *et al.* (1991) show that interesting relations seem to exist but that much remains to be explored. Such findings should stimulate promising future research in multitrophic relationships.

In conclusion

This latest, 8th symposium has again proven that the study of IPRs is a very fertile and attractive field of research. An important reason for this fact is that widely different disciplines ranging from evolutionary biology to molecular genetics are, and should be, involved. It is hoped that this interaction and co-operation between disciplines will flourish and still intensify in the future. This seems to be a recipe for progress in the applied aspects as well as in achieving a fundamental understanding of IPR in general.

Acknowledgements. I thank Drs R. Baur, S.B.J. Menken, J.A.A. Renwick, and J.H. Visser for discussions and improvements of the manuscript and the Schweiz. Nationalfonds, for financial support (Grant # 31-30059.90).

References

Bernays, E.A. (1991). General conclusions. In: Á. Szentesi & T. Jermy (eds), Proc. 7th Int. Symp. Insect-Plant Relationships 1989. *Symp. Biol. Hung.* **39**: 557-560.
Bernays, E.A. & D. Raubenheimer (1991). Dietary mixing in grasshoppers: changes in acceptability of different plant secondary compounds associated with low levels of dietary protein (Orthoptera, Acrididae). *J. Insect Behav.* **4**: 545-556.

Bernays, E.A., W.M. Blaney, R.F. Chapman & A.G. Cook (1975). The problems of perception of leaf-surface chemicals by locust contact chemoreceptors. In: D.A. Denton & J.P. Coghlan (eds), *Olfaction & Taste V*, pp. 227-229. New York: Academic Press.

Blaney, W.M., M.S.J. Simmonds & S.J. Simpson (1991). Dietary selection behaviour; comparison between locusts and caterpillars. In: Á. Szentesi & T. Jermy (eds), Proc. 7th Int. Symp. Insect-Plant Relationships 1989. *Symp. Biol. Hung.* **39**: 47-52.

Campbell, B.C. (1989). On the role of microbial symbiotes in herbivorous insects. In: E.A. Bernays (ed.), *Insect-Plant Interactions Vol. 1*, pp. 1-44. Boca Raton: CRC Press.

Christensen, T.A., H. Mustaparta & J.G. Hildebrand (1991). Chemical communication in Heliothine moths. 2. Central processing of intraspecific and interspecific olfactory messages in the male corn earworm moth *Helicoverpa zea. J. comp. Physiol. A* **169**: 259-274.

De Jong, R. & J.H. Visser (1988). Integration of olfactory information in the Colorado potato beetle brain. *Brain Res.* **447**: 10-17.

Dethier, V.G. (1987). Concluding remarks. In: V. Labeyrie, G. Fabres & D. Lachaise (eds), *Proc. 6th Int. Symp. Insect-Plant Relationships, Pau 1986*, pp. 429-435. Dordrecht: Junk.

Espelie, K.E., E.A. Bernays & J.J. Brown (1991). Plant and insect cuticular lipids serve as behavioral cues for insects. *Arch. Insect Biochem. Physiol.* **17**: 223-233.

Feeny, P. (1992). The evolution of chemical ecology: contributions from the study of herbivorous insects. In: M.R. Berenbaum & G.A. Rosenthal (eds), *Herbivores: Their Interaction with Secondary Plant Metabolites Vol. II*, pp. 1-44. New York: Academic Press.

Herrebout, W.M. (1991). Phylogeny and host plant specialization: small ermine moths (*Yponomeuta*) as an example. In: Á. Szentesi & T. Jermy (eds), Proc. 7th Int. Symp. Insect-Plant Relationships. *Symp. Biol. Hung.* **39**: 289-300.

Kogan, M. (1986). Plant defense strategies and host-plant resistance. In: M. Kogan. (ed.), *Ecological Theory and Integrated Pest Management Practice*, pp. 83-134. New York: Wiley & Sons.

Menken, S.B.J., W.M. Herrebout & J.T. Wiebes (1992). Small ermine moths (*Yponomeuta*): their host relations and evolution. *Annu. Rev. Entomol.* **37**: 41-66.

Rowell-Rahier, M., L. Witte, A. Ehmke, T. Hartmann & J.M. Pasteels (1991). Sequestration of plant pyrrolizidine alkaloids by chrysomelid beetles and selective transfer into defensive secretions. *Chemoecol.* **2**: 41-48.

Rueda, A.A., F. Slansky & G.S. Wheeler (1991). Compensatory feeding response of the slug *Sarasinula plebeia* to dietary dilution. *Oecologia* **88**: 181-188.

Schoonhoven, L.M. (1991). Insects and host plants: 100 years of "botanical instinct". In: Á. Szentesi & T. Jermy (eds), Proc. 7th Int. Symp. Insect-Plant Relationships. *Symp. Biol. Hung.* **39**: 3-14.

Simmonds, M.S.J., S.J. Simpson & W.M. Blaney (1992). Dietary selection behaviour in *Spodoptera littoralis* - The effects of conditioning diet and conditioning period on neural responsiveness and selection behaviour. *J. Exp. Biol.* **162**: 73-90.

Simpson, C.L., S.J. Simpson & J.D. Abisgold (1991). The role of various amino acids in the protein compensatory response of *Locusta migratoria*. In: Á. Szentesi & T. Jermy (eds), Proc. 7th Int. Symp. Insect-Plant Relationships 1989. *Symp. Biol. Hung.* **39**: 39-46.

Städler, E. & P. Roessingh (1991). Perception of surface chemicals by feeding and ovipositing insects. In: Á. Szentesi & T. Jermy (eds), Proc. 7th Int. Symp. Insect-Plant Relationships. *Symp. Biol. Hung.* **39**: 3-14.

List of participants

AUSTRALIA
R.E. *Jones*, James Cook University, Zoology Department, Townsville, QLD 4811

BELGIUM
L. *de Bruyn*, University of Antwerp RUCA, Department of Biology, Groenenborgerlaan 171, 2020 Antwerpen

J.M. *Pasteels*, Université Libre de Bruxelles, Laboratoire de Biologie Animale, 50 Avenue F.D. Roosevelt, 1050 Bruxelles

CANADA
C.J. *Bolter*, Agriculture Canada, 1400 Western Road, London, Ontario N6G 2V4

P. *Martel*, Agriculture Canada, Research Station, 430 Boulevard Gouin, St Jean Sur Richelieu, Quebec J3B 3E6

J.N. *McNeil*, Laval University, Department of Biology, Ste. Foy, P.Q. G1K 7P4

CHILI
A. *Givovich*, University of Chili, Faculty of Sciences, Department of Chemistry, Casilla 653, Santiago

CHINA
L.-E. *Luo*, Peking University, Department of Biology, Beijing 100871

CONGO
P.A. *Calatayud*, see France

CZECHOSLOVAKIA
J. *Harmatha*, Czechoslovak Academy of Science, Institute of Organic Chemistry and Biochemistry, Flemingovo Nám. 2, 166 10 Prague 6

DENMARK
A. *Kirkeby-Thomsen*, Danish Forest Research Institute, Skovbrynet 16, 2800 Lyngby

C. *Kjær-Pedersen*, National Environmental Research Institute, Department of Terrestrial Ecology, Vejlsovej 25, 8600 Silkeborg

J.K. *Nielsen*, Royal Veterinary and Agricultural University, Chemistry Department, Thorvaldsensvej 40, 1871 Frederiksberg C

FINLAND
M.L. *Helander*, University of Turku, Laboratory of Ecology and Animal Systematics, 20500 Turku 50

J. *Koricheva*, University of Turku, Department of Biology, Laboratory of Ecological Zoology, 20500 Turku 50

P. *Lyytikäinen*, Finnish Forest Research Institute, Department of Forest Ecology, P.O. Box 18, 01301 Vantaa

P. *Palokangas*, University of Turku, Department of Biology, 20500 Turku 50

H. *Roininen*, University of Joensuu, Department of Biology, P.O. Box 111, 80101 Joensuu

K. *Saikkonen*, University of Turku, Department of Biology, 20500 Turku 50
J. *Tahvanainen*, University of Joensuu, Department of Biology, P.O. Box 111, 80101 Joensuu

FRANCE

C.M. *Caillaud*, École Nationale Supérieure Agronomique de Rennes, Chaire de Zoologie, 65 Rue de Saint-Brieuc, 35042 Rennes

P.A. *Calatayud*, INSA Laboratoire de Biologie Appliquée, 20 Avenue Albert Einstein, Batiment 406, 69621 Villeurbanne

P. *Colyer*, European Science Foundation, 1 Quai Lezay-Marnesia, 67080 Strasbourg

S. *Derridj*, INRA Centre de Recherche de Versailles, Station de Zoologie, 78210 Versailles

G. *Febvay*, INRA Département de Zoologie, 20 Avenue Albert Einstein, INSA Batiment 406, 69621 Villeurbanne

J. *Huignard*, Université François Rabelais, Institut de Biocénotique Expérimentale des Agrosystèmes, URA CNRS 1298, Avenue Monge, Parc Grandmont, 37200 Tours

F. *Marion-Poll*, INRA Station de Phytopharmacie, Etoile de Choisy, Route de Saint-Cyr, 78026 Versailles

M.H. *Pham-Delègue*, INRA, CNRS Laboratoire de Neurobiologie Comparée des Invertébrés, P.O. Box 23, 91440 Bures sur Yvette

Y. *Rahbé*, INRA Départment de Zoologie, 20 Avenue Albert Einstein, INSA Batiment 406, 69621 Villeurbanne

E. *Thibout*, Université François Rabelais, Institut de Biocénotique Expérimentale des Agrosystèmes, URA CNRS 1298, Avenue Monge, Parc Grandmont, 37200 Tours

D. *Thiéry*, INRA, CNRS Laboratoire de Neurobiologie Comparée des Invertébrés, P.O. Box 23, 91440 Bures sur Yvette

GERMANY

A. *Biller*, Institut für Pharmazeutische Biologie der Technischen Universität, Mendelssohnstrasse 1, 3300 Braunschweig

M. *Boppré*, Forstzoologisches Institut der Universität Freiburg, Fohrenbühl 27, 7801 Stegen-Wittental

W. *Francke*, Universität Hamburg, Institut für Organische Chemie, Martin-Luther-King-Platz 6, 2000 Hamburg 13

H.-J. *Greiler*, Universität Karlsruhe, Zoologisches Institut, P.O. Box 6980, 7500 Karlsruhe 1

T. *Hartmann*, Institut für Pharmazeutische Biologie der Technischen Universität, Mendelssohnstrasse 1, 3300 Braunschweig

A. *Krüss*, Universität Karlsruhe, Zoologisches Institut, P.O. Box 6980, 7500 Karlsruhe 1

P. *Lösel*, Bayer A.G., Institut für Tierische Schädlinge, Pflanzenschutzzentrum Monheim, 5090 Leverkusen, Bayerwerk

P. *Ockenfels*, Zoologisches Forschungsinstitut und Museum A. Koenig, Adenauerallee 162, 5300 Bonn 1

HUNGARY

T. *Jermy*, Plant Proctection Institute, Hungarian Academy of Sciences, P.O. Box 102, 1525 Budapest

F. *Kozár*, Plant Proctection Institute, Hungarian Academy of Sciences, P.O. Box 102, 1525 Budapest

Á. *Szentesi*, Plant Proctection Institute, Hungarian Academy of Sciences, P.O. Box 102, 1525 Budapest

ISRAEL

M. *Burstein*, Tel Aviv University, Department of Zoology, 69978 Tel Aviv

D. *Wool*, Tel Aviv University, Department of Zoology, 69978 Tel Aviv

ITALY

C. *Coiutti*, Universita di Udine, Instituto di Difesa delle Plante, Localita Rizzi, Via Fagagna 208, 33100 Udine

G. *Giangiuliani*, University of Perugia, Faculty of Agriculture, Agricultural Entomology Institute, Borgo XX Giugno, 06121 Perugia

N. *Isidoro*, University of Perugia, Faculty of Agriculture, Agricultural Entomology Institute, Borgo XX Giugno, 06121 Perugia

L. *Mattiacci*, University of Perugia, Faculty of Agriculture, Agricultural Entomology Institute, Borgo XX Giugno, 06121 Perugia

F. *Nazzi*, Universita di Udine, Instituto di Difesa delle Plante, Localita Rizzi, Via Fagagna 208, 33100 Udine

JAPAN

T. *Ohgushi*, Faculty of Agriculture, Shiga Prefectural Junior College, Nishi Shibukawa 2, Kusatsu, 525 Shiga

THE NETHERLANDS

F.M. *Bakker*, University of Amsterdam, Department of Pure and Applied Ecology, Section Population Biology, Kruislaan 302, 1098 SM Amsterdam

A. *Bazelmans*, Rijk Zwaan Zaadteelt en Zaadhandel B.V., P.O. Box 40, 2678 ZG De Lier

L. *van den Berkmortel*, Bruinsma Seeds B.V., P.O. Box 24, 2670 AA Naaldwijk

K. *Booij*, Research Institute for Plant Protection IPO-DLO, P.O. Box 9060, 6700 GW Wageningen

J. *Bruin*, University of Amsterdam, Department of Pure and Applied Ecology, Section Population Biology, Kruislaan 302, 1098 SM Amsterdam

W. *Cappellen*, Bejo Zaden, P.O. Box 50, 1749 ZH Warmenhuizen

N.M. *van Dam*, University of Leiden, Department of Population Biology, P.O. Box 9516, 2300 RA Leiden

M. *Dicke*, Wageningen Agricultural University, Department of Entomology, P.O. Box 8031, 6700 EH Wageningen

M.J. *van Dijk*, University of Leiden, Department of Population Biology, P.O. Box 9516, 2300 RA Leiden

F.R. *van Dijken*, Centre for Plant Breeding and Reproduction Research CPRO-DLO, P.O. Box 16, 6700 AA Wageningen

H. *van Doorn*, Zaadunie B.V., P.O. Box 26, 1600 AA Enkhuizen

A. *van Eggermond*, De Ruiterzonen B.V., P.O. Box 4, 2665 ZG Bleiswijk

W.H. *Frentz*, Wageningen Agricultural University, Department of Entomology, P.O. Box 8031, 6700 EH Wageningen

B. *Gebala*, Centre for Plant Breeding and Reproduction Research CPRO-DLO, P.O. Box 16, 6700 AA Wageningen

F.C. *Griepink*, Research Institute for Plant Protection IPO-DLO, P.O. Box 9060, 6700 GW Wageningen

J.A. *Guldemond*, Research Institute for Plant Protection IPO-DLO, P.O. Box 9060, 6700 GW Wageningen

411

J.L. Harrewijn, Nickerson-Zwaan B.V., P.O. Box 4, 1747 ZG Tuitjenhorn

P. Harrewijn, Research Institute for Plant Protection IPO-DLO, P.O. Box 9060, 6700 GW Wageningen

M. van Helden, Wageningen Agricultural University, Department of Entomology, P.O. Box 8031, 6700 EH Wageningen

H. Inggamer, Centre for Plant Breeding and Reproduction Research CPRO-DLO, P.O. Box 16, 6700 AA Wageningen

C.M. de Jager, University of Leiden, Department of Population Biology, P.O. Box 9516, 2300 RA Leiden

C.R. Jager, University of Amsterdam, Institute of Taxonomic Zoology, P.O. Box 4766, 1009 AT Amsterdam

J. Janssen, Wageningen Agricultural University, Department of Entomology, P.O. Box 8031, 6700 EH Wageningen

R. de Jong, Wageningen Agricultural University, Department of Entomology, P.O. Box 8031, 6700 EH Wageningen

M.A. Jongsma, Centre for Plant Breeding and Reproduction Research CPRO-DLO, P.O. Box 16, 6700 AA Wageningen

J.W. Klijnstra, Plastics and Rubber Research Institute TNO, P.O. Box 6031, 2600 JA Delft

R.E. Kooi, University of Leiden, Department of Population Biology, P.O. Box 9516, 2300 RA Leiden

A.M.M. de Laat, Van der Have Research, P.O. Box 1, 4410 AA Rilland-Bath

J.C. van Lenteren, Wageningen Agricultural University, Department of Entomology, P.O. Box 8031, 6700 EH Wageningen

P. Lindhout, Centre for Plant Breeding and Reproduction Research CPRO-DLO, P.O. Box 16, 6700 AA Wageningen

J.J.A. van Loon, Wageningen Agricultural University, Department of Entomology, P.O. Box 8031, 6700 EH Wageningen

E. van der Meijden, University of Leiden, Department of Population Biology, P.O. Box 9516, 2300 RA Leiden

S.B.J. Menken, University of Amsterdam, Institute of Taxonomic Zoology, P.O. Box 4766, 1009 AT Amsterdam

L. Messchendorp, Wageningen Agricultural University, Department of Entomology, P.O. Box 8031, 6700 EH Wageningen

A.K. Minks, Research Institute for Plant Protection IPO-DLO, P.O. Box 9060, 6700 GW Wageningen

C. Mollema, Centre for Plant Breeding and Reproduction Research CPRO-DLO, P.O. Box 16, 6700 AA Wageningen

L. den Nijs, Research Institute for Plant Protection IPO-DLO, P.O. Box 9060, 6700 GW Wageningen

L.P.J.J. Noldus, Noldus Information Technology B.V., Vadaring 51, 6702 EA Wageningen

O.M.B. de Ponti, Nunhems Zaden BV, P.O. Box 4005, 6080 AA Haelen

L.E.L. Raijmann, University of Amsterdam, Institute of Taxonomic Zoology, P.O. Box 4766, 1009 AT Amsterdam

K. Reinink, Centre for Plant Breeding and Reproduction Research CPRO-DLO, P.O. Box 16, 6700 AA Wageningen

P.C.J. van Rijn, University of Amsterdam, Department of Pure and Applied Ecology, Section Population Biology, Kruislaan 302, 1098 SM Amsterdam

R.P.L.A. de Rooij, Research Institute for Plant Protection IPO-DLO, P.O. Box 9060, 6700 GW Wageningen

M.W. Sabelis, University of Amsterdam, Department of Pure and Applied Ecology, Section Population Biology, Kruislaan 302, 1098 SM Amsterdam

L.M. Schoonhoven, Wageningen Agricultural University, Department of Entomology, P.O. Box 8031, 6700 EH Wageningen

P. Scutareanu, University of Amsterdam, Department of Pure and Applied Ecology, Section Population Biology, Kruislaan 302, 1098 SM Amsterdam

M.M. Steenhuis-Broers, Centre for Plant Breeding and Reproduction Research CPRO-DLO, P.O. Box 16, 6700 AA Wageningen

D.C. Thomas, Wageningen Agricultural University, Department of Entomology, P.O. Box 8031, 6700 EH Wageningen

W.F. Tjallingii, Wageningen Agricultural University, Department of Entomology, P.O. Box 8031, 6700 EH Wageningen

J.H. Visser, Research Institute for Plant Protection IPO-DLO, P.O. Box 9060, 6700 GW Wageningen

K. Vrieling, University of Leiden, Department of Population Biology, P.O. Box 9516, 2300 RA Leiden

F. Wäckers, Wageningen Agricultural University, Department of Entomology, P.O. Box 8031, 6700 EH Wageningen

NIGERIA
N.A. Bosque-Pérez, International Institute of Tropical Agriculture, P.O. Box 5320, Ibadan

NORWAY
H. Mustaparta, Univeristy of Trondheim AVH, Department of Zoology, 7055 Dragvoll
A. Wibe, Univeristy of Trondheim AVH, Department of Zoology, 7055 Dragvoll

PERU
A. Golmirzaie, International Potato Center CIP, Apartado 5969, Lima

POLAND
A.P. Ciepiela, Agricultural and Pedagogic University, Institute of Biology, Department of Biochemistry, Ul. B. Prusa 12, 08-110 Siedlce

B. Gabrys, Agricultural University, Department of Agricultural Entomology, Cybulskiego 32, 50-205 Wroclaw

M. Kielkiewicz, Warsaw Agricultural University, Department of Applied Entomology, Ul. Nowoursynowska 166, 02-766 Warsaw

A. Tomczyk, Warsaw Agricultural University, Department of Applied Entomology, Ul. Nowoursynowska 166, 02-766 Warsaw

A. Urbanska, Agricultural and Pedagogic University, Institute of Biology, Department of Biochemistry, Ul. B. Prusa 12, 08-110 Siedlce

PORTUGAL
E.N. Barata, University of Evora, Department of Biology, Apartado 94, 7001 Evora
M.H. Carvalho Fernandes Bichao, Universtity of Evora, Department of Biology, Apartado 94, 7001 Evora

SPAIN
F. Camps, CID, CSIC Department of Biological Organic Chemistry, Jordi Girona 18-26, 08034 Barcelona
M.L. Gomez-Guillamon, CSIC Estacion Experimental La Mayora, Algarrobo-Costa, 29750 Malaga
C. Soria, CSIC Estacion Experimental La Mayora, Algarrobo-Costa, 29750 Malaga

SWEDEN
I. Åhman, Resistance Breeding Department, Svalöf AB, S-268 81 Svalöv
J. Sandström, Swedish University of Agricultural Sciences, Department of Plant and Forest Protection, P.O. Box 7044, 75007 Uppsala

SWITZERLAND
R. Baur, Swiss Federal Research Station, CH-8820 Wädenswil
B. Haegele, University of Basel, Zoological Institute, Rheinsprung 9, 4051 Basel
M. Rowell-Rahier, University of Basel, Zoological Institute, Rheinsprung 9, 4051 Basel
E. Städler, Swiss Federal Research Station, CH-8820 Wädenswil

UNITED KINGDOM
E. Bartlet, AFRC, IACR Rothamsted Experimental Station, Harpenden, Herts AL5 2JQ
E.A. Bell, Kings College, Department of Biochemistry, Strand, London WC2L 2RS
A.N.E. Birch, Scottish Crop Research Institute, Invergowrie, Dundee DD2 5DA
W.M. Blaney, Birkbeck College, Department of Biology, Malet Street, London WC1E 7HX
M.M. Blight, AFRC, IACR Rothamsted Experimental Station, Harpenden, Herts AL5 2JQ
P.G. Chambers, University of Oxford, Department of Zoology, South Parks Road, Oxford OX1 3PS
R.A. Cole, Horticulture Research International, Wellesbourne, Warwick CV35 9EF
K.R. De Souza, The Natural Resources Institute, Central Avenue, Chatham Maritime, Chatham, Kent ME4 4TB
L. Dinan, University of Exeter, Washington Singer Laboratories, Department of Biological Sciences, Perry Road, Exeter EX4 4QG
A.E. Douglas, University of Oxford, Department of Zoology, South Parks Road, Oxford OX1 3PS
K.A. Evans, The Scottish Agricultural College, Department of Crop Science and Technology, West Main Road, Edinburgh EH9 3JB
A.M.R. Gatehouse, University of Durham, Department of Biological Sciences, South Road, Durham DH1 3LE
P. Golob, The Natural Resources Institute, Central Avenue, Chatham Maritime, Chatham, Kent ME4 4TB
I. Gudrups, The Natural Resources Institute, Central Avenue, Chatham Maritime, Chatham, Kent ME4 4TB
R.J. Hopkins, Scottish Crop Research Institute, Invergowrie, Dundee DD2 5DA
R. Isaacs, Imperial College at Silwood Park, Department of Biology, Ascot, Berks SL5 7PY
A.J. Mordue, University of Aberdeen, Department of Zoology, Tillydrone Avenue, Aberdeen AB9 2TN
A.J. Nisbet, Scottish Crop Research Institute, Invergowrie, Dundee DD2 5DA
J. Orchard, The Natural Resources Institute, Central Avenue, Chatham Maritime, Chatham, Kent ME4 4TB

V. *Pike*, The Natural Resources Institute, Central Avenue, Chatham Maritime, Chatham, Kent ME4 4TB

G. *Powell*, AFRC Linked Research Group in Aphid Biology, Imperial College at Silwood Park, Ascot, Berks SL5 7PY

D. *Reavey*, University of York, Department of Biology, Heslington, York YO1 5DD

P. *Roessingh*, University of Oxford, Department of Zoology, South Parks Road, Oxford OX1 3PS

D.A. *Russell*, The Natural Resources Institute, Central Avenue, Chatham Maritime, Chatham, Kent ME4 4TB

M.S.J. *Simmonds*, Royal Botanic Gardens Kew, Jodrell Laboratory, Richmond, Surrey TW9 3AB

S.J. *Simpson*, University of Oxford, Department of Zoology, South Parks Road, Oxford OX1 3PS

S.E. *Trumper*, University of Oxford, Department of Zoology, South Parks Road, Oxford OX1 3PS

UNITED STATES OF AMERICA

M.D. *Abrams*, Pennsylvania State University, School of Forestry, 4 Ferguson Building, University Park, PA 16802

M. *Auerbach*, University of North Dakota, Department of Biology, P.O. Box 8238, University Station, Grand Forks, ND 58202

E.A. *Bernays*, University of Arizona, Department of Entomology, Tucson, AZ 85721

G. *de Boer*, University of Kansas, Department of Entomology, Haworth Hall, Lawrence, KA 66045

R.K. *Campbell*, Oklahoma State University, Department of Entomology, Stillwater, OK 74078

W.W. *Cantelo*, USDA, ARS Beltsville Agricultural Research Center, Building 470, BARC-East, Vegetable Laboratory, 10300 Baltimore Avenue, Beltsville, MD 20759

R.D. *Eikenbary*, Oklahoma State University, Department of Entomology, Stillwater, OK 74078

S.H. *Faeth*, Arizona State University, Department of Zoology, Tempe, AZ 85287

P.P. *Feeny*, Cornell University, Section of Ecology and Systematics, Corson Hall, Ithaca, NY 14853

R.S. *Fritz*, Vassar College, Department of Biology, P.O. Box 133, Poughkeepsie, NY 12601

D.J. *Futuyma*, State University of New York, Department of Ecology and Evolution, Stony Brook, NY 11794

W.A. *van Giessen*, USDA, ARS US Vegetable Laboratory, 2875 Savannah Highway, Charleston, SC 29414

R.A. *Haack*, USDA Forest Service, North Central Forest Experiment Station, 1407 South Harrison Road, East Lansing, MI 48823

F.E. *Hanson*, University of Maryland Baltimore County, Department of Biology, Baltimore, MD 21228

D.N. *Karowe*, Virginia Commonwealth University, Department of Biology, P.O. Box 2012, Richmond, VI 23284

W.J. *Lewis*, USDA, ARS Insect Biology Laboratory, P.O. Box 748, Tifton, GA 31793

S.M. *Louda*, University of Nebraska, School of Biological Sciences, 102 Lyman Hall, Lincoln, NE 68588

S.B. Malcolm, Western Michigan University, Department of Biological Sciences, Kalamazoo, MI 49008

W.L. Mechaber, Tufts University, Biology Department, Medford, MA 02155

J.R. Miller, Michigan State University, Department of Entomology and Pesticide Research Center, East Lansing, MI 48824

C. Montllor, ISK Mountain View Research Center Inc., 1195 West Fremont Avenue, Sunnyvale, CA 94087

D.M. Norris, University of Wisconsin, 642 Russell Laboratories, Madison, WI 53706

D.A. Potter, University of Kentucky, Department of Entomology, S-225 Agricultural Science Building North, Lexington, KE 40546

J.C. Reese, Kansas State University, Department of Entomology, Waters Hall, Manhattan, KA 66506

E.D. van der Reijden, Tufts University, Biology Department, Medford, MA 02155

J.A. Renwick, Boyce Thompson Institute at Cornell University, Tower Road, Ithaca, NY 14853

J.M. Scriber, Michigan State University, Department of Entomology, 243 Natural Science Building, East Lansing, MI 48824

E.L. Simms, University of Chicago, Department of Ecology and Evolution, 1101 East 57th Street, Chicago, IL 60637

M.T. Smith, USDA, ARS Southern Insect Management Laboratory, P.O. Box 346, Stoneville, MS 38776

K.A. Stoner, Connecticut Agricultural Experiment Station, P.O. Box 1106, New Haven, CT 06504

D.R. Strong, University of California, Bodega Marine Laboratory, P.O. Box 247, Bodega Bay, CA 94923

J.H. Tumlinson, USDA, ARS Insect Attractants, Behavior and Basic Biology Research Laboratory, P.O. Box 14565, Gainesville, FL 32604

T.C.J. Turlings, USDA, ARS Insect Attractants, Behavior and Basic Biology Research Laboratory, P.O. Box 14565, Gainesville, FL 32604

Index of authors

General index

acid rain 55, 347
Acrolepiopsis assectella 337
Acyrthosiphon pisum 117, 299, 301
aerial pollution 57
African armyworm 49
Agromyzidae 267
alarm pheromones 117
Allium 337
allozymes 201, 203, 209, 341
Amblyseiulus cucumeris 351
Ambrosia 191
American holly 39
Amphorophora idaei 275
anemotaxis 107, 109
ant predation 345
Anthonomus musculus 115
antibiosis 255, 265, 295, 304
antifeedants 153, 155, 159, 176, 179, 181,
 373, 383, 389
antixenosis 265, 316
aphid enzymes 277
aphid pectinases 289
Aphidius 349
aphids 33, 36, 61, 112, 117, 119, 179,
 181, 275, 277, 280, 283, 286, 289, 291,
 293, 295, 297, 299, 301, 304, 307, 313,
 395
Aphis fabae 112, 280, 283
Aphomia sociella 149
Apis mellifera ligustica 129
Arctiidae 83
Artemisia 191
Asclepias 43
associative learning 162, 403
assortative mating 203
Asteraceae 89, 191
attraction 103, 107, 365, 373
azadirachtin 176, 179
baculovirus 52
Barbarea vulgaris 205
behaviour 61, 71, 93, 105, 107, 112, 129,
 143, 151, 157, 159, 169, 181, 286, 293,
 295, 316, 359, 363, 365, 383
Betula 57, 345
Bicyclus anynana 65
biological control 297, 351, 363
biotypes 275, 289, 299
bird cherry-oat aphid 277

black bean aphid 112
black swallowtail butterfly 122
blackmargined aphid 61
Bostrichidae 325
Braconidae 297
Brassica 103, 105, 107, 129, 136, 141, 313,
 316, 319
Brassica juncea 136
Brassica napus 103, 105, 107, 129
Brassica oleracea 141, 313, 319
Brevicoryne brassicae 313, 319
Bruchidae 67
Burnet moth 125
cabbage root fly 141
cabbage seed weevil 103, 105
cabbage stem flea beetle 103
cabbage stem weevil 109
Cameraria 361
cardenolides 43, 155
cassava 235, 255, 353
cassava mealybug 255
Cecidomyiidae 271
cellulases 329
Cerambycidae 55, 133
cereal aphids 277, 349
Ceutorhynchus assimilis 103, 105, 107
Ceutorhynchus quadridens 109
chemical fingerprints 43
chemosensory data 171, 173
Chenopodiaceae 86
cherry fruit fly 143
Chloropidae 339
Chromolaena 89
Chrysanthemum 258, 263, 267
Chrysomelinae 341
circadian stability 165
Cirsium canescens 30
Coccoidea 46
Colorado potato beetle 253, 273
common bean 235
computer-aided analysis 171
contact chemoreceptors 139, 143
Coronilla varia 155
Cossidae 55
Cotesia glomerata 355
Cotesia marginiventris 363, 365
Cotesia rubecula 359
cotton 323